Geology
The Paradox of Earth and Man

Geology: The Paradox of Earth and Man

Keith Young University of Texas, Austin

Houghton Mifflin Company *Boston*
Atlanta Dallas Geneva, Illinois Hopewell, New Jersey Palo Alto London

To Marvin Young, my father,
and the Bighorn Mountains of Wyoming;
both taught me to love an ecosystem.

Contents

Preface

In spite of claims to the contrary, urbanites are characterized by a high degree of conformity. They work together, play together, panic together, and befoul the environment together.

Paul F. Hilpman, 1970

Man either alters his geologic environment, remains ignorant of it, or accepts it as fate. Since many problems can be avoided, or resolved, an enlightened planning is appropriate. Knowing all these things, man's behavior toward his earth is still one of destruction and ill-conceived despoilage. This is the paradox so apparent in man's behavior toward his only resources.

Air pollutants, removed and buried in the ground, even if only temporary before polluting the nearest aquifer or the nearest stream, become geologic problems. Water contaminants, when removed, most often return to the stream from which they were removed or enter some underground aquifer or other stream; they also become geologic problems.

The garbage of man is a thermal pollutant of the ground, a liquifier of the substrate, and is often discharged by ground water into streams and underground aquifers. The common belief that such pollutants are purified in the ground is based on a minimum input that is now all too often exceeded. These are also geologic problems.

Human waste, if not neutralized, is a health hazard that increases with increasing population. If neutralized, its fantastic fertility can upset ecologic balance by eutrophication. In or on the ground it is a geologic problem.

Partly because of man's instinct to bury his poisons and partly because of natural cycles, rocks and soils are involved in nearly all pollution problems.

Geologic hazards include earthquakes, volcanoes, floods, land movements, and climatic changes. These are also part of man's world and a danger to some of his activities.

The essential outline of this book was first published in Milwaukee in 1970 in the American Geological Institute's Short Course on Environmental Geology. At that time it had already been taught for five semesters, with considerable success. Since then it has been taught another seven semesters.

The rationale of this book is environmental—to treat those aspects of geology that interface directly with man. In this respect it is to be distinguished from texts dealing with classical geology as well as from those dealing with the theoretical aspects of geology on which the activities of man do not directly impinge. Occasionally it has been deemed necessary to present background information. Some of the background information, where it would have produced non sequiturs in the text, has been included in the appendices.

The book is philosophically divided into three sections. The first eight chapters are concerned with geological aspects of ecosystems, including a consideration of the principles from biological, geological, and conservational aspects (Chapter 1). Sucessive chapters then deal with resources (2), atmosphere (3), oceans (4), water resources (5), erosion (6), soil (7), and the environmental health aspects of the biosphere-lithosphere interface (8). The dominant, though not exclusive, thrust of these chapters is concerned with man's effects on the habitable earth.

The second part of the book, Chapters 9 through 13, is concerned with geologic hazards—earthquakes, volcanoes, landslides, subsidence, permafrost, floods, and climatic changes. The dominant, though again not exclusive, thrust of this section is concerned with earth-initiated effects on man.

The third section of this book consists of Chapters 14 through 16—planning, population, and evolution, in that order. These chapters are primarily concerned with behavior. Land use planning is a potential behavior of man that can be developed for the benefit of ecosystems, which in turn is like casting "bread upon the waters"; in other words there is a beneficial return to society by treating ecosystems optimally. Population is a behavioral aspect of man, which, if allowed to proceed unchecked, negates all other beneficial activities. Chapter 16 is concerned with those aspects of man's behavior that may be inherited and that are behavioral adaptations to situations that no longer exist. Such behavior impedes man's ability to adapt to a rapidly changing society and culture.

The purpose of this book is to gain insight into those activities of man that degrade the earth and those earth phenomena that are potentially and actually harmful to man. Prevention, cures, and amelioration are considered, but the primary implication throughout is that proper understanding followed by proper planning prevent damage and death, and are less costly than prescription and cure.

The author's course has consisted of a mixed group of students—upper division, lower division, students with prior geology, students without geology, and students with earth science experience in secondary schools. However, the book is aimed at first-year college students with or without secondary school earth science experience. Technical language is held at a minimum

Acknowledgments

and usually explained upon first usage. The glossary is concerned as much with non-technical language that may be outside of the student's experience as with technical language.

In order to hold the book to a reasonable length there have been many compromises over inclusions and omissions. The author sympathizes with the user who cannot find a pet interest, since many of his own pet areas were left out at the suggestion of editors and reviewers. Yet, the book is so constructed that any teacher can find time to cover subjects of special interest.

Keith Young

I am grateful to many colleagues, too numerous to mention, who have inspired my interest in a broad range of geological phenomena affecting man. More specifically, for their help as reviewers at various stages of the manuscript, I would like to thank Robert E. DeMar, University of Illinois, Chicago Circle; John G. Dennis, California State University, Long Beach; Laurence Lattman, University of Cincinnati; George R. McCormick, University of Iowa; and Jack F. Schindler, St. Petersburg Junior College. In addition, I would like to acknowledge the valuable input and ideas of Geoffrey Smith, Ohio University, and O. T. Hayward, Baylor University.

None of this would have been possible without the help and encouragement of my wife, Ann S. Young, who has suffered through months of my preoccupation and absence of communication.

Chapter 1

Geology in the Ecosystem

To all practical purposes, Western man remains obdurately pre-Copernican, believing that he bestrides the earth round which the sun, the galaxy, and the very cosmos revolve. This delusion has fueled our ignorance in time past and is directly responsible for the prodigal destruction of nature and for the encapsulating burrows that are the dysgenic city.
Ian McHarg, *The Fitness of Man's Environment,* 1968

Primitive man almost certainly considered his environment an enemy, or at least a stubborn contestant from which he had to wring a precarious existence. Men of the Paleolithic or Old Stone Age, dependent on crude tools and weapons for gathering and hunting, were particularly vulnerable to their environment. Although improved technologies gave men of the Neolithic or New Stone Age an advantage over their Paleolithic progenitors, and eventually gave them primitive agriculture and domesticated animals, the environment was still fought, dreaded, and placated in times of want, but enjoyed in years of plenty. Just staying alive was a constant struggle, and everything useful was presumed to be there for man's taking and, hence, to exist for his benefit.

Man the Unnatural Animal

As man evolved, he gradually developed more sophisticated attitudes toward his environment and the entire world in which he lived. The thought that everything on the earth existed for the benefit of man was passed from one culture to another—Hebrew, Phoenician, Greek, and Roman—and to succeeding cultures. Early man didn't think about "spoiling" his environment; such a thought was beyond his comprehension. Man's adjustment to nature was so necessary to his survival that living in balance with nature was a continuing part of his existence and was selected for through evolution. As soon as developing technology and culture no longer required this balance, either biologically or culturally, it disappeared.

Prehistoric overkill

The introduction of Neolithic technology produced what Professor Paul S. Martin of the University of Arizona has called "prehistoric overkill." Professor Martin believes that whereas the culture of Paleolithic man did not allow him to upset the balance of nature, that of Neolithic man did, with its superior weapons and hunting techniques. If Martin is correct, man's destruction of his environment began long before the appearance of the cultures we consider characteristically human, and even long before the agricultural revolution that immediately preceded the beginning of historic times.

Prehistoric overkill is thought to have begun in Africa and Southeast Asia just over 40,000 years ago (Table 1-1). Even though animal life in modern Africa is more diverse than on any of the other continents, probably as many as one-third of the larger species that lived in Africa were destroyed as man developed his ability to kill. Subsequent extinctions coincide with the migrations of man. Between 20,000 and 8000 years ago, most of the larger mammals in Europe and South America, and about 32 percent of the larger mammals in North America, became extinct.

Between 13,000 and 8000 years ago, about half of the larger marsupials in Aus-

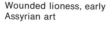
Wounded lioness, early
Assyrian art

Table 1-1

Prehistoric overkill in years B.P. (before the present). Data from Paul S. Martin, "Prehistoric Overkill." In *Pleistocene Extinctions* (New Haven: Yale University Press, 1967).

Prehistoric Overkill

Area	Years B.P.
Africa	about 40,000
North America	20,000 to 8,000
South America	20,000 to 8,000
Europe	20,000 to 8,000
Australia	13,000 to 8,000
New Zealand	4,000 to 400
Malagasay	4,000 to 400

tralia became extinct. Man did not migrate into New Zealand and Malagasay until about 4000 years ago. Between 4000 and 400 years ago, nearly fifty varieties of large birds disappeared from the islands of New Zealand as man killed them for food. When New Zealanders ran out of birds, they started eating each other, thus stabilizing a population that was overreaching a diminishing food supply. The giant birds of Malagasay were exterminated by human predation about the same time as those of New Zealand. Since that time, man has continued to destroy many varieties of animals. Dramatic examples are European wild cattle and the Asiatic lion, about 400 years ago, the Australian wombat and the American bison in the last century, and even now the Siberian and Indian tigers.

Earth and agriculture

Neither as a hunter nor as a food-gatherer—not even when he began to domesticate animals, and his herds overgrazed the land—did ancient man seriously alter his surroundings. Indeed, he did little more than rodents and insects in stirring up and loosening the soil.

When man began to cultivate the soil in order to grow food—when he began to destroy the original ground cover by turning the earth and exposing it unprotected to wind and water—then, for the first time he began seriously to alter his environment geologically. Primitive agriculture did little harm to the land surface, but as man's agricultural instruments improved—from a stick to a flint hoe to a metal plow—the amount of erosion substantially increased. In many areas of intensive agriculture 4000 to 5000 years ago, salt deposition from the irrigation of soils was more detrimental than erosion. Although man's assaults on his geological environment may not exceed nature at its worst, per unit of time, man tears up his soil continually, whereas nature tears it up only during the worst years of climatic cycles. Nature rehabilitates the land between such times, but man never lets it rest long enough for natural restoration.

Although man continually improved his technology for the production of food, his attitudes toward his environment didn't change, remaining primitive. The Accadian, ancient Egyptian, and Phoenician attitudes remained anthropocentric (man-centered). The Hebraic tradition of "be fertile and increase, fill the earth and subdue it" is representative of all of these cultures. The Hellenes in the Golden Age of Greece particularly emphasized that the earth and everything on it was for the benefit of man. Today most men have still not altered their environmental attitudes, which our culture has inherited from the Hebraic culture via

Christianity and from the Greeks via the Romans.

The frontier tradition in America

As recently as the settlement of the American West, these primitive concepts not only held but were even strengthened. During this massive population movement, that continued well into the present century, man was thought of as a conqueror subduing a hostile nature as he would subdue any other enemy. Nature was more likely to be conquered if one took from it everything it had to offer. This required hard work, and work was a major virtue. It is no wonder, then, that there were outcries of indignation from the West when, as the frontier dwindled, "Eastern conservation-

ists" tried to tie up the natural park lands, the forests, and all the other natural treasures that until the 1890's had been free for the taking. The agricultural frontier followed frontier tradition: one should strive for the greatest harvest per acre whether or not it is good for the land. Impending famine is not likely to change this attitude.

Industrialization

Of all the many aspects of human culture, industrialization has had the most serious effects on the geological environment. Neolithic man did not remove enough tar from the Autun area in eastern France to affect the environment. Likewise, primitive methods of producing oil in Bangladesh and Burma a thousand years ago left most of the re-

Lumbering white pine in the American Northwest

serves in the ground. Nor did the American Indians take enough oil from the Pennsylvania tar pits or enough copper from the Michigan peninsula to reduce the reserves by any appreciable amount.

Supplies of some minerals, metals, and other natural resources, however, were dramatically reduced and even exhausted quite early in man's industrial history. The mining area just north of the Gulf of Aqaba (Figure 1-1) at Ezion-Geber may have been depleted of copper and iron ores in the time of King Solomon's technicians, but new methods allow modern Israel to rework some of these old deposits that had not been mined for nearly three thousand years. Tin deposits in Tartessus, now Spain, were so depleted in early Roman times that the Phoenicians opened the tin deposits in Cornwall, which in turn were depleted early

Figure 1–1 (a and b)

Ezion-Geber, on the Gulf of Aqaba, reactivated copper and iron mining district.

(a)

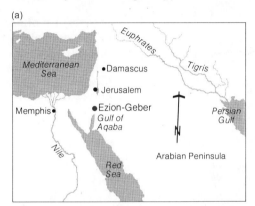

(b)

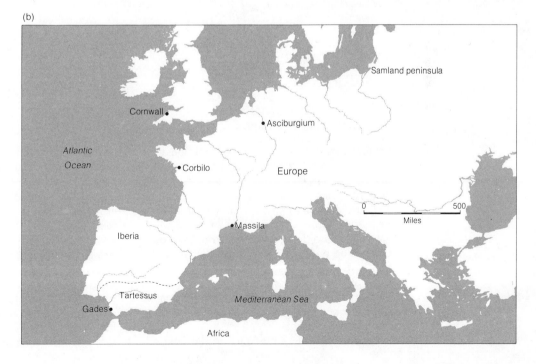

Figure 1-2

Iron and copper consumption since 1800 compared to population growth.

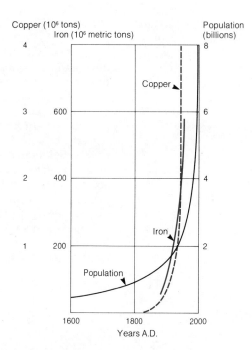

in the period of the Christian culture. The best deposits of Baltic amber were apparently depleted by Roman times.

With the steady improvement of technology, the use of nonrenewable resources expanded rapidly. The industrial revolution brought a fantastic increase in the rate with which natural resources were used. Indeed, since that time the acceleration curves for the use of many resources have risen even more sharply than the population curves (Figure 1-2).

Urbanization and pollution

Urbanization interacts with the geological environment by crowding large numbers of people into areas of probable hazards, such as earthquake, flood, tidal wave, and landslide areas. A city can be built over natural resources before they are known to exist. When they are discovered, those resources can not be produced economically because that land is already so expensive. The huddling of large numbers of people in small areas produces huge local concentrations of garbage, trash, human waste, and other manmade pollutants.

Even in Neolithic times men in the Orkney Islands, Cornwall, and the Gironde, France, dug trenches under their dwellings to drain sewage to a beach or into the nearest stream. The Romans were well acquainted with methods of sewage disposal; the Cloaca Maxima, the great sewer of Rome, dates from the sixth century B.C. During the Dark Ages this technology disappeared; the streets of medieval cities—as well as many modern cities throughout the world—were the constant receptacle of slops, refuse, and human waste. Systematic sewage systems were not built again until the nineteenth century. It is perhaps no exaggeration

Scene of old London

Air shaft of an ancient coal mine

to say that the greatest single contribution of technology to the quality of life is plumbing.

The first pollution (as we now generally use the term) to affect the geological environment was air pollution in the valley of the Thames River near London in the thirteenth and fourteenth centuries, resulting from the introduction of coal as a residential and an industrial fuel. There can be little doubt that the widespread pollution of air, earth, and water—of which we

Air pollution from industry—
before federal regulation

are now so acutely aware—would not have developed so swiftly had it not been for the disastrous population explosion of the last several centuries. The fact remains that once man ceased to be a part of nature, he began to disturb its balance so irretrievably that the quality of his own existence is now threatened.

Figure 1-3

Organization of animals and plants into a community.

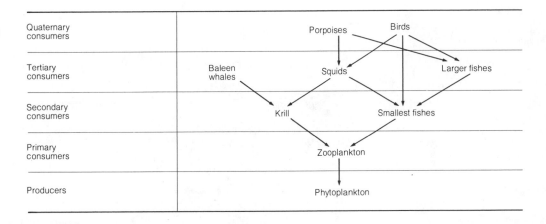

Quaternary consumers		
Tertiary consumers		
Secondary consumers		
Primary consumers		
Producers		

Porpoises   Birds

Baleen whales   Squids   Larger fishes

Krill   Smallest fishes

Zooplankton

Phytoplankton

Ecology and Ecosystems

The study of the relationships of organisms to their environments is called ecology. This study includes all forms of life and everything that affects life—air, water, soil, rocks, minerals, and other forms of life. For purposes of discussion, the total environment is usually divided into the organic environment and the physical environment, though the two are so interdependent that they should be thought of as one interacting entity. After all, a bird that depends almost entirely on earthworms as its major source of food is restricted to areas with some soil cover hospitable to earthworms.

Community and biome

In many similar ways geology enters into the ecological balance of the community A community is a place where all the populations of organisms occupy a given area, or all the populations of organisms live in some kind of interdependency, provided that "interdependency" is not defined too broadly (Figure 1-3). Man is a part of a community. Major terrestrial communities are called biomes. They are largely defined according to the climatically controlled distribution of plants. This makes definition easier and the concept of biomes more usable than if they were named and defined according to more mobile animal groups. Major biomes of North America include tundra, boreal forest, deciduous forest, grassland, desert, and tropical forest (Figure 1-4).[1] Man's own way of living varies tremendously from biome to biome. Communities are less formal units than biomes and can be named for plants or animals.

Ecologists usually use the terms "biome" and "community" to refer to the organic environment. These and larger categories

[1] There is no rule governing the naming of biomes, and ecologists are not in agreement as to their number or their names.

Figure 1–4

Major biomes of North America.

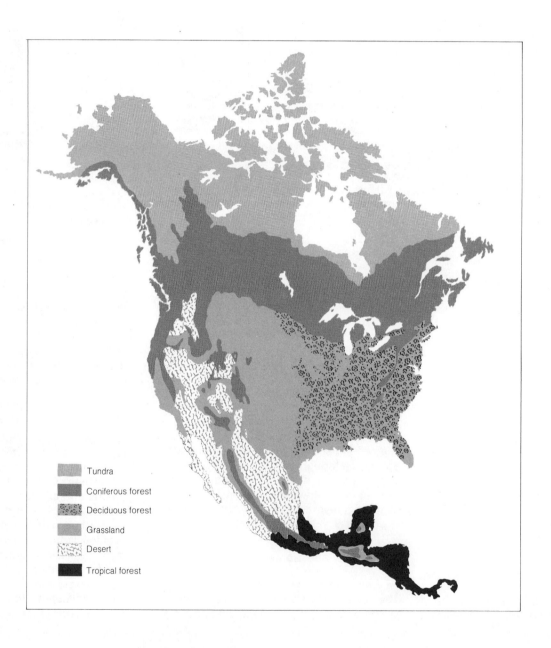

Tundra

Coniferous forest

Deciduous forest

Grassland

Desert

Tropical forest

of nature can also be called ecosystems, if they include living organisms interacting with nonliving materials. It is through the ecosystem that geology becomes a real part of ecology, for it is at this level that there is an exchange of earth materials with living materials. Air and water are earth materials, as are minerals necessary for the development and well-being of the living. Also a part of ecology are earth systems that cleanse the ecosystem and neutralize pollutants.

There is also a great deal to be learned by looking at urban problems from the ecological point of view, and studying man as part of the biological communities and subcommunities that are subject to ecological principles. Eugene P. Odum, outstanding American ecologist, has long advocated an approach to environmental problems through ecosystem studies which include man as a part of the system—a distinct reversal of cultural trends of the last few thousand years.

Climax and disclimax

A biome is usually named for the dominant type of vegetation, but not always—as in the desert biome. These dominant types of vegetation are known as climax and represent the end product of a series of steps or stages (called seres) in the development of the community. When, for example, a mature forest is destroyed by a forest fire, the types of plants in the forest before it was burned do not return immediately, but are preceded by a succession of seres. Not only do the animals and plants of a biome evolve through time, but as they do so, they affect the survival of every other plant and animal in the biome, so that the entire system evolves as a single unit over millions of years. In addition, each of the seres evolves as the biome evolves.

A simple kind of forest fire climax cycle is that of Labrador, where the climax forest is spruce. After a forest fire the area appears dead for a long time, until the return of the first sere, composed of mosses and lichens. Deciduous trees follow the lichens as the second sere, then, after another long period, the spruce reappears and the climax has returned.

Before man acquired fire, lightning was the primary cause of forest fires, as it still is in primitive areas. If fire does not come too often, each forested area returns to the climax community. A simple cycle occurs in the eastern Rocky Mountains in the white pine community of the coniferous forest biome. After a forest fire the first plants to appear are wild strawberries, fireweed, Indian paintbrush, and various coarse grasses (Figure 1-5). Almost immediately following this community, the young jack pines appear. This sere gives way—in seventy or eighty years or more—to another mature white pine forest (climax), which when burned starts the cycle all over again. But, in many areas forest fires occur so often that the young pines never completely mature. Consequently, community after community is burned off in the jack pine stage. Such a phenomenon in which the mature climax vegetation never develops is called a disclimax. Other climax cycles, as in some of the communities of the deciduous forest biome, are much more complex and may take hundreds of years.

Figure 1–5

Simple climax cycle in the white pine community of the Bighorn Mountains, Wyoming.

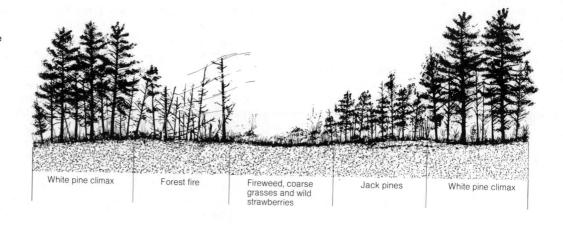

| White pine climax | Forest fire | Fireweed, coarse grasses and wild strawberries | Jack pines | White pine climax |

They, too, are essentially a series of successive seres leading to the climax.

The biome is so constructed and the ecosystem so interrelated, that different animal communities are represented in different seres. Therefore, the sere and the climax can be recognized by the animals as well as by the plants.

The Ecosphere

The earth is a giant ecosystem, for which Lamont Cole has coined the term ecosphere. As an ecosphere the earth may be compared to a giant spaceship; this implies that the earth receives energy only from without. Like a spaceship its capacity to receive poisons is limited. Life on a spaceship does not survive without fresh water; neither can life on earth. Life on a spaceship cannot augment its food supply from without; neither can life on earth. Life on a spaceship can probably discard its wastes; life on earth cannot yet do this

and must learn to live with its metabolic and cultural by-products.

Earth's energy

Since man is annually consuming energy that took over one million years to accumulate geologically, it behooves us to look at the earth's energy budget, upon which depend all ecosystems. The earth receives approximately 250,000 calories per square centimeter per year from the sun. So much of this is sent back to outerspace by reflected light or radiated heat that only 130,000 calories, or about 56 percent, reach the land or water surface. Of this 130,000 calories, 55,000 are returned as radiated heat and 5,000 are immediately reflected, leaving 70,000 effective calories per square centimeter per year. According to these estimates, the earth gives back to space, as heat and reflected light, about the same amount of energy as it receives from the sun, since most of the 70,000 calories

Example of man's pollution,
even in the Sahara Desert

power the hydrologic cycle. Only between 0.15 and 0.3 percent of the available solar energy is used in photosynthesis.

The energy from the earth's interior amounts to from 30 to 50 calories a square centimeter a year, or about one-tenth of the solar energy used in photosynthesis. This probably is not even as great as the margin of error in the measurement of solar energy, and is insignificant from the standpoint of the ecosphere, except that it is the escaping part of the energy that powers the rock cycle. The earth then uses internal energy for continental drift and mountain building, and solar energy for erosion, sediment transport, deposition, and other aspects of the hydrologic cycle.

The hydrologic cycle

Of the 70,000 effective calories per square centimeter a year received by the earth, most are used in evaporating water and

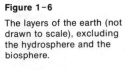

Figure 1–6

The layers of the earth (not drawn to scale), excluding the hydrosphere and the biosphere.

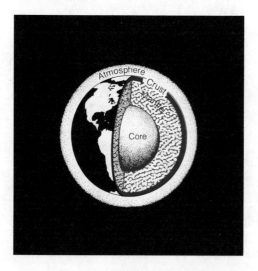

thus escape into space as heat. But in this process they power the hydrologic or water cycle (Chapter 5). In this cycle water evaporates from various bodies of water and from soils, transpires from the leaves of plants, and is distributed through the atmosphere by air currents. It is returned to the earth's surface in the form of rain, snow, hail, sleet, and dew. Of the precipitation that falls on land, some is consumed by plants; some evaporates; some joins the existing groundwater; and some runs off, eventually to the oceans, carrying with it fragments of earth to produce sediments in the ocean. The hydrologic cycle aids in distributing water and mineral nutrients to all organisms and through each ecosystem.

The Lithosphere

The layered earth is almost a sphere; it is also the solid part of the ecosphere. It is enclosed in a discontinuous layer of water called the hydrosphere (Chapters 4 and 5), and is completely enclosed by the gaseous envelope popularly called the atmosphere (Chapter 3). Scattered through the outer few feet of rock and soil and through the hydrosphere and lower atmosphere are many organisms, collectively called the biosphere. The soil and rock within the other spheres constitute the lithosphere The outer part of the lithosphere is the earth's crust. At the center of the earth is the core, and between the crust and the core is an intervening mantle (Figure 1-6). The rocks of the crust are forever changing and are much more varied than they look. There are actually two kinds of crust: continental crust and oceanic crust Beneath the continents the crust is very

Figure 1–7

Principal components of the earth's crust.

thick, averaging about 35 kilometers, and thickening even more under the central parts of mountain ranges. The composition of the upper part of the continental crust is sialic, that is, rich in silicon and aluminum. At greater depths beneath the continents the rocks are rich in silicon, magnesium, and iron, and are called simatic. The sialic layer appears to be absent from the crust beneath the ocean basins, so that the thinner oceanic crust is essentially simatic (Figure 1-7).

Types of rock

The rocks that make up the outermost part of the earth's crust are an important part of the environment. Those rocks at the surface produce soils and sediments through weathering, and erosion of these rocks shapes the landscape we see around us. All of these processes contribute to ecosystems. Classified by origin, these rocks are of three main groups: igneous, sedimentary, and metamorphic.

Igneous rocks are those that have crystallized from a body of molten rock called magma. If such rock solidifies beneath the surface of the earth, it is called intrusive, but if it breaks through the surface and cools as lava, it is called extrusive. The composition of igneous rock varies greatly. Granite is the most abundant intrusive igneous rock and is most common in the cores of great mountain chains. Basalt, the most abundant extrusive igneous rock, occurs widely distributed along the midoceanic ridges and in large plateaus on the continents.

Sedimentary rocks are formed from older rocks that have been broken down by weathering and resolidified by compaction and cementation to form a new rock. Although sedimentary rocks differ in origin, composition, and appearance, they can be grouped into four major categories: claystones, sandstones, limestones, and evaporites.

Sandstones are rocks made up of discrete, visible grains, usually less than 2 millimeters in diameter. The chief mineral is usually quartz, although feldspar, basalt fragments, and other minerals are known in sandstones. Sandstones are as stable as their mineral constituents.

Claystones make up about 75 percent of the sedimentary rocks of the continents and are usually composed of clay minerals. If finely laminated, they are called shales Nearly all claystones are unstable to some extent, but some with a capacity to take up large amounts of water, like montmorillonite, are more unstable than others.

Limestones consist of either calcite or dolomite. They may be formed by precipitation from solution, or they may be constructed of whole shells of fossils or minute parts of these shells. Limestones are found throughout much of the United States, and they are used commonly as building stone and construction aggregate. Many spectacular caverns, such as those of Kentucky, New Mexico, and West Virginia, result from the ready solution of some types of limestone.

Evaporite deposits, although not as abundant as other types of sedimentary rock, are sometimes of important economic value. These rocks form by precipitation of various minerals from water. Most deposits of salt were formed in this way. Gypsum, a material used extensively in the building trades, is also precipitated from water.

Metamorphic is the third major rock type, and it refers to the alteration of any other type of rock by temperature, pressure, or chemical fluids. Quartz sandstone by metamorphism becomes metaquartzite, claystone becomes schist or slate, and limestone is metamorphosed to marble. Igneous rocks high in quartz and feldspar become gneiss, and those high in iron and magnesium minerals become schist.

Not only do many of these rocks have unique environmental qualities, but through the following cycles each participates as part of ecosystems in many environmental phenomena.

Mineral Cycles

There are many mineral cycles in the complex of ecosystems called earth. The rock cycle can be used as a basis on which mineral cycles of ecosystems are constructed, especially if one realizes that the rock cycle relates the crust to the earth's interior.

The rock cycle

All rocks are subject to weathering, during which they are broken into small pieces or taken into solution by water. When weathered particles or dissolved chemicals are deposited, or are precipitated from air or water, they produce sediments. Sediments are lithified (made into rock) by compaction or by cementation, that is, by the addition of precipitated chemicals that serve as a binder to hold the fragments together. It is by lithification that sediments become sedimentary rocks. By heat, pressure, and chemical change within the earth, sedimentary and igenous rocks are altered to metamorphic rocks. Any of these may be melted and in turn produce igneous rocks upon recrystallization. The entire sequence from erosion through deposition, metamorphism, melting, crystallizing, and back to erosion is known as the rock cycle. As can be seen from Figure 1-8, parts of the cycle can be short-circuited. Rocks from different stages of the rock cycle have unique properties of environmental importance.

Figure 1-8

The rock cycle illustrates the interchange of materials of the lithosphere through the crust of the earth.

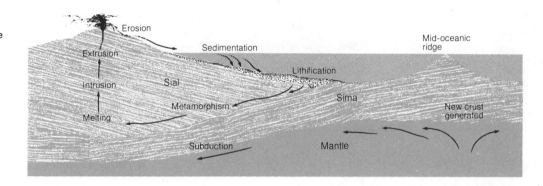

New mineral matter is constantly being added to the hydrosphere and the crust along the midocean ridges by the upwelling of molten rock from the earth's mantle in a process known as seafloor spreading. (See Appendix B, Continental Drift, page 440.) The new minerals added to the hydrosphere, and to the newly formed portions of the earth's crust along these ridges, are distributed through the ecosystems via the hydrologic and rock cycles.

The rock cycle also interchanges with the hydrosphere and the atmosphere through the hydrologic cycle and through igneous activity, including volcanism. When gases are emitted into the atmosphere by volcanism they become available to plants and animals. They are dissolved by falling rain and can be obtained by plants and animals in the direct consumption of water. All surface rocks in time become soil, from which minerals may be dissolved by water and directly consumed by plants and animals. Plants can extract minerals from soil by a chemical process known as ionic exchange, and in this manner provide min-

erals for herbivores. The food cycle begins with these basic minerals, originally derived from the rock cycle. Many animals receive all their mineral requirements from eating other living things, but most land animals get some of their mineral requirements directly from the earth—at salt licks, or by eating dirt, or by consuming the minute amounts of soil particles adhering to the leaves of plants as dust.

Different parts of ecosystem cycles have different time values, many of which are unknown. Carbon in the atmosphere has a residence time of about four years, but in the ocean it has a residence time of about 400 years. Dust has a residence time of one year in the atmosphere. The residence time of elements in soils depends on the climate and the solubility of the individual elements. The rock cycle is completed much more slowly—from 100 to 600 million years.

The nitrogen cycle

Figure 1-9 illustrates the nitrogen cycle Plants obtain nitrogen from soil minerals

Figure 1-9

The nitrogen cycle.

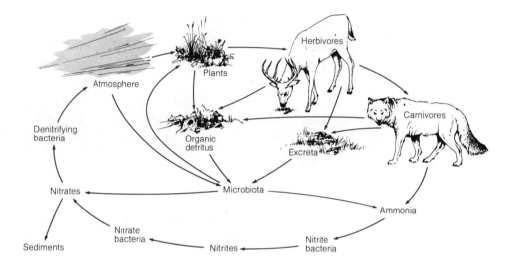

(soluble nitrates). Nitrates in the soil are fixed by bacteria and blue-green algae or are a part of the rock mineral, either naturally or as applied in fertilizer. It has recently been demonstrated that some plants can also obtain nitrogen directly from the atmosphere by absorption of ammonia (NH_4) through their leaves. Herbivores obtain their nitrogen from plants, and carnivores in turn obtain theirs from herbivores.

The atmosphere receives its nitrogen from volcanism, denitrifying bacteria, and animal wastes. Nitrogen can also be obtained from nitrates in sediments through bacterial action and by protein synthesis. Some nitrates are lost to the oceans by deposition as sediments and can only be recovered after hundreds of millions of years via the rock cycle. Presumably, this loss is balanced by nitrogen obtained from volcanic activity and by weathering of nitrates from marine rocks that have been exposed as a result of mountain building.

The phosphorus and carbon cycles

The phosphorus cycle (Figure 1-10) is another important cycle to animals and plants. The route of phosphorus parallels that of nitrogen, but is less complex because there is no counterpart of the nitrite state, and no major intermediate stage between phosphate and protein synthesis. The carbon cycle (Figure 1-11) is another example of a well studied cycle from rocks to air to life and back to rocks again.

Some ecologists fear that we will use up all of our sulfur, phosphate, and other fertilizers because they will be washed out of the soil and deposited in the oceans. Residence time of these products in rock before reappearance on the earth's surface is from 100 to 600 million years.

Figure 1–10 (left)

The phosphorus cycle represents interchange of organic and inorganic materials.

Figure 1–11 (right)

The carbon cycle includes photosynthesis and combustion of fossil fuels.

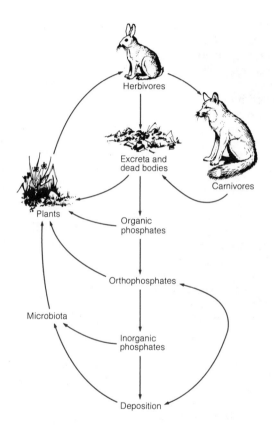

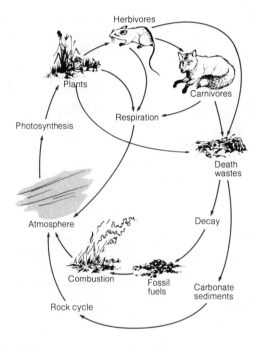

disclimaxes, because man never allows the biological habitat to develop normally and fully. Because of technology and increased population the disclimax is a continually changing entity. Unfortunately, under the hand of man the change is usually detrimental to the ecosystem.

2. Man must use nearly all of his habitat, especially with increasing population. This means he is either the sole or the major determinant of the direction of change of habitat.

3. Man is able to analyze his habitat systems, to reason from cause to effect, and to be aware of the consequences of his own actions.

4. Man (a) uses resources and materially

The Conservation Ethic

In a thoughtful article on the need for an intelligent conservation policy designed to conserve the environment and make the best use of it, the New Zealand ecologist G. L. Kesteven[2] has made the following points:

1. "Habitats change *irreversibly.*" By the destruction of certain aspects of the living environment all of man's habitats become

[2]G.L. Kesteven, A policy for conservationists, *Science* 160:857–860, 1968.

changes his environment, (b) has an esthetic appreciation of his environment, and (c) has an ethical responsibility to maintain his environment.

5. Conservation represents the overall human strategy toward habitat—scientific, managerial, and ethical.

6. Man's extraction of resources usually is directed toward only part of the habitat, thus producing an imbalance requiring some reaction on the part of man or the habitat to set up a new balance (usually a disclimax).

7. Man has an especially heavy responsibility as a guardian of all gene pools.

Kesteven is a biologist, but his first six points are applicable to all areas of conservation. Concerning statement 7, the upsetting of a geological balance may affect the gene pools, and thus protection of gene pools may become a geological problem. Kesteven then adds three policy statements:

1. No plant or animal (as individual, stock, or species) should be destroyed unless its continued existence is not essential, its removal does no harm to the natural community to which it belongs, and distinct benefit accrues to the human community at large through the benefit that accrues to those who remove the organism. Alternative grounds for destruction could be that the organism is noxious to man, either directly or indirectly through the natural community to which it belongs, and that the consequences of its removal are, on balance, beneficial.

2. There should be no further modification of, or interference with, water, air, soil, substrate, rock, or biotope unless the immediate and necessary benefits of that modification are accompanied by long-term benefits.

3. There should be neither deliberate nor careless acts in contravention of the above, either directly, as in hunting and mining, or indirectly, as in pollution and the use of insecticides.

C. K. Leith[3], former University of Wisconsin geologist, formulated the following concept of conservation: "Conservation is the effort to insure to society the maximum present and future benefit from the use of natural resources."

The primary concepts of biologist and geologist in relation to conservation do not greatly differ. But their statements, perhaps naively, do not consider recent cultural and population trends. R. O. Slatyer[4], an Australian ecologist, has also beautifully implored man to behave as though he were an intrinsic part of the ecosystem with which he evolved. Slatyer closes his plea with three imperatives, concerned with cultural and population trends, that man must follow if he is to survive:

First, he must regulate his numbers. Unless population stability is achieved, everything else must ultimately fail.

Secondly, he must conserve and recycle the basic materials he uses to the greatest possible degree.

Thirdly, he must ensure that his food supply is adequate for his regulated numbers, and that the means of its production are not in themselves leading to environmental deterioration.

Kesteven, Leith, and Slatyer are saying the same thing, but each with a different emphasis. Kesteven emphasizes the biological aspect of the ecosphere, Leith stresses the resource aspect of the ecosphere, and

[3]C.K. Leith, Conservation of minerals, *Science* 82: 109–117, 1935.
[4]R.O. Slatyer, Man's use of the environment—the need for ecological guidelines, *Australian Journal of Science* 32(4):146–153, 1969.

Slatyer argues for that part of the ecosphere that deals with the production of food. All recognize that immediate circumstances, if important to survival of large numbers of individuals, may necessitate the compromise of a general rule.

Conservation involves several aspects of behavior toward natural resources, no matter whether the resources being considered are valuable gems, industrial minerals, food, esthetic values, or recreational areas. Resources must not only be judiciously used, but some must be saved. There is a need for economizing and for eliminating or at least restricting waste. Many commodities should be recycled and substitutes extensively developed, provided that they in turn are not environmentally detrimental.

In conclusion, we must incorporate into our environmental ethic the concepts that (1) man's only habitat for the present and the near future is the earth; (2) man is responsible only to man for what he does to the earth, and (3) man has been irresponsible for such a long, long time that his very survival, for very many more centuries, depends on his conservation of air, water, and other natural resources. Even as exploration and technology develop further, there may not be enough lead time to develop the resources for the teeming millions who are already on the way or are projected to be on the way.

Selected Readings

Buckholtz, Curtis W. 1969. William Logan and Glacier National Park. Montana 19:2–17, Summer.

Dansereau, Pierre. 1957. Man's impact on the landscape. In Dansereau, Pierre, Biogeography: An ecological perspective. New York: Ronald Press, pp. 258–293.

Glacken, Clarence J. 1967. Traces on the Rhodian Shore. Berkeley and Los Angeles: Univ. of Calif. Press.

Hart, R. A. 1973. Geochemical and geophysical implications of the reaction between seawater and the oceanic crust. Nature 243:76–78.

Ilic, Miloje. 1971. O "novoj globalnoj tektonici." Zbornik Radova, Univ. u Beogradu, SV 14, pp. 9–39.

Kesteven, G. L. 1968. A policy for conservationists. Science 160:857–860.

Malin, James C. 1948. The grassland of North America: Prolegomena to its history. Lawrence, Kansas: James C. Malin.

Martin, Paul S. 1967. Prehistoric overkill. In Martin, Paul S. and H. E. Wright, Jr., eds., Pleistocene Extinctions: A Search for a Cause; vol. 6, Proceedings VII Cong. of Internat. Assoc. for Quaternary Res., New Haven: Yale Univ. Press, pp. 75–120.

Martin, Paul S. 1973. The discovery of America. Science 179:969–972.

Odum, Eugene P. 1959. Fundamentals of Ecology. Philadelphia: W. B. Saunders.

Odum, Eugene P. 1969. The strategy of ecosystem development. Science 164:262–270.

Rienow, Robert, and Leona Train Rienow. 1967. Moment in the Sun. New York: Dial Press.

Slatyer, R. O. 1969. Man's use of the environment—the need for ecological guidelines. Australian Journal of Science 32:146–153.

Smith, Henry Nash. 1947. Rain follows the plow: The notion of increased rainfall for the Great Plains. Huntington Library Quarterly 10:159–194.

Stevenson, F. J. 1972. Nitrogen cycle. In Fairbridge, Rhodes H., ed., The Encyclopedia of Geochemistry and Environmental Sciences, vol. IVA of the Encyclopedia of Earth Sciences Series. New York: Van Nostrand Reinhold, pp. 801–806.

White, Lynn, Jr. 1967. Historical roots of our ecological crises. Science 155:1203–1207.

Chapter 2

Is the United States a Have-not Nation?

I think that the energy problem we are facing right now, although it seemingly is a surprise to many, is not something that has come upon us overnight.

Hollis M. Dole, Assistant Secretary of the Interior, 1972

As Arnold Toynbee has pointed out, cultures have declined for many reasons; the depletion of resources, or the lack of a technology capable of using existing resources, has contributed to the waning of civilizations. One of the major problems facing the British Isles today is the decreased value and desirability of coal as a source of energy, and the growing difficulty of mining poorer and poorer coal from thinner and thinner deposits. In our own Southwest, ores once profitably mined by the Spanish with slave labor are no longer economical to extract, even with our greatly advanced technology. The United States is rapidly becoming a have-not nation, partly because we have wasted our resources, and partly because advancing technology is not planned around resources "in pocket." Indeed, technology consumes increasing quantities of resources of which the United States—and often the entire world—is in short supply. Figure 2-1 shows that we import substantial percentages of vital minerals, and Figure 2-2 shows that our known supplies of others will be exhausted in a relatively few years.

Until recently, it was thought that our resources were inexhaustible. Indeed, some economists and geographers still believe that we need not worry about our supplies of natural resources because (a) as economically workable deposits are depleted, new technology will make the use of lower-grade deposits practicable, or (b) we will find satisfactory substitutes, or (c) our expanding economy will enable us to exploit deposits of poorer grade. These propositions constitute the doctrine of *laissez faire* capitalism that may be labelled "the

Figure 2–1

Imports of some minerals vital to the United States, shown in percentages of national consumption.

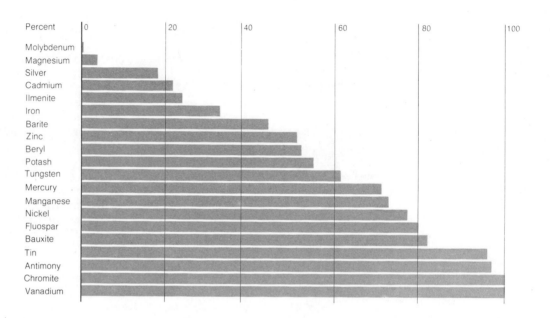

Landsberg School," after Hans H. Landsberg, one of its principal spokesmen, who as recently as 1970 restated the main points of the position.

In contrast to this comfortable optimism is the view of geologist and geophysicist M. King Hubbert, who has surveyed the world's energy resources several times in the past two decades. Hubbert believes that the epoch of fossil fuel will be a relatively brief period in human history. Indeed, he predicted that the United States would reach the peak of its hydrocarbon reserves before 1970, and that reserves would subsequently shrink steadily. Hubbert's dates may be in error because of recent discoveries of deposits such as those on the North Slope of Alaska. On the other hand, these new supplies may be offset by a more rapid acceleration of demand than Hubbert anticipated. Such discoveries seem likely to stave off the inevitable for only a few years. They will not cure the shortage. Hubbert's position has recently been supported by numerous geologists.

Variety of Shortages

The transformation of the United States from a self-sufficient nation to a have-not nation has been extremely rapid. Most of the resources used in the United States must be imported. This abrupt change, which took place within a few decades, is the result of two factors: (1) the accelerating rate at which we have depleted and are continuing to deplete our own resources, and (2) the ever-increasing variety of materials needed to produce one manufactured unit. For example, an axe or a skillet is no longer made from iron alone—several scarcer metals are added to improve the

Figure 2-2

Reserve life of selected known and hypothetical reserves in the United States.

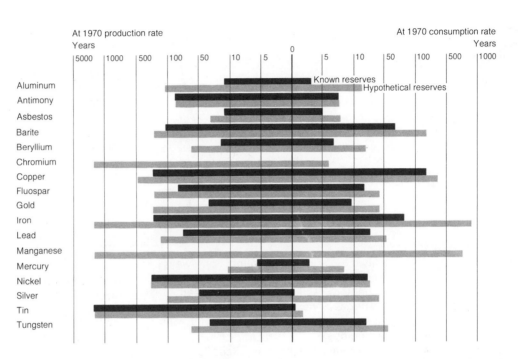

At 1970 production rate
Years

At 1970 consumption rate
Years

products' quality. If we stop to think about the variety of rare earths and metals (niobium, tantalum, columbium, zirconium, germanium, and so on), other rare commodities (quartz crystals), and the great variety of plastics and alloys used in modern electronics alone, we may get some sense of the complex logistics necessary to maintain our present style of life.

The rate of consumption of resources is a major concern. The use of many natural resources is rising even more steeply than is the population curve. If production is a measure of affluence, then this phenomenon is also reflected in a graph of the gross national product and the U.S. population (Figure 2-3). Many of the resources we are using are exhaustible. One commodity in surprisingly short supply is sand and gravel, essential in all construction. Many areas no longer have sand and gravel, shell, or whatever aggregate they once used. The use of a technological replacement—crushed stone—has caused a rapid rise in the cost of housing, industrial construction, highways, railroads, and all other construction for which aggregate is a major commodity.

Future Resources

Obviously, the future supply of resources depends on the substitution of technologically produced replacements for depleted resources, and the discovery of new deposits previously hidden by complex geological and geographical conditions or economically inaccessible. Substitution is

Figure 2–3

Comparison of population and GNP growth.

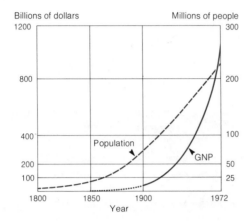

limited by strict requirements not always recognized by economists. Much new technology requires larger and larger quantities of rarer and rarer resources—a factor in the increasing cost of improving technology.

A typical example is superconductors, which are of great importance to future long-distance power transmission. The metals involved in research with superconductors are lead, mercury, tin, technetium, tantalum, vanadium, aluminum, zinc, titanium, combinations of niobium-aluminum-germanium, and combinations of carbide-niobium nitride. From the point of view of resource economics, this list raises massive impediments to the extensive use of superconductors. Mercury, tantalum, niobium, and germanium will never be plentiful enough for extensive use. Whereas sufficient quantities of these might be collected to help ward off radiation from spacecraft, such a use would put these rare commodities in even shorter supply for other purposes. Of the metals listed, only aluminum and perhaps titanium can be

produced in quantities sufficient to exceed present demand. Superconductors for extensive use will probably be restricted to these two metals.

The substitutes most likely to succeed will be those developed from abundant substances, though curious researchers often prefer to work with rare commodities and sometimes make valuable discoveries while doing so. The most successful substitutes will probably be derived from organic materials, provided that large quantities of rare metals are not required. Aluminum substitutes have the great advantage of ready availability, again provided that rare alloys are not required.

Future discoveries in already explored areas

Some estimates of reserves attempt to take future discoveries into account. Their accuracy depends in part on the ability to estimate both future discoveries, based on the rate of decline in discoveries, and the thoroughness of present exploration. Considering how little one or two really large discoveries alter predictions of reserves of common resources, it is evident that future planning should take such predictions into account.

Improved technology

Technology will undoubtedly come up with substitutes and innovations. Thus with copper (Figure 2-4)—as with iron, manganese, and aluminum—tonnage of available ore rises as its copper content decreases. For this reason a slight increase in price per pound of copper, or a slight technological advance, brings much more ore within

Figure 2-4

Decline of percentage of copper in mined ores in the United States. (After U.S. Bureau of Mines)

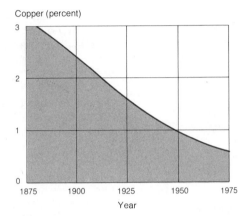

Copper (percent)

Altered economic demands

The primary economic stimuli to renewed production or new development of lower-grade ores are inflation and increased demand. Some geologists and economists believe that a number of mineral deposits have recently been opened or reopened purely because inflated dollars allowed a price increase. Yet increases in the prices of mercury and silver in the last decade far exceed those that could be attributed to economic growth. Such increases have also forced users such as the United States Mint, which cannot compete at such high prices, to turn to substitutes or do without. Improved technology and higher prices have failed to produce desired amounts of silver (Figure 2-5) and mercury, and prices do not rise by this magnitude when demand can be met. On the other hand, research and development can improve technology, and improved technology increases economic growth even when there is no growth in population.

economic reach. On the other hand, such commodities as beryllium, natural quartz crystal, and commercial-grade feldspar do not have gradual decline curves for the percentage of marketable mineral in a potential ore. They occur only in rare mineral aggregates. When depleted, these ores are either gone forever or there is a long wait for deposits of much less concentration—now economically and technologically unavailable—to be developed.

The continental shelves

Because most placer deposits are still economically unavailable, and many are still technologically unavailable, most of the resources of the continental shelves do not appear in reserve estimates. Except for certain types of dredging operations for tin and diamonds, there has been little submarine mining.

Even the extraction of hydrocarbons from offshore deposits is unlikely at 1973 prices. Drilling costs must be halved, or the price of oil doubled, before hydrocarbons can be economically extracted from 600 feet or more of water. Neverthe-

Figure 2-5

World silver production and consumption.

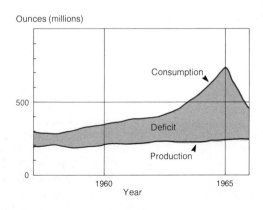

Ounces (millions)

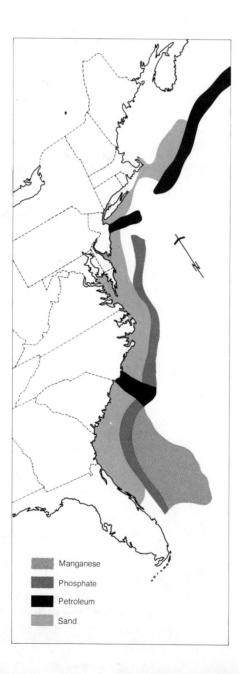

Figure 2-6

Areas of exploration for specific resources off the Atlantic coast of the United States. (From *Resources and Man: A Study and Recommendations* by the Committee on Resources and Man of the Division of Earth Sciences, National Academy of Sciences-National Research Council, with the cooperation of the Division of Biology and Agriculture. W. H. Freeman and Company. Copyright © 1969.)

Manganese

Phosphate

Petroleum

Sand

less, there has been great interest in the possibility of submarine mining, as evidenced by the leasing of concessions for tin off Tasmania, Indonesia, Malaysia, and Thailand; monazite off South America; sulfur off Louisiana (until Canada flooded the sulfur market); and so on. Except for the dredging of placer deposits, many of these are still economically inaccessible, and many are still technologically impossible.

Different continental shelves yield different minerals, depending on the type of shoreline and the age and kind of rock beneath them. The kind of rock inland from the shelf may also be important, and so may the type of erosion the sediment has undergone. For example, diamonds dredged off the Atlantic coast of Southwest Africa have been carried there by inland streams. Near-shore tin placers in Indonesia likewise result from stream erosion of tin from inland pegmatite deposits, concentrated by deposition in sand or gravel. Along the Atlantic coast of North America, good deposits of coarse aggregate occur only where glacial ice has deposited them on the continental shelf. Petroleum deposits on the continental shelf may occur in basins of deposition with the right source beds and permeable trapping beds. Phosphorites are found in areas of winnowing at the outer edges of continental shelves. Manganese occurs in offshore areas of little or no deposition (Figures 2-6 and 2-7).

Before continental shelves and shallow seas can be completely exploited, geological mapping similar to that in progress for the English Channel and the North Sea is necessary. Such mapping is a slow process, and requires expensive equipment such as special core-barrels, special winches, drag

Figure 2-7

Section across a continental margin, showing the most likely localities for potential resources.

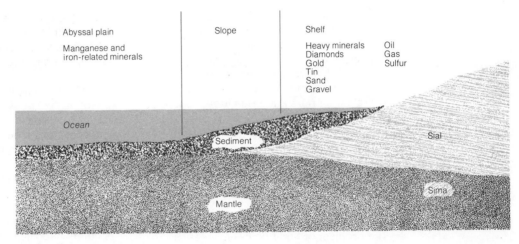

Abyssal plain

Manganese and iron-related minerals

Slope

Shelf

Heavy minerals   Oil
Diamonds   Gas
Gold   Sulfur
Tin
Sand
Gravel

Ocean

Sediment

Sial

Mantle

Sima

buckets, submarine diving gear, and even small ships. However, resources such as the placer gold on the shelf around Alaska and eastern Siberia are attractive to the potential producer. The relationship of some placers to the coastline suggests that rich deposits will someday be worked (Figure 2-8).

So far the offshore area has added little to the total economy of the world, and it is not certain that great mineral riches are there to be found, except for hydrocarbons and diamonds. The belief that there is mineral wealth on the continental shelves stems from a general knowledge of their geology and of the types of mineral deposits associated with similar geology on land.

The deeper ocean basins

The deeper ocean basins are composed of rocks that do not commonly contain ore deposits. Furthermore one would not expect to find ore deposits of the kinds associated with granitic rocks or pegmatites in the deep ocean basins, where only basaltic rocks are known. It has been reported that large quantities of deep ocean clays are of the correct composition for direct kiln-

Figure 2-8

Energy used by man, changed through time and technology.

United States, 1967 A.D.

United States, 1850 A.D.

Roman Empire, 50 A.D.

Coal

Wood

Oil and natural gas

Nuclear, hydro, etc.

Water, wind, animal, man

ing to cement, and dredged manganese nodules from the deeper basins might someday produce low-grade sources of cobalt, copper, nickel, and other metals.

Optimists hope that the oceans will solve our problems, but the economic and technological barriers are tremendous, and it will be many years before many possibilities can be realized. We must wait until technology and the economic climate lower the cost curve, or until scarcity forces the price curve up to the point of practicality.

Energy

The problem of finding new sources of energy is important enough to consider separately. In the next decade shortages of energy loom larger than those of other mineral products. One measure of increased affluence in the United States during the last century was the replacement of manual and wood energy with other energy sources (see Figure 2-8). Long the major energy source of the world, coal has been replaced by gas and oil in many economies. Coal has not always been able to compete with hydrocarbons because it costs more to produce, and other energy sources pollute the environment less. At present only a very small amount of the

coal in the ground is economical to mine. For example, 14-inch beds of coal at depths of 3,000 feet have been included in some reserve estimates, but no one is now going to mine such unrewarding deposits. Abundant coal in some nations will not economically help others that have never had coal or have depleted their supplies. Still, coal can be burned cleanly by the use of such techniques as gasification, recovery of sulfur, combustion at high pressures, and collection and use of fly ash. As other energy sources diminish, coal—of which we have more than of any other source of energy now being used—will regain its importance.

Someday our descendants, if they survive the environment we leave them, will complain about our wasteful use of fossil and nuclear fuels. The fuel we burn in a day was deposited over thousands of years of geologic time; it has been estimated that in slightly over 2000 years man will have used up fossil fuels it took 600 million years of history to create.

For the past several hundred years, the geography of energy has been one controlling factor in the distribution of wealth. Energy is distributed quite unevenly over the world, and some nations—including the United States, Canada, Germany, France,

Figure 2-9

Estimates of world resources of minable coal and lignite.

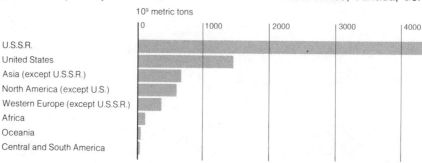

10^9 metric tons

Figure 2-10

Gross national product versus fuel and mineral consumption for the United States.

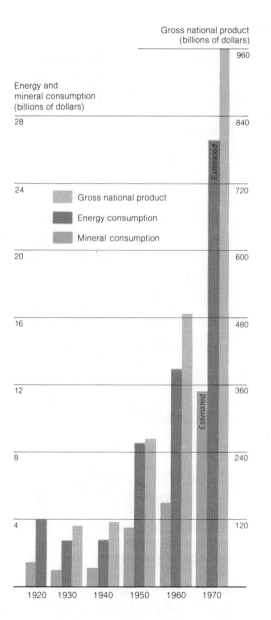

England, and the U.S.S.R.—have had large reserves of energy fuels within their boundaries. France and England are now suffering from the depletion of their mainstay, coal, and have only scant reserves of oil, gas, and nuclear fuels. The United States has depleted its hard coal, but has large reserves of soft coal (Figure 2-9), some of which are not now economical to mine.

The southern hemisphere is deplorably lacking in coal, as Figure 2-9 shows, and is not much better supplied with oil. Coal and oil occur in basins of deposition and were formed in nonpolar climates. The part of the southern hemisphere that was Gondwana has few basins, because the continental blocks formed before the evolution of coal-forming plants and because Gondwana was spread around the pole for a long period. Most of the undeveloped nations are associated with Gondwana continents and are largely without the two most important sources of industrial energy, coal and hydrocarbons. Since these nations seem unlikely to develop either source, many of the newer nations of the world are at an economic disadvantage.

In his book *Affluence in Jeopardy,* the economic geologist Charles F. Park has emphasized that the United States' gross national product depends heavily on energy consumption (Figure 2-10), and that affluence, as measured by the GNP, cannot exist without natural resources. Most of the developed nations, except the U.S.S.R., are feeling the same pinch. Can the present standard of living be maintained if energy consumption levels off? What is the probability that undeveloped nations can become affluent if developed nations are forced to lower their levels of affluence due to a shortage of resources?

Even for a large nation in the southern hemisphere like India, the major barriers to development are the size of the population and lack of energy. To meet the goals of India's fourth five-year plan, coal production would have had to expand from 64 to 66 million tons a year in 1965 to about 120 million tons a year by 1970—an increase of almost 100 percent. Coking coal would have had to expand from .17 to 18 million tons in 1965 to 40 million tons in 1970. But the coal mined in India was already deteriorating in quality in 1965.

Moreover, India does not have enough oil and gas, even though it imports more petroleum than any other commodity. In the absence of hydrocarbons, India will have to depend on coal for more than 60 percent of the fuel it requires for electric power. Because of energy deficiencies, India's fourth five-year plan was doomed to failure before it started.

Coal

The United States has enough soft coal to last a few centuries (see Figure 2-10), if the reserve figures are reasonably accurate.

Oil derricks off the coast of Texas

Technology should be able to develop ways of using this coal for many purposes, including coking.

Hydrocarbons

In 1962 M. King Hubbert caused considerable comment when he presented his fore-cast of energy reserves to a committee of the National Academy of Sciences. The American public, with the exception of experts in mineral exploration, had been rather well indoctrinated by those who believed resources to be inexhaustible. Hubbert showed that hydrocarbon reserves would peak out before 1970 (Figure 2-11a), and that gas production could peak out in the early 1970s (Figure 2-11b); both would probably be depleted early in the 21st century. The decline curve for world production of crude oil is almost as steep and short (Figure 2-12). Based on Hubbert's predictions, hydrocarbon reserves in the United States can no longer sustain increased yearly withdrawals.

Although reserves have increased in the last decade, a large 30-billion-barrel oil field lasts less than five years given the present United States consumption level of seven billion barrels of oil a year. The 18-billion-barrel potential of the Alaskan North Slope would be consumed in less than three years if it were our only source of hydrocarbons. Hubbert's predictions have already been upheld by production and reserve declines in two of our major oil-producing states, Louisiana and Texas. The ratio of natural gas reserves to production is steadily declining at a time when demand is rapidly accelerating. Many parts of the country are now experiencing such a shortage that new residential gas connections cannot be obtained, existing connections are cancelled, and gas for industry and residences is rationed. By the early 1980s we may not be able to depend on hydrocarbons as our primary source of energy, even if their products can be cleaned up for the preservation of the environment. They

Figure 2-11a

Complete cycle of U.S. crude oil production as first projected by Hubbert in 1956.

(a)

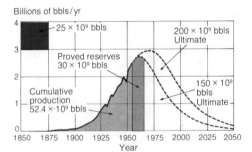

Figure 2-11b

Cycle of natural gas production in the United States.

(b)

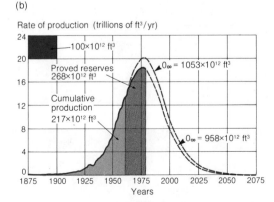

must be supplemented by coal, including synthetic gas from coal, and by nuclear fuel.

Figure 2-12

Complete cycle of world crude oil production.

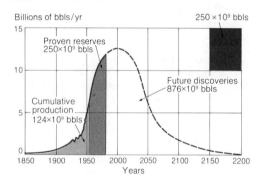

The United States imports large quantities of hydrocarbons: 1.6 million barrels per day or about one-seventh of its consumption in 1971. Considering our present shortage and its impending increase, could future national security best be served by importing hydrocarbons now and preserving our own supplies? This question should have been asked 20 years ago, because it looks now as though we have waited too long. Given the world political situation, we will not get more than our share of the Middle East's oil, and even that much imported oil might move Hubbert's curve only one more decade to the right (Figures 2-11a

Figure 2-13

Parts of the world where oil is found.

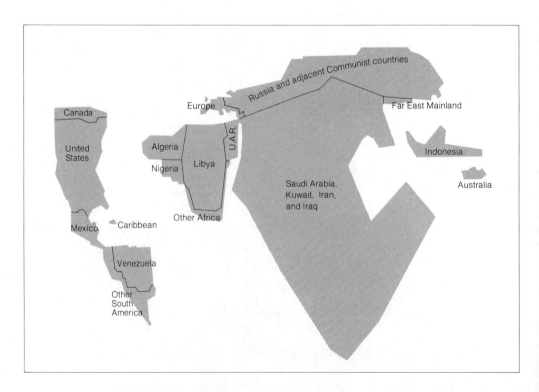

and 2-13). Much of the United States' hydrocarbon imports comes from Venezuela. Venezuela is a member of OPEC (Oil Producing Economic Community), which is dominated by Middle Eastern countries. OPEC has increased hydrocarbon prices until they can no longer be regarded as low. Considering our tenuous political relations with the Middle Eastern countries, a secure source of hydrocarbons is more important than a cheap source. This situation makes the North American Arctic Province and the Athabasca tar sands more attractive.

Tar sands

Tar sands are outcropping sandstones full of tar. In the geologic past, the volatiles and liquids escaped as the migrating hydrocarbons reached the outcrop, leaving the solids as a tar residue in the sandstone. These tars can be heated, and gases and liquids extracted from them. The Athabasca tar sands cover an area of 12–15,000 square miles in Alberta, Canada, and contain an estimated 300 billion barrels of oil. As they are developed, the United States will undoubtedly use all that Canada will allow us.

Oil shale

We will probably also be able to develop economically feasible ways to extract oil from deposits known as oil shales (Figure 2-14). Certain shale formations contain large amounts of oil or organic matter from which oil can be produced. Three methods have been suggested for extracting oil from oil shale: 1. One method is surface retorting, in which oil shale is heated to high temperatures, the oil and gas driven off and collected, and the spent shale used as fuel to heat the next batch. 2. Another method is to drill wells in the oil shale, pump air down one or more wells, and retort the shale by burning it in the ground. Volatilized and liquified hydrocarbons would then be forced to the surface in other nearby wells. 3. A third method is to fracture the oil shale by a nuclear explosion, which would in turn volatilize and liquify hydrocarbons. Production would use conventional drilling methods.

Although the U.S. Bureau of Mines has produced usable petroleum by experimental surface retorting, the Bureau's cost estimates failed to include a profit or the cost of rehabilitating the land. To extract one-tenth of the crude oil consumed by the United States from oil shale by the Bureau of Mines' surface retorting method would require the burning of enough oil shale

Figure 2-14

Basins of the United States that may eventually produce commercially from Eocene oil shale.

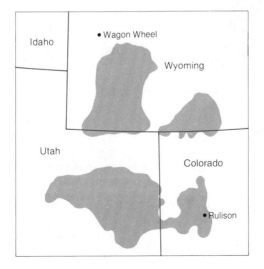

Figure 2–15

Technique of stimulating hydrocarbon production in oil-bearing strata with nuclear explosions.

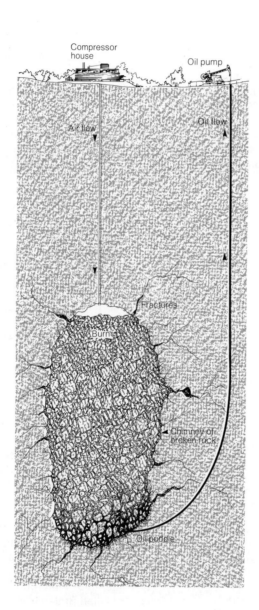

daily to cover one square mile to the depth of 2.5 feet with ashes and cinders, or about 10 square miles to a depth of 80 feet a year. There may not be enough dirt-moving equipment in the United States to handle this much material. What would the Sierra Club say to such a project? Reclamation costs, which would be as high as production costs or higher, were not considered in the pilot studies.

The economics of retorting oil shale at depth have not been worked out but, since the material is neither very porous nor very permeable, recovery must be less than with the use of similar techniques in oil fields. Another limiting economic factor is the question of how much of the recovered fuel must be used to pump air down to the fire in the producing horizon.

Nuclear-aided hydrocarbon production

Two attempts have been made to recover hydrocarbons by means of underground nuclear explosions, though not in oil shale. These are the Gasbuggy and Rulison projects (Figures 2-14, 2-15, and 2-16). After the strata are fractured by nuclear explosion, the oil is allowed to accumulate, and enough time is allowed to reduce radiation to safe levels. These areas can then be drilled by conventional methods. This procedure would be much cheaper than the fantastic earth-moving and reclamation costs of surface retorting, though the cost of nuclear fuel would have to be added to the conventional costs of production. Furthermore, the waiting time required for radiation to subside appears to be much longer than was first projected. The proponents of this method insist that its eco-

Figure 2-16

Geologic conditions at the site of the Rulison nuclear explosion. (After *World Oil*)

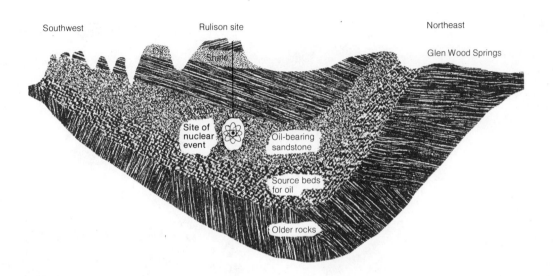

Southwest

Rulison site

Northeast

Glen Wood Springs

Site of nuclear event

Oil-bearing sandstone

Source beds for oil

Older rocks

nomic outlook is good; antagonists say it is impossible. We can only wait and see. The production of usable liquids and gases from oil shale requires types of innovation of which industry is instinctively wary, and the lead-time will be much longer than originally anticipated. In the meantime we need additional hydrocarbon reserves immediately.

The Gasbuggy and Rulison explosions were set off in types of sandstone that produce oil in other areas. These sandstones are more amenable to production than are oil shales, but after several years the gas at both Rulison and Gasbuggy is still too radioactive to produce. If these sites do eventually give a satisfactory yield, the next attempt may be made in a good oil shale formation. Project Wagon Wheel and Project Wasp in west-central Wyoming, still in the planning stage, are similar to Gasbuggy and Rulison. Project Thunder-

bird is a plan for a nuclear detonation in a 200-foot-thick coal bed at depth in the Powder River Basin of northeast Wyoming. It is expected that the heat and fracturing will produce accumulations of coal gas that can be drilled by conventional methods. Whether by the Thunderbird plan or by conventional mining and treatment, the production of gas from coal looks more economical than the production of gas from oil shale.

Nuclear power

Because of other energy shortages, Great Britain embarked on a nuclear power program earlier than did most other nations, but some reports on her progress in developing nuclear electric power are much too favorable. Furthermore, the discovery of gas in the North Sea has raised other expectations and slowed Britain's nuclear

Nuclear power plant

efforts. France and the smaller nations of Europe had planned to follow Great Britain's lead, but the prospect of North Sea and Siberian gas is slowing down nuclear development in these nations also.

The United States had started or ordered 125 nuclear power plants as of October 1971; 49 were on order, 55 under construction (one in Puerto Rico), and 21 in operation, though only a few if any were operating at full potential (Figure 2-17). In 1968 there was much talk about our entrance into the nuclear age. But in 1969 all discussion of nuclear-electric plants suddenly disappeared from the public press.

Various reasons have been given for the slowdown in nuclear electric construction, including disagreements over thermal pollution and the alienation of nearby communities. A more likely explanation is that

Figure 2-17

Status of the nuclear electric program in the United States, October 1971.

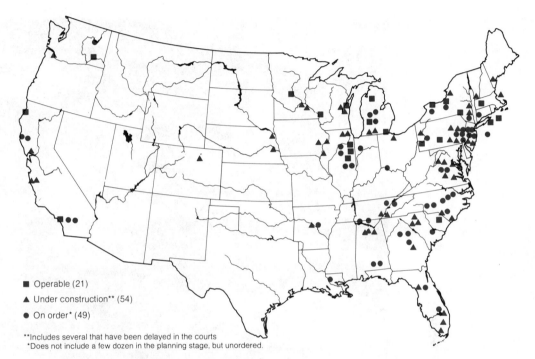

■ Operable (21)

▲ Under construction** (54)

● On order* (49)

**Includes several that have been delayed in the courts
*Does not include a few dozen in the planning stage, but unordered.

uranium ores were in short supply at quoted prices. In 1969 it was thought that ore at $8 a pound was competitive with coal, and that at $10 a pound it was not. An eight-year supply of fuel should be under contract when a nuclear plant begins to operate; an eight-year supply at $8 could not be guaranteed after 1969, nor could an eight-year supply at $10 be guaranteed after 1974. Most plants had been projected on the basis of $8 ore. Hence the slowdown in nuclear construction in 1969.

But planning for the construction of nuclear electric generating plants showed a new spurt in 1970. The cost of coal to the electric generating industry had gone up, partly as a result of wage increases and partly due to increased costs resulting from the Mine Safety Act of 1969. The shortage of natural gas gave further impetus to the planning of nuclear electric installations. Thus, while $10 fuel is projected only to 1990 (provided we double known reserves), the cost of nuclear fuel is a much smaller item in the total cost of electricity than is the cost of coal or gas.

Nuclear reactors called fast breeder reactors, which produce more nuclear fuel than they use, are now being designed. A pilot plant is projected for the United States in the early 1980s. If this type of plant is successful, nuclear fuel reserves will cease to be a problem.

The future of nuclear energy production

Cooling towers of an atomic
power plant

in the world is little different from that in the United States, and uranium and thorium ores are so unevenly distributed that most nations have none at all. A true international market for nuclear fuels is essential to nations without them. Yet national fears, prides, and jealousies are not likely to abate in time for an international nuclear fuel market to be useful.

We are now beginning to hear about energy from controlled fusion—energy released when two atoms unite. The atoms of deuterium, tritium, and lithium are susceptible to fusion. But even if experiments continue as planned, a pilot plant cannot be expected before 1995, nor commercial use before 2005.

Until the engineering of fast breeder reactors is made completely safe, the future of nuclear energy is not bright, despite opportunistic magazine articles and newspaper headlines.

Hydroelectric power

Much of the world's water power has already been developed. Areas near the mountains of central Europe owe much of their early economic development to water power. Central Africa, central Asia, and central South America still have tremendous

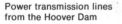
Power transmission lines from the Hoover Dam

Source of hydroelectric
power, T.V.A. Wilson Dam,
Alabama

undeveloped potential. Hydroelectric production of power can be more than doubled
in the United States; however, many of
the best hydroelectric sites in the United
States are in wilderness areas, such as
Hells Canyon, on the boundary between
Idaho and Oregon, and the Salmon River
in Idaho. Few people really want such
scenic areas "developed." Hydroelectric
power represents a declining percentage of
the total energy consumed in the U.S.

At a few places in the world, unusually
high tides funnel through narrow entrances
to sizeable bays: Passamaquoddy Bay in
Maine, the Bay of Fundy in Nova Scotia,
and the Rance River Estuary near Brest,
France, are the most notable examples. The
French site has the only operating tidal-
electric plant in the world, though a joint
U.S.-Canadian plant for the Passamaquoddy
has been discussed. The potential of tidal
energy is not overwhelming, and some projects have been disappointing, usually because of faulty design. Yet the world may
someday need every watt of clean energy it
can produce.

Solar energy

Since photosynthesis consumes less than 0.3 percent of the solar energy that reaches the earth, and about 28 percent is sufficient to operate the earth's heat machine, almost 72 percent of our solar energy—or about 180,000 calories per square centimeter per year—is returned to space. This is a tremendous reservoir of energy that can be drawn upon in the future without disturbing the ecosphere. Whether it will eventually be converted to human use by means of earth sites or space stations is still uncertain. A pilot plant has started operation near Tucson, Arizona, to test an earth site. In addition, experiments to test the practicality of units to collect the sun's heat for

Solar power generator in the French Pyrenees

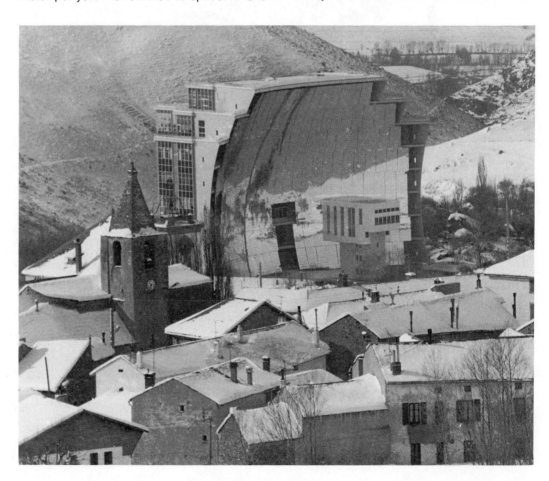

Rooftop television antennas and solar water boilers, Israel

Geothermal power generated from volcanic steam, Baja California, Mexico

individual residences and other buildings are being conducted at Pennsylvania State University and in Washington, D.C.

Geothermal energy

All geothermal energy is produced by steam, from sources that may be shallow or deep. Shallow sources give rise to live steam and geysers at the surface, and steam can be tapped by drilling shallow wells. Many geyser areas—including Italy, Iceland, New Zealand, Japan, Kamchatka, and Sonoma County, California—are now making use of steam energy or planning to do so.

There are many areas where geothermal potential is not evident on the surface, but may still be found accessible by drilling, usually to greater depths than is necessary

Figure 2-18

Model of a geothermal operation.

Figure 2-19

Increase in daily energy consumption from 1970 to 1980.

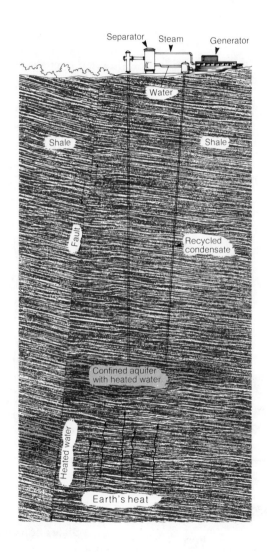

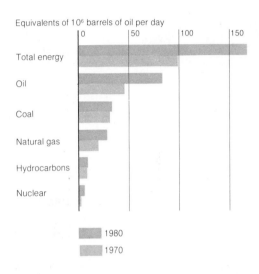

Equivalents of 10^6 barrels of oil per day

water hot enough to produce steam, then energy can be produced. Two holes are drilled to the formation (Figure 2-18). One yields steam, and the other returns the spent water to the producing formation after its heat has been used for power. The recycled water thus provides pressure so that steam can be produced continuously, and the water is recycled so that the supply is not depleted. Even in developed areas of favorable potential, geothermal energy will probably not supply more than 5 percent of total energy needs.

Summary

With regard to many commodities, the United States has become a have-not nation. Our balance of trade suffers increasingly from the need to purchase and import foreign raw materials. The more this deficit is met by recycling, the sounder will be our foreign exchange position. The

in geyser areas. In fact these areas may be a richer source of thermal energy than known geyser and steam areas. If it is possible at economically feasible depths to drill to a permeable bed that contains

Figure 2-20

Source of energy during the decade 1971 to 1980.

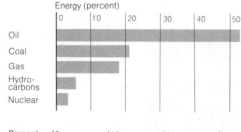

Figure 2-21

Percentages of energy consumed by selected parts of the world.

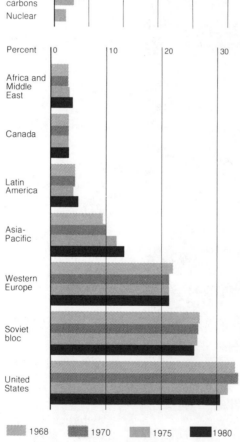

The overall supply of energy from known sources should last for several generations, but the next two to four decades will be critical. Figures 2-19 and 2-20 indicate a tenfold increase in nuclear energy produced in the next decade, but by 1980 nuclear sources will still supply only a small fraction of the world's power.

Although liquid hydrocarbons cannot be included in long-term planning, they will continue to be the primary source of energy through the next decade or two, even though many areas, including the United States, will experience shortages. Synthetic gas, tar sand, and shale oil technologies have not developed rapidly enough to bridge this gap within the next ten years. As yet there has been little planning to counter the even greater decline of liquid hydrocarbons in the period to follow.

Already the Federal Power Commission has been asked by one northeastern state to divide the remaining gas fairly among the states. This would mean that citizens of oil-producing states, who have experienced pollution for years, would be required to endure a hydrocarbon shortage in order to send gas to states whose administrators will not allow drilling because of potential pollution.

The percentage of the world's energy consumed by the United States is decreasing, and will decrease further by 1980. In certain years since World War Two, the United States consumed as much as 55 percent of the world's production of energy. By 1970 this figure had dropped to 42 percent. And it is estimated that by 1980 the United States and Canada together will consume only about 33 percent (Figure

shortage of energy is our most severe problem, but many other commodities are almost equally scarce.

2-21). Since our actual consumption has continually increased and will continue to increase, the consumption of the remainder of the world must be rising even more rapidly than ours. This means that our standard of living, even if it rises, will be more nearly in line with that of the rest of the world in the 1980s.

If we are to prevent mankind from lapsing into a new Stone Age, and at the same time preserve the environment, the most desirable sources of power are those derived from the sun and from nuclear fusion. Both are clean. But the extensive use of solar energy is still two to four decades in the future. And energy from fusion appears to be at least four decades away, even if the engineering problems can be solved. In the meantime, we shall no doubt have to go on using all the sources of energy available to us, even though they are not always desirable. Hydroelectric and geothermal sources are not large. Fossil fuels produce dangerous sulfur products and toxic metals. And nuclear fuels can produce damaging radiation. But until better sources are developed, they are all we have.

Selected Readings

Abelson, Philip M. 1968. The inexorable exponential. *Science* 162:11.

Barnea, Joseph. 1972. Geothermal power. *Scientific American* 225(1):70–77.

Charlier, Roger H. 1969a. Harnessing the energies of the ocean, part I. *Mar. Tech. Soc. Jour.* 3(3):13–32.

———. 1969b. Harnessing the energies of the ocean, part II. *Mar. Tech. Soc. Jour.* 3(4):59–81.

Cloud, Preston E., Jr., et al. 1969. *Resources and Man.* San Francisco: W. H. Freeman.

Coppi, Bruno and Jan Rem. 1972. The tokamak approach in fusion research. *Scientific American* 227(1):65–75.

Hilderbrandt, A. F.; G. M. Haas; W. R. Jenkins; and J. P. Colaco. 1972. Large-scale concentration and conversion of solar energy. *Eos*–:684–692.

Lovering, T. S. 1968. Non-fuel mineral resources in the next century. *Texas Quarterly* 11(2):127–147.

Manners, Gerald. 1964. *The Geography of Energy.* Chicago: Aldine.

McKetta, John J. 1972. U.S. warned about energy chaos. *World Oil* June:76–78.

National Petroleum Council. 1972. U.S. energy outlook; a summary report of the National Petroleum Council. Washington: National Petroleum Council.

Owings, M. J. 1972. Petroleum liquids in energy supply and demand—some significant influences. *Journal of Petroleum Technology* May:521–529.

Risser, Hubert E. 1971. Fossil fuel supplies and future energy needs. *Eos* 35:763–767.

Sherwin, R. J. 1972. Energy—major sources and consumption. *APEA Journal* –:102–107.

Singleton, Arthur L. 1968. Sources of nuclear fuel. U.S. Atomic Energy Commission, Division of Technologic Information.

Weinberg, Alvin M. 1972. Social institutions and nuclear energy. *Science* 177:27–34.

Chapter 3
The Air We Breathe

Hell is a city much like London—a populous and smoky city.
Percy Bysshe Shelley, 1815

Polluted air is nothing new. In the time of Christ, the poet Ovid complained of the sooty smoke that filled the city of Rome. In the summer of 1257 Queen Eleanor of England left Nottingham, where air was so foul with coal smoke that she declared it uninhabitable. By the end of the thirteenth century, smog had become so severe in London that Edward I issued a proclamation drastically curtailing the use of coal—the first attempt by an English government to control pollution.

In the next two centuries, at least three attempts at controlling pollution were made by Parliament, but all were unsuccessful. Killer smogs were known in London in the seventeenth century; hundreds of people died in the foggy winter of 1873–1874, and again in 1952. The adjacent mainland has also been affected; on December 3–5, 1930, hundreds were killed in the Meuse Valley in Belgium. Before the cause of these deaths was ascertained, rumors attributed them to a new plague. During the first 600 years in the history of London smog, no effective controls were enacted. And, although Parliament passed several bills in the nineteenth century, none was effective.

In time the New World experienced the same blight. Early sailing ships traveling from Europe to North America knew when they were approaching land by a pink tint in the sea foam, produced by pollen carried to sea by the prevailing westerly winds. In the present half-century, an even stronger sign of approaching land is the plumes of black smoke tailing downwind from our cities, carried to sea by the prevailing westerlies. Thirty years ago the Atlantic coastal plain of the United States was recognizable from

View of New York City
enveloped in smog

the air because only there was the ground
obscured by smog at 5000 feet. Today the
whole country, except for the mountains,
is hard to see from the air unless the atmosphere has recently been cleansed by rain
or snow.

Origin of the Earth's Atmosphere

It is now widely assumed that the solar
system condensed as a result of mutual
attraction among particles within a vast
cloud of interstellar dust. As the matter
in the cloud became more and more dense,
gravitational forces produced the sun and,
together with the great centrifugal force
generated by the rotation that began when
the sun's substances collided, at least some
of the planets. These events probably did
not all happen concurrently throughout the
solar system, nor were they necessarily
identical for the sun and all the planets.
In the sun the pressure of condensation
generated such intense heat that it became,
and remains, a gaseous and molten mass.
The outer planets probably never achieved
the density of the sun, or even of the inner
planets, and are still composed largely of
gases.

According to this theory, about 4.5 billion years ago Earth, like the outer planets,
evolved an atmosphere of methane (CH_4),
ammonia (NH_4), helium (He_2), and hydrogen
(H_2), with traces of the inert gases argon,
neon, and krypton, in proportions similar
to those in interstellar space. The fact
that the ratios of these gases in the earth's
atmosphere are now quite different from
those of the atmospheres of the outer
planets and interstellar space has been
used to argue that earth at some point lost
its original atmosphere and later developed a new one. Since the atmospheres of
Mars and Venus also differ greatly from
those of the outer planets, it is thought
that those planets too may have lost their
original atmospheres.

One explanation for the loss of earth's
original atmosphere is that the forces of
gravity that the materials of the young
planet exerted on each other were so great
that earth melted from the resulting heat.
When it melted, the heavier elements—
such as iron, nickel, cobalt, and chromium
—sank to earth's center, while the lighter

Figure 3-1

Major steps in the evolution of the atmosphere.

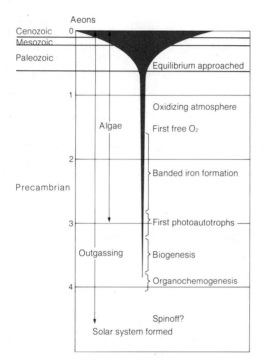

Indeed, it is thought that one or more meltings or partial meltings may have taken place since the original thermal event. This process, known as outgassing—the emission of gases generated by liquid or solidifying magma following any stage of melting—is thought to have produced earth's present atmosphere. The process is still going on.

The hypothesis that outgassing produced earth's present atmosphere is about 100 years old, and while it is more widely accepted today than in earlier decades, there are still objections to it. Some scientists question whether the centrifugal force generated by increased rotation was sufficient to let lighter elements escape from earth's field of attraction. Proponents of this view propose instead that sunlight broke down water vapor, and that the hydrogen thus released escaped into space, leaving free oxygen. For whatever reason, our atmosphere is different from those of the outer planets, and seems to have changed in composition over time. Gases important in our atmosphere, such as water vapor (H_2O), sulfur dioxide (SO_2), carbon dioxide (CO_2), carbon monoxide (CO), nitrogen (N_2), hydrogen chloride (HCl), and a few rarer ones, can be obtained from hot or molten rock materials within the earth, and are emitted by volcanoes. The best available evidence, combined with the best geochemical theory, indicates that earth's atmosphere was produced by just such gaseous emissions.

Evolution of the Earth's Atmosphere

Earth's atmosphere, then, has evolved over a period of about 4.5 billion years. During that time, one—and perhaps more—atmo-

elements moved outward, forming layers of varying density. As this process continued, the speed of rotation would have increased, for the same reason that a spinning ice-skater revolves more rapidly as he brings his arms inward toward his body, applying more "spin" to a less widely distributed mass. According to this theory, the increased speed of rotation would provide enough centrifugal force to spin off the light gases of the original atmosphere. The earth would thus have been without an atmosphere for a time (Figure 3-1).

The heat generated as earth became more compact, and as the heavier and lighter elements in it separated out into "layers," initiated intense thermal activity, which persists in the form of volcanism.

Table 3-1

Constant components of the atmosphere (dry, by volume) and variable components of the atmosphere. Constant components are from Brian Mason, *Principles of Geochemistry*, 3rd ed. (New York: John Wiley & Sons, 1966, but converted to percentages. Variable components are compiled mostly from Bernard D. Tebbens, "Gaseous Pollutants in the Air." In Arthur C. Stern, ed., *Air Pollution*, vol. 1 (New York: Academic Press, 1968).

Constant Components of the Atmosphere

Substance	Symbol	Percent
Nitrogen	N_2	78.09
Oxygen	O_2	20.95
Argon	Ar	0.93
Carbon dioxide	CO_2	0.03
Neon	Ne	0.0018
Helium	He	0.00052
Methane	CH_4	0.00015
Krypton	Kr	0.00010
Hydrogen	H_2	0.00005
Nitrous oxide	N_2O	0.00005
Xenon	Xe	0.000008

Variable Components of the Atmosphere

Substance	Symbol	Amount
Water vapor	H_2O	1–3 percent
Hydrogen peroxide	H_2O_2	trace
Ozone	O_3	0–1.0 ppm
Ammonia	NH_4	0–3.0 ppm
Sulfuretted hydrogen	H_2S	trace to 1 ppm
Sulfur dioxide	SO_2	0–3.2 ppm
Sulfur trioxide	SO_3	trace
Carbon monoxide	CO	0–65 ppm
Hydrocarbons		0–4.6 ppm
NO and NO_2		0–3.0 ppm
PAN		0–0.00035 ppm
F and HF		0–0.018 ppm

spheres were lost entirely, and one or more were created, probably by a combination of outgassing and photosynthesis, a process performed by green plants which will be discussed shortly. It is generally assumed that between 0.6 and 1.0 billion years ago—some time before the Cambrian—the atmosphere reached its present composition, and has remained in a state of near-equilibrium ever since.

The primary components of the present atmosphere are nitrogen, oxygen, argon, and carbon dioxide (Table 3-1), which make up 99.99 percent of its total mass. The nitrogen and carbon dioxide are probably products of outgassing, and the free oxygen was probably produced by living organisms.

The Atmosphere and Living Things

Biogenesis, the generation of living things, took place nearly three and a half billion years ago. It is possible that the first photoautotrophs, organisms chemically able to transform light into energy, appeared nearly three billion years ago. If these organisms were not direct ancestors of green algae, and thus of green plants, they at least began the gradual alteration of the atmosphere by producing oxygen. These primordial forms probably preceded the algae, the next level of photosynthesizers, by three to four hundred million years.

The process of photosynthesis can be expressed by the formula:

$$6\ CO_2 + 12\ H_2O = C_6H_{12}O_6 + 6\ O_2 + 6\ H_2O.$$

A plant is a chemical factory that transforms carbon dioxide from the atmosphere, and water from its own substance, into (1) glucose, a carbohydrate that becomes the main source of energy for all forms of animal life; (2) oxygen, which it releases into the atmosphere; and (3) more water. No chemical reaction can take place without energy; the energy that plants use in photosynthesis is radiant energy, which they obtain from sunlight.

But this reaction is reversible, and if the

oxygen were not removed, the plant would oxidize its own cells—that is, the oxygen and glucose would be transformed back into carbon dioxide and water, and the whole process would end in no net gain. Modern plants possess an oxygen-mediating enzyme, a substance that makes it possible for a chemical process or transfer to take place. This enzyme traps the oxygen, transfers it to the atmosphere, and prevents self-oxidation in plants.

Some scientists believe that primitive photosynthesizers did not have an oxygen-mediating enzyme, but were able to get rid of the oxygen resulting from photosynthesis because of its strong affinity for iron. Evidence supporting this belief is found in a phenomenon known as banded iron formation (BIF). Apparently, iron in the environment united with and fixed the oxygen, making it unavailable for further reactions. The earliest known banded iron formation, found in South Africa, has been dated at more than three billion years old, and may mark the appearance of the first organisms capable of photosynthesis. BIF is restricted to the Precambrian shield areas of the world (Figure 3-2). The end of banded iron formation could similarly be taken to mark the evolution of an oxygen-mediating enzyme, indicating that green plants no longer depended on iron to relieve them of their oxygen, and could release it directly into the atmosphere. This development in turn marked the beginning of the modern atmosphere, with its large oxygen component. Thus, for almost two billion years the free oxygen produced by photosynthesis may have been consumed by ferrous iron to produce BIF.

About 1.5 to 1.8 billion years ago the deposition of BIF ceased, except for isolated local occurrences, and oxygen began accumulating in the atmosphere. The amount of carbon dioxide decreased relative to oxygen, and ozone (O_3) increased.

Figure 3-2

Shield areas of the world.

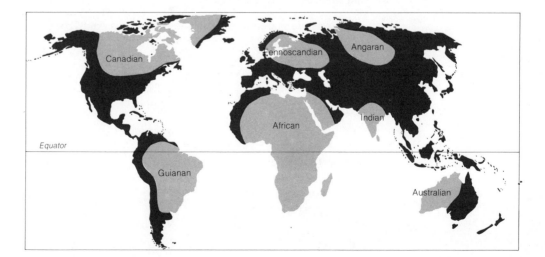

Red beds, colored by iron more thoroughly oxidized than in BIF, were deposited, and oxygen became available for the deposition of carbonate rocks $CaMg(CO_3)_2$ and $CaCO_3$, which require the addition of oxygen to carbon dioxide to form the carbonate radical CO_3. All these phenomena are evidence of the addition of free oxygen to the atmosphere. Free oxygen also made possible the development of animal metabolism, for while plants take in carbon monoxide and release oxygen as a waste product, animals consume oxygen and give off carbon dioxide.

Thus our atmosphere evolved over a period of some 4.5 billion years. Earth's present atmosphere has remained in a state of near-equilibrium since its evolution some time before the Cambrian, except for variations in carbon dioxide content of plus or minus 25 percent (e.g., between 0.025 and 0.04 percent of atmosphere in dry weight).

Can Man Destroy the Equilibrium?

Earth's atmosphere appears to have been reasonably stable for the past half billion to billion years; man is today pouring harmful substances into it in alarming quantities. To the chief present components—nitrogen, oxygen, argon, and carbon dioxide—we are now adding at unprecedented rates: carbon dioxide, nitrous oxide, ozone, hydrogen peroxide, ammonia, hydrogen sulfide, sulfur dioxide, sulfur trioxide, carbon monoxide, hydrocarbons, aldehydes, nitric oxide, PAN (peroxyacetylnitrate), fluorine, hydrogen fluoride, the products of chemical and biological warfare agents, and many artificial radionuclides. Some sources of major pollutants are listed in Table 3-2.

Effects on health

Many pollutants, particularly the oxides of nitrogen, the oxides of sulfur, carbon monoxide, the hydrocarbons, and the aldehydes, are serious health hazards in our major cities, especially to tobacco smokers, cardiovascular patients, and people who already suffer from diseases of the respiratory system. Ozone and oxides of nitrogen react

Table 3-2

Sources of some major atmospheric pollutants.

Source	Particulate	Ammonia	Oxides of nitrogen	Oxides of sulfur	Carbon dioxide	Carbon monoxide	Hydrochloric acid	Hydrofluoric acid
Inorganic chemistry industry		X	X	X	X	X	X	X
Stationary combustible sources				X	X	X		
Incinerators			X			X		
Gasoline and diesel engines			X		X	X		
Fly-ash (coal-fired plants)	X			X				
Blast furnaces	X							
Metallurgical industries	X			X				
Pulp and paper industries	X			X				
Refineries			X		X			
Plant and animal waste		X	X					

Air pollution in industrial
area of West Germany

with hydrocarbons and aldehydes to form
organic compounds that are more injurious
to the health of animals and plants than are
the separate compounds. Other air pollutants may also damage animals and plants.

Some Specific Pollutants

Carbon monoxide

Carbon monoxide is a highly poisonous
gas emitted in large quantities by automobile engines and composed of equal parts
of carbon and oxygen. There are about
530×10^6 tons of CO in the atmosphere,
amounting to approximately 8×10^{-12} percent of the atmosphere by weight. The
amount of CO added to the atmosphere
annually by man is about 270×10^6 tons,
and added by nature is between 6×10^9 and
7×10^9 tons. That is, the production of CO
by nature is over 20 times that by man.

Carbon monoxide is unstable and oxidizes,
when oxygen is available, to form carbon

dioxide (CO_2). Thus the residence time of CO in the atmosphere is only about one month. This is why billions of tons of CO can be added to the atmosphere annually, and yet the atmospheric content will still remain at only a little over one-half billion tons. Because of the short residence time there is little danger of man overloading the atmosphere with carbon monoxide.

The residence time of carbon monoxide in the atmosphere is long enough, however, for local concentrations to constitute a health hazard to man and animals. Sources of high local concentrations of CO are shown in Table 3-2. Published evidence shows a considerable decrease in CO in downtown New York City since 1900. In that city there were 379 accidental deaths from CO poisoning in 1945, and only three in 1970. This finding has been used as evidence for a continuing decline in CO. The decrease in CO, and in accidental deaths, is correlated with the decrease in the use of anthracite (hard coal), which burns less completely in stoves and furnaces. Although the suggested maximum exposure to CO varies according to the reporting organization, it is generally suggested that after exposure to 30 ppm (parts per million) or more for as long as 8 hours, one should stay away from CO pollution for the next 24 to 36 hours.

Carbon monoxide occasionally reaches dangerous levels in some cities, and it is said that CO concentrations along heavily travelled highways are high enough to produce adverse physical effects. These include delayed reaction times after several hours of driving, increasing the probability of accidents.

Attempts to control CO emission by the use of antipollution devices on automobiles cannot be completely satisfactory, because frequent tuneups are necessary to keep recyclers and catalytic converters efficient, and many people cannot afford to have their cars tuned up every two or three months.

Sulfur

As an atmospheric pollutant, sulfur appears both in hydrogen sulfide (H_2S) and in oxides of sulfur (SO_2, SO_3), although SO_3 is rare. By reacting with oxygen, H_2S becomes sulfuric acid (H_2SO_4); by reacting with atmospheric moisture (H_2O) and oxygen, sulfur dioxide (SO_2) also becomes sulfuric acid. Sulfuric acid is highly corrosive, and damages stone buildings, gravestones, statuary, and paint. The corrosive action of sulfur is causing ancient and Renaissance marble statuary in Italy to deteriorate at an alarming rate.

Sources of atmospheric H_2S and SO_2 are volcanoes, decaying organic matter, organic wastes, and the burning of fossil fuels. SO_2 constitutes about 95 percent of sulfur emissions into the atmosphere.

Total sulfur entering the atmosphere is about 550 million tons per year, over two-thirds of which is in the northern hemisphere. About 70 percent is natural and 30 percent manmade.

Massive destruction of trees near Butte, Montana, and Ducktown, Tennessee, was caused by oxides of sulfur, in both areas from stack pollutants given off by copper smelters. It is said that only two conifer trees in St. Louis, Missouri, predate the use of coal as a residential fuel. All others

Bronze statues eroded by
air pollution, Venice, Italy

Paris covered with a smog layer

were killed by sulfur oxide fumes from the chimneys of the city. Conifers are much less resistant to sulfur than are deciduous trees. The mass deaths of men, plants, and hundreds of thousands of animals during the blue haze volcanic eruption of 1783 in Iceland are thought to have been due to SO_2. Many animals not killed by the volcanic fumes died of starvation because the plant life on which they fed had been destroyed.

The effect of sulfur on man is insidious. In SO_2 pollution events, people with bronchial and respiratory diseases are the first to suffer and die. Cardiovascular patients are next. Although such deaths are often attributed to the primary disease, it is obvious that the incidence of death among the elderly and ill is greatly increased by SO_2 pollution.

Sulfur compounds enter the air from ocean dispersal, volcanoes, decaying organic matter, and the burning of fossil fuels. The atmosphere is cleansed primarily through the hydrologic cycle, to be discussed in Chapter 5. Rainfall washes sulfur compounds out of the air into streams and eventually into the oceans. From the oceans they can again be dispersed into the atmosphere, or deposited as sediment to enter the rock and crustal cycles.

As fossil fuels, especially coal, take on increasing importance as a result of the energy shortage of the next two decades, sulfur pollution will become more severe.

The total effect of sulfur on the ecosphere is slight. Fortunately, the residence time for sulfur in the atmosphere is short and, from the standpoint of a healthy ecosphere, the atmosphere is the best sink for sulfur pollutants. Yet local health hazards to animals and plants cannot be ignored. This is why the development of some 500,000 megawatts of coal-fired electric power in the next two decades is planned for relatively unpopulated areas.

Most sulfur can be removed from fossil fuel stack gases and smokes, but there is a serious question about what to do with it. Already, by removing sulfur from hydrocarbons, Canada and France alone have flooded the market and depressed the price of sulfur from $40 a ton in 1969 to $6 a ton in 1972. As we recover more sulfur, we will not be able to give it away. What does one do with unwanted sulfur? To leave it lying on the ground is to allow it to pollute the surface or ground water in far greater quantities than it would if processed into the atmosphere. Someday we may be forced to bury excess sulfur in impermeable shales where it cannot escape into the ecosphere.

Lead

Young child peeling lead-based paint

From 200,000 to 300,000 children in the United States suffer from mental deficiency because of lead poisoning. The usual explanation for this tragedy is that hungry young children eat leaded paints peeling from the walls of houses in poorer neighborhoods.

It has recently been shown that less than 2 percent of lead that is eaten, and about 2 percent of lead that is consumed with liquids, actually enters the human system, but that 40 percent of lead that is inhaled enters the system (Figure 3-3). Thus a number of experts believe that aromatic leaded ethyl gasoline from automobile engines plays a larger role in lead poisoning than had been supposed. The poorer areas of cities usually have the highest automobile pollution. Futhermore, in such areas children play in the streets, near the pavement, where the heavier leaded ethyl gases are likely to accumulate.

Figure 3-3

Percent of consumed lead absorbed by the human body.

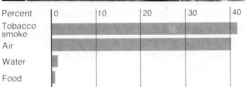

Percent	0	10	20	30	40
Tobacco smoke					
Air					
Water					
Food					

Ethylated lead is added to gasoline to make it fire better in the automobile engine. Fossil fuels also contain some natural lead and cadmium, which are released into the atmosphere. The quantities of these substances have not been closely studied, but they would probably be insignificant if leaded ethyl were removed from motor fuels. If not, the increment from fossil fuels may be the amount that alters the lead content in some localities from tolerable to critical. The Environmental Protection Agency and automobile manufacturers are now attempting to force the removal of lead from gasoline. This will be a tremendous improvement, but will require alterations in refineries around the world that will cost hundreds of millions of dollars.

Mercury

Mercury is another product of fossil fuels. Illinois coals contain from 100 to 500 ppb (parts per billion) of mercury, and average about 300 ppb. Coals from the western states contain less mercury than do Illinois coals, and have a low sulfur content. Conservative scientists disclaim knowledge of the state of mercury in pollutants emitted by fossil fuel installations, but burning mercury with carbon and hydrogen is most likely to produce methylated mercury (CH_3HgCH_3), which enters animal systems much more easily than do other mercuric compounds. Lead and mercury are discussed further in Chapter 8.

The Economics of Air Pollution

Many other pollutants in addition to sulfur dioxide damage or kill plants. Sulfur dioxide kills cotton and damages wheat and barley. Ozone damages spinach, parsley, alfalfa, rye, barley, wheat, and carnations. Fluorine damages chrysanthemums, gladioli, apricots, and citrus fruits. Hydrogen chloride and hydrogen fluoride damage trees, and ammonia damages cotton. In Los Angeles, 40–60 percent of the roses, carnations, and orchids are affected by pollution. As many as 25,000 acres of pine trees in San Bernardino County have been damaged by smog, and the economic loss from air pollution damage to citrus orchards in California alone amounts to more than $10 million a year. The total economic loss from pollution of plants in the United States is well over $500 million a year, of which one-fourth is in California.

Except for damage to plants and hazards to health, the economic consequences of air pollution have not yet been fully examined. Many of the oxides of nitrogen and sulfur react with water to produce corrosive acids. These react with metals, paints, stone, brick facings, and various other surface coverings.

The corrosive and erosive effects of man-made acids, excluding those that occur in nature, cost hundreds of millions of dollars annually in the United States alone. Although much of the native and artificial stone used in buildings weathers under natural conditions, weathering is much more rapid and severe if there are small amounts of sulfuric, nitric, hydrofluoric, or hydrochloric acid in the air.

An often cited example of the effect of climate in weathering is the obelisk referred to as Cleopatra's Needle. This obelisk had stood in the dry climate of Alexandria, Egypt, for about sixteen centuries before being transferred in 1878 to New York City's Central Park, where in slightly less than one hundred years it has weathered far more than it did in the previous sixteen centuries.

In recent decades the deterioration of statuary and buildings has increased, especially in industrial areas, and local atmospheric pollution has been blamed. The problems are both economic and aesthetic. Although organic acids may be partly responsible, the primary corrosive agents are carbonic acid (H_2CO_3), nitric acid (HNO_3), and sulfuric acid (H_2SO_4), derived respectively from the reaction of water (H_2O) with carbon dioxide (CO_2), nitrogen dioxide (NO_2), and sulfur dioxide (SO_2) plus oxygen (O_2). Hydrochloric acid (HCl) and hydrofluoric acid (HF) are also potent atmospheric weathering agents.

Limestone is the building material most susceptible to these acids, and especially to the most abundant of them, carbonic acid. But even if the carbon dioxide in the atmosphere has increased by 10 percent in the past century, such a small increase cannot account for the sudden spectacular increase in the rate of weathering. It seems more likely that sulfur dioxide from fossil fuels is the primary agent responsible for defacing limestone buildings and Renaissance and Hellenic statuary in Italy:

$$(2\ SO_2 + O_2 = 2\ SO_3 + 2\ H_2O = 2\ H_2SO_4)$$
$$(H_2SO_4 + CaCO_3 = H^+ + HCO_3^- + Ca^{++} + SO_4^{--})$$

The latter four ions wash away in rainwater. In other words, sulfur dioxide, water, and oxygen produce sulfuric acid, a very corrosive acid, which in turn etches away the limestone and marble of structures and statuary. Similar reactions occur with the other acids. There is even a fad among some architects to use those limestones that weather most rapidly. Such buildings lose their "newness" more quickly, adding to the owner's prestige and, in the opinion of some, the buildings' beauty.

Some air pollutants, particularly hydrogen sulfide, are harmful to paints, particularly lead paints with lead carbonate or lead sulfate as white pigments. Lead sulfide is black, and lead paints turn an ugly gray or black under reducing environments in the presence of sulfur and some sulfur compounds. This phenomenon can still be observed on houses painted with lead paints and located downwind from pulp mills or other sources of sulfur pollution. This and the problem of lead poisoning have led to a search for substitutes for lead paint. Replacements equally resistant to general weathering were difficult to develop, but titanium oxide pigment is now used more successfully than other substitutes to replace white lead pigments.

During the 1930s combinations of sulfuric, nitric, and organic acids contributed to the obsolescence of rayon, a synthetic fabric, for general use. In heavy smog, rayon reacted with these acids to give way and even to melt. Women found their stockings and blouses deteriorating before their eyes. Nylon much more effectively resists strong acids.

Hydrochloric, nitric, and hydrofluoric

acids are also very corrosive weathering agents. No reliable estimate exists of the economic damage to buildings and other structures, or the costs of repainting, resulting from weathering by air pollutants. But it must be tremendous. The aesthetic damage to statuary, art, buildings, and even ornamental walls and fences can be appreciated by all who see Cleopatra's Needle in New York City, the deteriorating statuary in several Italian cities, or structures and statuary in other heavily polluted areas.

Long-range Geologic Effects of Air Pollution

The long-range geologic effects of atmospheric pollution are thought to be (1) diminution of the supply of oxygen and (2) climatic change. Some people have feared that we might add so much carbon to the atmosphere and use up so much oxygen as to endanger animal life. But it seems unlikely that we will deplete the supply of oxygen. There are several thousand times as much oxygen as carbon in the atmosphere; even if the most dire predictions came true, there would be several thousand times as much oxygen as carbon dioxide

in the atmosphere even after several centuries of burning fossil fuels at present rates. Climatic change, however, is a remote possibility.

The greenhouse effect

When solar energy of short wavelength strikes earth, some of it is reflected back in longer wavelengths. Energy of longer wavelength does not readily escape the lower atmosphere, and some of it is thus reflected back to earth and retained near the surface as heat energy. Clouds, carbon dioxide, or even smog can scatter some long-wave terrestrial radiation and retain it near the earth (Figure 3-4). The greater the density of the atmosphere, the more energy will be reflected back to earth, and the greater will be the accumulation of heat.

Many scientists have maintained that the continued addition of carbon dioxide to the atmosphere will increase its density and allow less and less radiant energy to escape from the atmosphere, thus raising the temperature on earth. This has been termed the greenhouse effect, since the glass in a greenhouse lets the sun's rays in but doesn't let all of the heat back out again.

Figure 3–4

Atmosphere with a higher density of heavy molecules has a greater retention of heat energy near the surface of the earth. (Copyright © 1971 by W. W. Norton & Company, Inc.)

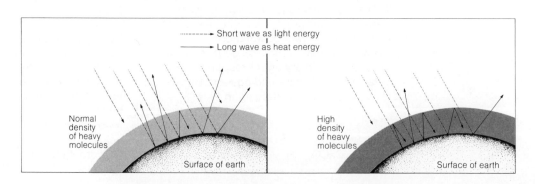

Figure 3-5

Winter temperature increases for different parts of Europe since 1750. Each division represents one degree centigrade.

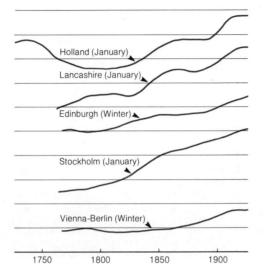

Holland (January)

Lancashire (January)

Edinburgh (Winter)

Stockholm (January)

Vienna-Berlin (Winter)

1750   1800   1850   1900

It has been postulated that the 4–5 degree rise in temperature recorded during the last century at Ottawa, Pittsburgh, in North Africa, and in other parts of the world is the result of increased CO_2 in the atmosphere since the beginning of the Industrial Revolution. That most of this increase in temperature is local now seems assured, although Emmanuel LeRoy Ladurie, in a study of Europe's climate over the last thousand years, demonstrates some general warming in the last century and a half (Figure 3-5). It has also been pointed out that the warming effect of increased carbon dioxide in the atmosphere is offset by the cooling effect of increasing pollution, by droplets or particles of less than twenty-five thousandths of an inch in diameter, collectively called aerosols

The amount of CO_2 that will be added to the atmosphere in each of the next dozen years can be estimated by comparing the amount of fossil fuel used in any recent year with the estimate for 1985, assuming that the amount of CO_2 produced by burning fuel is one-tenth of its weight. The mass of the atmosphere is about 5.7×10^{16} tons. The amount of carbon dioxide in the atmosphere is about 1.88×10^{13} tons. In 1960 we added about 31×10^8 tons of carbon dioxide to the atmosphere, not counting that added by the burning of wood for fuel. Thus in 1985 we will add about 45×10^8 tons of carbon dioxide to the atmosphere. If this amount remained steady, in the 100 years from 1985 to 2085 we would add 45×10^{10} tons to the atmosphere. This is an increase of about 0.024 percent per century, an infinitesimal change. Over many centuries such an increment might become serious, but it now looks as though we would run out of fossil fuel, or kill off mankind with some other pollutant, long before our supply of oxygen is depleted because of excess carbon monoxide or carbon dioxide.

Other writers have arrived at more frightening estimates. One estimates a 10 percent increase in carbon dioxide in the atmosphere between 1860 and 1960. Another concludes a 30 percent increase between 1890 and 2000. These estimates are based on the assumption of a higher base—that there was more CO_2 in the atmosphere in 1970 than 100 years ago. This is probably true. But how valid is a measurement or estimate made 100 years ago?

These two estimates are at extreme odds. Either the few measurements we have represent local rather than general conditions, or estimates of total carbon dioxide production by the combustion of fossil fuels are completely in error. Since there are many ways to check the amount of carbon diox-

City incinerator, Philadelphia, one method of solid waste disposal

ide produced by a unit of fossil fuel, one suspects that the measurements we have do not reflect the average carbon dioxide content of the atmosphere. The records suggest that a single volcanic eruption may have a greater effect on atmospheric temperature than several decades of burning fossil fuels, though in the opposite direction. The volcanic dust thrown into the atmosphere reflects incoming solar radiation back into outer space, and thus has a marked cooling effect. There is every reason to believe that the presence of other types of pollutant particles in the atmosphere would have the same effect.

For all these reasons one may suspect that air pollution will not create a greenhouse effect, but may even reduce the temperature at the earth's surface. The increase in temperature since about 1750 (see Figure 3-5) is probably related to the "1400-year climatic cycle" or to some other natural climatic cycle (to be discussed in Chapter 13). Both the increase in temperature since 1750, and the decline in temperature in the middle United States from 1930 to 1970, can be explained by climatic curves developed from oxygen isotope studies. In other words, we cannot definitely attribute any specific climatic changes to the greenhouse effect, particulate matter, or short-term climatic effects of atmospheric pollution. Considering how much we do not know, we can only suggest that man conserve his atmosphere carefully until enough significant data can be collected to reach valid conclusions on climatic change.

Cleansing the Atmosphere

Today, at a time when many people are conscious of pollution, we hear a great deal about curing and preventing air pollution, but little about the natural cleansing of the atmosphere. There are several ecocycles that do cleanse the atmosphere, and we should be aware of their beneficial effects. These are the hydrologic and rock cycles (discussed in Chapter 1) and many of the mineral or element cycles, such as sulfur, nitrogen, and phosphorus, some of which are also involved in the biochemical cycles. All these cycles cooperate to remove pollutants from the atmosphere.

The hydrologic cycle

Through part of the hydrologic cycle, chemicals and particles are washed out of the air by solution into or absorption into raindrops. One result of this phenomenon is the 17 parts per million of sulphate in rainwater falling on Washington, D.C., where SO_2 in the atmosphere is picked up by falling raindrops and carried to the ground.

Monterrey, Mexico, is one of the smoggiest cities in the world. But after a rain one can see the beautiful mountains surrounding the city, because the particulates have been washed out of the air, even though on the previous day the mountains may have been totally invisible. The same effect is observable at Denver, Washington, D.C., and a number of other cities where air is often heavily laden with particulates. Particles deposited directly into the oceans by rainfall immediately become part of the rock cycle. Particles that fall on land join the soil or are carried by streams to the oceans, where they too join the rock cycle as ocean sediment. The average residence time for dust in the atmosphere is about one year. This estimate is borne out by temperature charts following major volcanic eruptions, which show that the temperature effects of atmospheric volcanic dust are of sufficiently short duration to be easily explained by a residence time of one year. Since this residence time is so short, enough particulates may accumulate over a short period to alter climate locally, but there is little likelihood of long-term accumulations of dust in the atmosphere.

One might assume that gaseous pollutants would be washed out of the air by the hydrologic cycle as rapidly as are particles, but apparently they are not. If the figures on carbon dioxide in the atmosphere for 1895 and 1970 are correct, the residence time in the atmosphere for carbon dioxide is greater than for particles (dust). Some experts think it is about four years.

Gases are taken out of the atmosphere by direct exchange across the ocean surface, but the amounts are apparently insignificant for short periods. Although little is known about the behavior of gases at the ocean surface, it is known that carbon dioxide passes from atmosphere to ocean in some areas, and from ocean to atmosphere in other areas. But the total balance of carbon dioxide crossing the ocean surface in either direction is not known. We know even less about the behavior of other gases at the ocean surface, and almost nothing about the behavior of common pollutants, such as sulfur dioxide, ammonia, and nitrous oxides.

The pH is a measure of acidity in which a value of 7 is neutral, lower values are acid, and higher values are alkaline. The pH for the mild carbonic acid, $H_2O + CO_2 = H_2CO_3$, in a saturated solution is 3.8. The greater the amount of dissolved CO_2, the greater the amount of carbonic acid, and the more acid the rainwater. The pH of normal rainwater is about 5.7—that is, only slightly acid. This means that rainwater can take up more carbon dioxide. Rainwater may thus be able to wash even greater amounts of carbon dioxide from the atmosphere, unless falling rain can only dissolve the amount of carbon dioxide that produces a pH of 5.7.

Since the concentrations of other gaseous pollutants of the atmosphere are much lower than that of carbon dioxide, rainwater may not be saturated with these gases either. This means that the hydrologic cycle may be able to cleanse the atmosphere of even greater quantities of pollutants and still maintain their present residence time in the lower atmosphere. One is led to suspect that, except for biochemical cycles, exchange across the ocean surface is not significant and that the hydrologic cycle is the primary cleanser of the atmosphere.

Biochemical cycles

Important as it is, the hydrologic cycle is not the only sink for atmospheric pollutants. Biochemical cycles also cleanse the atmosphere. Land plants use carbon dioxide in photosynthesis. Some botanists have emphasized that in deserts lack of water is the factor that limits plant growth, but that in nondesert areas, where water is plentiful, the limiting factor may be carbon dioxide. Although these botanists have not considered trace nutrients as limiters of plant growth, experiments demonstrate that plant growth is increased remarkably in laboratories when up to 30 percent more carbon dioxide is added to an otherwise natural atmosphere. There are no estimates of lag time between increases in carbon

dioxide and plant growth in nondesert areas. However, young plants and young forests, or early seres in the climax cycle, are much more efficient at removing carbon dioxide and manufacturing oxygen than is the climax sere. Consequently, the lag time should not be great. Some think that additional plant growth will help retain the present residence time of carbon dioxide in the atmosphere by utilizing greater amounts of it for plant growth. Marine plants also use carbon dioxide, but their source is marine water; most marine plants do not have access to air. Some carbon monoxide is taken up directly by soil processes.

Plants also take up nitrogen through biochemical cycles, but nitrogen is not a serious global pollutant and this is not an environmental problem. The nitrogen oxides are probably cleansed only through the hydrologic cycle, although some may enter the soil from the air and be converted to soluble minerals for plant use. Except for ammonia, most of the gaseous pollutants must either break down chemically in the atmosphere or be washed out of the air by precipitation. Ammonia can be washed out of the air, and direct uptake of ammonia by plants has recently been demonstrated. This is an additional sink for ammonia and nitrogen.

Other biochemical cycles require that gaseous pollutants be cycled through the hydrosphere or the soil before again becoming part of the biosphere. The hydrologic cycle must transfer chemicals from the atmosphere to the hydrosphere or the soil before they can enter plants or animals.

Although some gaseous pollutants seem to reside in the atmosphere longer than dust, we cannot yet estimate how long.

The ocean surface is a poor transmitter of gases. Consequently, the ecosphere depends almost entirely on the hydrologic cycle and the carbon and nitrogen cycles to cleanse the atmosphere.

Summary

The earth's present atmosphere probably evolved as a result of outgassing and of the addition of oxygen through photosynthesis. The amount of carbon dioxide in the atmosphere has varied within broad limits for the past billion years, and these limits can probably safely accommodate the amounts man has introduced.

As to how significantly man has altered the total atmosphere, the evidence is inconclusive. This is a good reason for approaching the question of atmospheric pollution with care. But local damage to biological systems is widespread. Many people, zoo animals, and plants become sick or die of the effects of sulfur, nitrogen, lead, hydrocarbons, and other atmospheric pollutants from industrial, residential, and transportation sources. Public health and welfare are at present the most compelling reasons for controlling atmospheric pollution.

Such specific substances as lead, sulfur, nitrogen, and unburned hydrocarbons and aldehydes need to be removed from pollutants and neutralized chemically or buried in impermeable claystone formations to prevent them from reinvading the ecosystem.

Selected Readings

Bray, J. R. 1971. Climatic change and atmospheric pollution. *New Zealand Ecological Society* 18:38–41.

Cloud, Preston E., Jr. 1968. Atmospheric and hydrospheric evolution of the primitive earth. *Science* 160: 729–736.

Johnsen, S. J.; W. Dansgaard; and H. B. Clausen. 1970. Climatic oscillations 1200–2000 A.D. *Nature* 227: 482–483.

Kellogg, W. W., et al. 1972. The sulfur cycle. *Science* 175: 587–596.

Landsberg, Helmut E. 1970. Man-made climatic changes. *Science* 170:1265–1274.

Lave, Lester B. and Eugene P. Seskin. 1970. Air pollution and human health. *Science* 169:723–733.

Rodin, L. E. and N. J. Basilevic. 1968. World distribution of plant biomass. In *Functioning of terrestrial ecosystems at the primary production level,* UNESCO, Proceedings of the Copenhagen Symposium, pp. 45–52.

Rohrman, F. A., B. J. Steigerwald, and J. H. Ludwig. 1967. Industrial emissions of carbon dioxide in the United States: a projection. *Science* 156:931–932.

Salinger, Richard and Wallace West. 1968. Conserving our waters and clearing the air: Study unit for science and social studies class. Student manual. New York: Am. Petroleum Inst.

Singer, S. Fred. 1970. Human energy production as a process in the biosphere. *Scientific American* 223(3): 174–176ff.

Stern, Arthur C., ed. 1968. *Sources of air pollution and their control,* 2nd ed. *Air pollution,* vol. 3. New York: Academic Press.

Wagner, Richard D. 1971. *Environment and man.* New York: Norton.

Wassink, E. C. 1968. Light conversion in photosynthesis and growth of plants. In *Functioning of terrestrial ecosystems at the primary production level,* UNESCO, Proceedings of the Copenhagen Symposium, pp. 53–66.

Weinstock, Bernard and Hiromi Niki. 1972. Carbon monoxide balance in nature. *Science* 176:290–292.

Wise, William. 1968. Killer smog, the world's worst air pollution disaster. Chicago: Rand McNally.

Woodford, A. O. 1972. Origin and history of oxygen in the air. *Journal of Geological Education* 20:276–280.

Chapter 4

The Oceans: The Final Receptacle of Pollutants

The ocean is the earth's greatest storehouse of minerals.

Rachel Carson, 1951

Extent of the Hydrosphere

The hydrosphere—the waters of the earth, comprising its oceans, lakes, and streams—is vital to mankind and to all life as we know it. It is unique within our solar system; there are no oceans on other planets. The hydrosphere is responsible for many physical and climatic features which give our earth its most significant characteristics. Water is the critical fluid that circulates through the hydrologic cycle and runs the earth's "heat engine." The hydrologic cycle thus affects the earth's climate and weather. It was the waters of the ocean which, about 3.5 billion years ago, contained the peculiar chemical mixtures from which life on earth emerged. Today the oceans are the home of billions of algae and other one-celled marine organisms (phytoplankton) that produce about 90 percent of the oxygen in the earth's atmosphere.

In terms of actual and potential economic wealth, the oceans cover vast deposits of natural resources which man will probably need in the future. They contain vast stores of resources—mineral, plant, and animal—which can improve the quality of man's life (Chapters 2 and 15); through the hydrologic cycle they aid in cleansing the atmosphere; and they become the final repository of many pollutants from air and land—to the extent that their ability to cleanse themselves, particularly near continental shores, is sometimes in doubt. It has generally been assumed that the main body of the oceans was in no immediate danger of pollution. The Norwegian anthropologist and explorer Thor Heyerdahl found, however, a large ac-

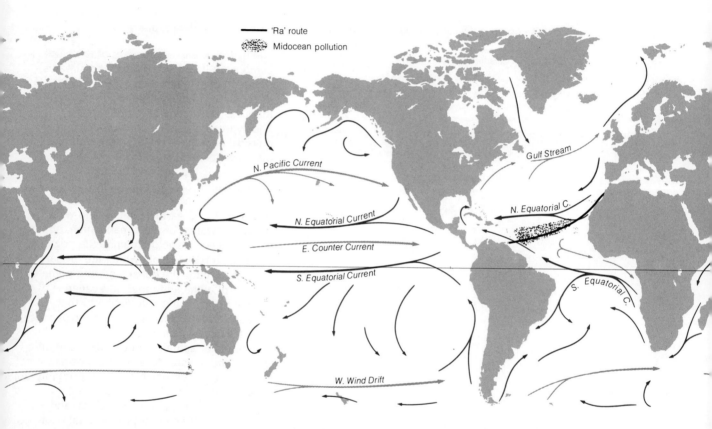

——— 'Ra' route
⬚⬚⬚ Midocean pollution

Gulf Stream

N. Pacific Current

N. Equatorial Current

N. Equatorial C.

E. Counter Current

S. Equatorial Current

S. Equatorial C.

W. Wind Drift

Figure 4–1

Major ocean currents of the world.

cumulation of cultural debris in the Sargasso Sea (Figure 4-1). The oceanographer Jacques Yves Cousteau has also emphasized man's increasing pollution of the open ocean.

Origin of the Hydrosphere

The origin of the earth's water closely parallels the origin of its atmosphere. It is probable that before the earth lost its original atmosphere it also possessed some sort of light fluids. It is thought that these were spun off with the original atmosphere at the time of the first major thermal event, when the earth melted and its heavier elements—such as iron, nickel, and other metals—sank toward the center to become part of the earth's core.

The earth's present oceanic water, like its present atmosphere, is believed to be the product of volcanism and outgassing,

Figure 4−2

Content of juvenile emanations from various types of rocks and volcanic processes.

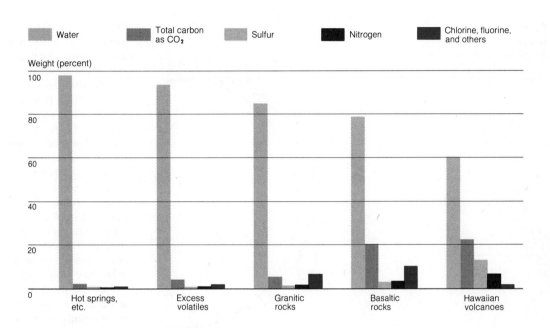

which took place extensively immediately following spin-off, and has continued throughout geologic time at a lesser rate This water that appeared on the earth's surface for the first time is called juvenile or magmatic water. The hydrosphere has been accumulating this water ever since. Water is recycled from the hydrosphere back to earth through the rock cycle.

Just how long ago the greater portion of the waters of the world was formed is a matter of conjecture, but it is known that the water of the oceans is older than the modern ocean basins. Nowhere in modern ocean basins are there rocks dated at more than 180 million years old. Since life and water are known to have existed on the earth for well over three billion years, the ocean basins must have been distributed differently in the pre-Mesozoic era than now (Appendix B).

Different kinds of igneous rocks contribute different percentages of water and other chemicals to the hydrosphere (Figure 4-2). Basaltic rocks add far greater amounts of carbon and sulfur than do granitic rocks, whereas the latter contribute greater amounts of chlorine and fluorine.

All the geologic evidence demonstrates that most of the liquids of the ocean systems are old. Indeed, the composition of the ocean waters does not seem to have changed much in the past billion years. For a billion or more years before that, as free oxygen was gradually becoming available, the amount of carbonates, sulfates, nitrates,

Figure 4-3

Routes of flow of shallow and deeper ocean currents.

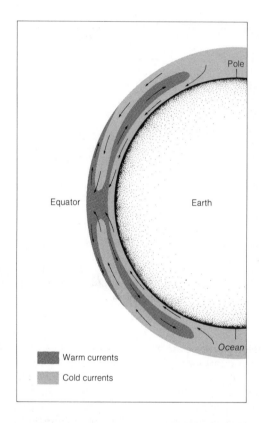

the tropics more directly than they strike the polar regions, and they strike the temperate belts at an angle somewhere between the other two. The amount of solar energy reflected increases as this angle departs from ninety degrees.

Hence the amount of heat received and retained by the earth's surface increases steadily as one moves away from the poles toward the equator. Tropical waters are therefore warmer than polar waters. Moreover, colder and heavier polar waters tend to sink, initiating a slow oceanic circulation (Figure 4-3). At the same time, surface water moves toward the poles to replace the sinking polar waters. The circulation keeps the polar climates warmer than they would be without it.

Tropical oceans also warm the overlying atmosphere. The warm air rises and then moves both north and south toward the polar regions, warming even the polar atmosphere. Thus, air and ocean together, powered by solar energy, provide a more hospitable environment for life everywhere on earth. The importance of oceanic circulation—and of its absence—is demonstrated by Antarctica, where there is little or no polar circulation of warm water (see Figure 4-1), and the great ice cap remains.

Local variations in energy patterns of ocean and atmosphere produce local variations in climatic conditions. They are all nevertheless triggered by the earth's "heat engine." Tropical storms—hurricanes and typhoons—are thought to result from the tremendous energy released in limited areas of ocean which have been differentially heated by solar energy. Such local warming creates areas off the east coasts of continental masses especially

and other oxidized compounds in the ocean was gradually increasing, until a few hundred million years before the Cambrian its composition, like that of the atmosphere, may have become stabilized within certain broad limits.

The Earth's "Heat Engine"

The ocean and the atmosphere cooperate in an energy cycle which is the major factor in controlling and equalizing the temperature of the earth's surface. The sun's rays strike

favorable to the genesis of tropical storms. Furthermore, other air masses that modify climate may be maintained and modified by localized heat flow from the oceans.

Most of our climate, over both the short and the long range, results from the combined effects of the earth, ocean, and atmosphere. This aspect of our environment is being intensively studied, and great strides in understanding it have been made through the use of weather satellites. In the future we may better understand the impact of man, if any, on the delicate balance of the earth's heat engine. If the heat engine proves to be sensitive, we can expect attempts at climate engineering. Climate engineering, if and when it is possible, will have to be monitored closely for major environmental changes.

Composition of the Oceans

The oceans contain vast quantities of various substances that arrived in the oceans in five ways. The first is outgassing, the process that brought forth the hydrosphere itself. The second, closely associated with the first, is the addition of mineral matter from new rocks appearing at the spreading zones along the midocean ridges. Third, most of the minerals in the oceans are probably dissolved from rocks by surface and ground water and brought to the oceans by streams and rivers. A fourth source is the atmosphere, and a fifth source of ocean materials is life. By both metabolism and decay, organisms produce unique organic molecules which were not in the oceans before the appearance of living things. Through the hydrologic and biogeochemical cycles, the atmosphere is cleaned of natural and artificial substances; whether pollutants or not, these become part of the oceans.

After oxygen and hydrogen (the components of water) the most abundant elements in the oceans, in order of decreasing amount, are chlorine, sodium, magnesium, sulfur, potassium, calcium, bromine, carbon, boron, strontium, and fluorine (Table 4-1).

Resources beneath the ocean were discussed in Chapter 2. The oceans contain a number of elements that will be needed in the future. In World War II, magnesium was obtained from sea water. Iodine is obtained by harvesting certain marine organisms. Our needs for such commodities as scandium, germanium, rubidium, niobium, and other rare elements will soon exceed the shortage of land-based supplies. It is now uneconomical to produce these elements from sea water, but in the future, as by-products of desalination, their economic production may be possible. Research is being done at the present time on the use of lithium as a fuel for fusion. The most likely source of lithium in sufficient amounts will be the ocean. The other fuels being investigated as sources of energy by fusion are deuterium and tritium, the isotopes of hydrogen. The ocean would also be a primary source of these. The most likely heat collector in fusion reactors is now thought to be a blanket of niobium. Niobium is also rare on land, and in the future may best be obtained from the ocean. The necessity of processing tremendous quantities of water to acquire deuterium and tritium may eventually lead to the processing of other ocean materials as by-products. Although

Table 4–1

Composition of the ocean for the 55 most abundant elements, excepting hydrogen and oxygen (4700 tons per cubic mile is approximately one part per million).

Element	Tons/cu. mi.	Element	Tons/cu. mi.	Element	Tons/cu. mi.	Element	Tons/cu. mi.
Chlorine	91.2×10^6	Phosphorus	330	Vanadium	9	Xenon	0.5
Sodium	29.4×10^6	Iodine	280	Manganese	9	Germanium	0.3
Magnesium	6.48×10^6	Barium	140	Titanium	5	Chromium	0.2
Sulfur	4.248×10^6	Indium	94	Antimony	2	Thorium	0.2
Potassium	1.824×10^6	Zinc	47	Cobalt	2	Scandium	0.2
Calcium	1.72×10^6	Iron	47	Cesium	2	Lead	0.1
Bromine	312,000	Aluminum	47	Cerium	2	Mercury	0.1
Carbon	134,400	Molybdenum	47	Yttrium	1	Gallium	0.1
Boron	58,880	Selenium	19	Silver	1	Bismuth	0.1
Strontium	38,400	Tin	14	Lanthanum	1	Niobium	0.05
Fluorine	6,240	Copper	14	Krypton	1	Thallium	0.05
Nitrogen	2,400	Arsenic	14	Neon	0.5	Helium	0.03
Lithium	800	Uranium	14	Cadmium	0.5	Gold	0.02
Rubidium	570	Nickel	9	Tungsten	0.5		

Garbage barge in New York harbor

all the resource possibilities of sea water will not be discussed, these few examples demonstrate their potential.

Pollution in the Oceans

The oceans can be polluted from the atmosphere, from the land through streams, or by man through direct dumping. Natural and artificial atmospheric pollutants settle into the ocean from the atmosphere. From streams the ocean receives all kinds of sediment and any pollutants, either natural or man-made, that are poured into the streams before they enter the sea. Some of man's wastes go into the ocean directly. Examples are sewage poured directly into bays, estuaries, or the ocean, and garbage, which until recently was barged from New York City onto the continental shelf and dumped. All these pollutants come from the land; the ocean produces few pollutants of its own.

Thus far man, in the infinite variety of his activities, has not greatly affected the earth-ocean-atmosphere machine. Whether he will some day fill the air and the water with enough pollutants to alter this machine is questionable.

For example, it is doubtful that the harmful products of fossil fuels will permanently taint the total ocean any more than they will the total atmosphere. Carbon dioxide poured into the oceans will be deposited with lime mud. Nitrates and phosphates, even if washed into the ocean as soluble fertilizers, will be precipitated as sediment, or consumed by the organic world. The same can be said for other nonparticulate pollutants. Particle pollutants will be washed out of the air by rain and deposited on the sea bottom as a fine-grained sediment. But, can

the rock cycle cleanse the water fast enough to keep it clean enough for the more sensitive animals and plants? We cannot answer this question without more data than we have now.

Atmospheric pollutants probably do not endanger the total ocean. Even human waste is not that serious, since organisms will consume that portion of it which can be converted into food, and the remainder will settle to the bottom as sediment. Some people worry that man may put enough nutrients into the ocean to pollute it by the overproduction of minute plants, but this danger seems remote. However, the above statements do not refer to toxicants, nor to the fringe areas or margins of the oceans.

Toxicants

Chlorinated Hydrocarbons. Some of the most insidious pollutants are chlorinated hydrocarbons, also called organochlorides The most widespread of these are DDT and polychlorinated byphenyls (PCBs). Sweden and the United States have banned the use of DDT, except for emergencies, and Great Britain is in the process of abandoning its DDT programs. But other organochloride pesticides, some of them more toxic and more persistent than DDT, have not been included in these restrictive orders. These compounds have adverse effects, as discussed below.

They are such fine dusts that they are easily windborne. It has been estimated that windborne organochloride pesticides, primarily from Africa and India, are accumulating in the Bay of Bengal at the rate of three tons a year.

Figure 4-4

Decline in photosynthetic activity of four species of phytoplankton with increase of DDT in the water.

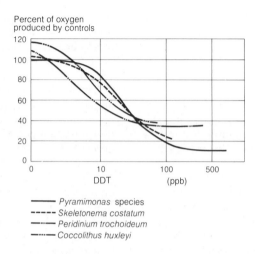

Percent of oxygen produced by controls

Pyramimonas species
Skeletonema costatum
Peridinium trochoideum
Coccolithus huxleyi

DDT accumulates in organic tissues and becomes more concentrated at each step higher in the food cycle. It is highly concentrated in tuna, which is near the top of the food cycle.

Because DDT builds up in the food cycle so easily and is then consumed by scavengers, it has been spread worldwide by wind and by wide-ranging marine carnivores. DDT has been found in soils, run-off water, rainwater, air, and most animals, even in the Antarctic.

Perhaps of more importance and greater danger, DDT, in concentrations as low as a few parts per billion, inhibits photosynthesis by some phytoplankton (Figure 4-4). PCBs and mercury have also been found to be toxic to some phytoplankton. PCBs are as widespread in the ocean as DDT. Phytoplankton are at the base of the marine food cycle, and any decrease in their photosynthetic processes decreases the entire food supply up to and including man. Figure 4-5 shows diatoms as the type of phytoplankton which are the primary producers or photosynthesizers in the cold waters of the Antarctic.

The main primary consumers, the second link in the chain, are krill—small, shrimp-like animals. At the third level, feeding on the krill, are the balleen whales (which have mesh-like strainers instead of teeth), the blue whale, flying birds, small fishes, some penguins, and the crab-eater seal. Larger birds, fishes, and other seals are on the fourth level (tertiary consumers), and the leopard seal and the killer whale are the top carnivores in this particular chain. Man is more omnivorous, and uses all levels above the krill for food. Because of their effect on phytoplankton, continued use of organochloride pesticides could greatly reduce the food resources of the ocean. Since these same phytoplankton take up carbon dioxide and release oxygen to the atmosphere, such pesticides would also reduce the amount of carbon dioxide cleansed from the ocean-atmosphere cycle. The total ecological effects of organochlorides on phytoplankton are not known, and caution in their use is important until these effects can be determined.

The Bureau of Commercial Fisheries has reported that all marine fisheries are contaminated with organochlorides, not enough to be a health hazard to human beings, but enough to damage commercial fisheries. It might be asked how the population of commercial fishes can be reduced by organochlorides without endangering the people who are eating those fishes not yet destroyed. In any event, we are now alert to the danger created by these pesticides,

Figure 4–5

The food cycle in Antarctic waters, from phytoplankton through krill, and consumer levels up to the killer whale.

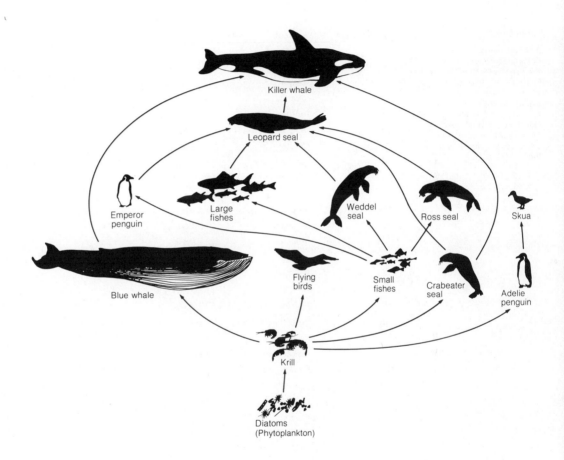

Killer whale

Leopard seal

Emperor penguin

Large fishes

Weddel seal

Ross seal

Skua

Blue whale

Flying birds

Small fishes

Crabeater seal

Adelie penguin

Krill

Diatoms (Phytoplankton)

and have begun to take steps to reduce the damage they do.

Radioactive Pollutants.   Geochemists can isolate man-introduced radioactivity in any fifty gallons of sea water taken from any place in the ocean. These harmful substances are radioactive forms of atoms, sometimes called radioisotopes, or more commonly radionuclides, of the standard elements. Artificial radionuclides are from two sources. (1) Some are products of nuclear fission—produced by a nuclear explosion such as that of an atomic bomb. (2) Others are induced—produced as a by-product of a generating plant powered by nuclear energy. The oceans also contain radionuclides of natural origin (Table 4-2).

In fish and other animals the highest con-

Table 4-2

Artificial and naturally oc-
curring radionuclides in
the sea. Data from Theodore
R. Rice and Douglas A.
Wolfe, ''Radioactivity—
Chemical and Biological
Aspects.'' In Donald W.
Hood, ed., *Impingement of
Man on the Oceans* (New
York: John Wiley & Sons,
Inc., 1971), p. 328.

Radionuclide	Source	Half-life	Radionuclide	Source	Half-life
^{3}H	Natural	1.2×10^1 years	^{110m}Ag	Induced	249 days
^{10}Br	Natural	2.7×10^6 years	^{113m}Cd	Induced	14 years
^{14}C	Natural	5.5×10^3 years	^{129m}Te	Fission product	33 days
^{32}P	Induced	14.3 days	^{129}Te	Fission product	74 minutes
^{32}Si	Natural	7.1×10^2 years	^{137}Cs	Fission product	30 years
^{35}S	Induced	87.1 days	^{137m}Ba	Fission product	2.6 minutes
^{40}K	Natural	1.3×10^9 years	^{128}I	Fission product	1.6×10^7 years
^{51}Cr	Induced	27.8 days	^{141}Ce	Fission product	32.5 days
^{54}Mn	Induced	300 days	^{144}Ce	Fission product	290 days
^{55}Fe	Induced	2.94 years	^{144}Pr	Fission product	17.5 minutes
^{59}Fe	Induced	45.1 days	^{144}Nd	Fission product	2.4×10^{15} years
^{57}Co	Induced	270 days	^{147}Pm	Fission product	2.5 years
^{58}Co	Induced	72 days	^{147}Sm	Fission product	1.3×10^{11} years
^{60}Co	Induced	5.27 years	^{131}I	Fission product	8.05 days
^{65}Zn	Induced	245 days	^{140}Ba	Fission product	12.8 days
^{87}Rb	Natural	5.0×10^{10} years	^{185}W	Induced	74 days
^{89}Sr	Fission product	50.4 days	^{187}W	Induced	24 hours
^{90}Sr	Fission product	28 years	^{226}Ra	Natural	1.6×10^3 years
^{90}Y	Fission product	64.4 hours	^{228}Th	Natural	1.9 years
^{91}Y	Fission product	58 days	^{228}Ra	Natural	6.7 years
^{95}Zr	Fission product	63.3 days	^{230}Th	Natural	8.0×10^4 years
^{95}Nb	Fission product	35 days	^{231}Pa	Natural	3.2×10^4 years
^{103}Ru	Fission product	41.0 days	^{232}Th	Natural	1.4×10^{10} years
^{103m}Rh	Fission product	54 minutes	^{235}U	Natural	7.1×10^8 years
^{106}Ru	Fission product	1.0 year	^{238}U	Natural	4.5×10^9 years
^{106}Rh	Fission product	30 seconds			

centration of radionuclides occurs in the
liver, bones, and stomach. Thus processed
seafoods from which these organs are not
first removed contain higher proportions of
radionuclides than fresh fish from which
the internal organs have been removed.
Clams and crabs are particularly high in
radionuclides. Fortunately, the amount of
radionuclides present in marine foods is
well below the level prescribed by health
standards, but it has been suggested that
an exclusive seafood diet would substantially increase the dosage which individuals are now receiving from both natural
and artificial sources, and bring the radioactive isotope level in the bones of the
consumer up by many percentage points.
It must be emphasized that natural radionuclides are as dangerous as artificial,
but their level is one to which life evolved
and adapted. Artificial radionuclides are
not, individually, any more dangerous than

Fulton fish market, New York

test explosion on the Aleutian Island of Amchitka in spite of widespread protest against the experiment, including strong statements by the governments of Canada and Japan. As it happens, no radiation has yet escaped into the waters of the ocean from this operation (code-named Cannikin) but at the time it was widely feared that both land and water could be polluted, and there was such fear of a tidal wave that a worldwide watch was conducted for 24 hours after the explosion.

While the ocean is a widespread carrier of pollution by radionuclides, it is also a cleansing agent. It receives these substances from the atmosphere by way of the hydrologic cycle, and although some radionuclides concentrate in living organisms, others are carried to the bottom along with the clay and organic sediments.

For several years concrete casks of spent radioactive fuel were dumped at sea. Because of objections to this indiscriminate practice, a committee of the U.S. National Academy of Sciences investigated the handling of radioactive wastes and recommended against any type of disposal at sea.

natural, but they do raise the total radiation level. Therein lies their danger to life.

In spite of the widespread publicity which has been given to the pollution of marine food supplies by radiation, there have been some interesting cases of indifference to the general hazards of radiation in the waters of the ocean. Thus a proposed nuclear excavation of the harbor at Cape Keraudren, West Australia, was called off not for fear of radiation pollution, but for fear of earthquakes. And late in 1971 the United States Atomic Energy Commission carried through an underground nuclear

Chemical Warfare Agents. As late as 1969, a full decade after the warning from the National Academy, it was decided to dump excess and outdated chemical and biological warfare products in the Atlantic Ocean. Both CBW and radioactive wastes were placed in what were thought to be nonbreakable and noncorrosive casks. But a medium with the force of the ocean—which can move a 2600 ton reinforced-concrete jetty out to sea overnight, never to be found again, or toss a boulder through the roof of a lighthouse 100 feet above sea level—is not to be treated lightly. Suppose the casks were shifted until they rolled, banged, and bumped down the continental slope with a mud slide. Rock-boring pelecypods have been known to drill holes through submarine oil lines made of iron pipe, and into the steel platforms of drilling rigs. Such platforms have also been corroded and etched through by organic acids emitted at the point of encrustation of marine animals adhering to these structures. Marine pelecypods and gastropods bore into the concrete, limestone, and granite of coastal jetties. Casks of toxic wastes dumped into the oceans are not only subject to the furies of tropical storms, but must withstand chance turbidity currents or submarine landslides. If none of these damage the casks and release the contents, the casks are certain to be bored into by marine organisms or corroded by organic acids and sooner or later opened, allowing their toxic contents to invade the open ocean. For these reasons toxicants of all kinds should be kept from the marine environment. Instead, at the time of manufacture, later neutralization should be designed and planned, and facilities should be constructed.

Other Ocean Dumping. The United States has not been the only country guilty of indiscriminate dumping of wastes in the oceans; many other nations have participated in this practice. After World War II excess mustard gas (a chemical warfare agent) was dumped into the Baltic Sea and into the Bay of Biscay. Because of mustard gas leaking from the depths of the Baltic Sea, many swimmers have suffered from sore eyes, and there have been reported instances of blindness. The release of mustard gas from containers dumped in the Bay of Biscay has recently resulted in the closing of certain oyster fisheries in that Bay.

Other intentional dumpings since 1945 include:
1. Norway: Reportedly dumps 18,000 tons of formalin annually into the North Atlantic.
2. Germany: Reportedly dumps 1,200 tons of sulfuric acid daily into the North Sea.
3. Great Britain: The Royal Navy dumped ferric chloride into the English channel in July 1970.

Mercury. Recently mercury has been found in tuna, swordfish, and other marine foods. The quantity is well below the danger level as prescribed by the federal government before August 1972, but one would not want a steady diet of such meats. Mercury appears to be ubiquitous in marine animals and environments. Analysis of marine organisms indicates that their mercury content is not the result of pollution by man, but is natural. In other words, marine animals and plants which have been in museums for many decades contain as much mercury as marine animals and plants recently collected from the oceans. On the other hand, the mercury content of land animals and plants, and of the land environment itself, has steadily increased for the last 60 or 70 years because of mercury pollution.

Mercury has an affinity for clay minerals and for organic molecules which attach themselves to clay minerals. Most of these sediments are trapped in lakes, ponds, and reservoirs for various lengths of time on their way to the oceans. For these reasons much of modern mercury pollution may not yet have reached the ocean. Moreover, the volume of the ocean is so great that a century of pollution may not be enough to produce recognizable increases.

The Disturbance and Destruction of Natural Environments

The most sensitive component of the natural environment—and hence the first to be damaged—is the organic community. Not that all members of the community are equally sensitive to minute changes, but enough of them are so sensitive that change in a community's structure is the first observable effect of pollution. Certain organisms can be used as pollution indicators. Fish are most sensitive to some pollutants; hermit crabs are extremely tolerant of most pollutants; different species of algae tolerate different degrees of pollution. Furthermore, since species vary in their sensitivity to pollutants, the degree of pollution is also a measure of community structure. From the human viewpoint, it is desirable that pollution not be allowed to alter the community to any great degree, because altered communities do not produce a full food cycle and are of less benefit to man.

Pollution by human waste

Of primary importance to the maintenance of any community is food. Altered biologic communities mean altered food cycles and the production of less food by and for the marine community. Overpollution, particularly by human waste or toxicants, can so completely befoul the environment that

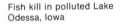
Fish kill in polluted Lake Odessa, Iowa

Family of swans showing the effects of an oil spill in the Thames River near Groton, Connecticut

beaches may create a health problem. The *Escherichia coli* count in bathing waters along some British beaches is above public health standards. *E. coli,* a common intestinal bacterium, is itself harmless, but it indicates the probable presence of harmful bacteria. Relatively high concentrations of *E. coli* have been found on the sea bottom 110 miles off shore from New York City. They are spread widely and range as far as 80 miles from New York City's outfall for raw sewage.

Red tides and similar phenomena are sudden blooms of microorganisms (plant or animal) that kill literally millions of animals, and sometimes kill man if he eats the flesh of animals that carry them. Certain mass mortality events are called red tides because the organisms that produce them give the water a pinkish tinge. Similar phenomena, not producing a tint in the water and actually nontoxic, can be produced when blooming microorganisms deplete the waters of oxygen and thus cause the death of as many animals as if they were indeed poisonous. A sudden supply of mineral foods is usually required for the production of such toxic organisms in sufficient numbers to produce mass mortalities, either by red tides or by asphyxiation. Effluent from human or agricultural wastes can provide the food for blooms of such microorganisms.

In the face of all the evidence that exists for the deterioration of the ocean's fringes from the addition of toxicants and human and agricultural wastes, it is surprising that the oceans should be suggested as a final repository for human waste from the great megalopolis on the east coast of the United States. The human component may not be

animals cannot survive. Moreover, overpollution by human waste can provide so much food for phytoplankton and other marine plants that their growth depletes the water of oxygen at night, and prevents or destroys the growth of marine animals. Water can also be depleted of oxygen by exothermic chemical reactions—that is, those which give off heat—when it is polluted with some contaminants. Bays, lagoons, and estuaries can be partly or totally depauperated by untreated or improperly treated effluent from our cities. Raw sewage which concentrates along

Sign along the Potomac
River

serious, although changes in currents before sewage is completely diluted could result in asphyxiation kills through oxygen depletion. That waste is seemingly ubiquitous was observed by oceanographers in submersibles at depths of 2000 feet off Miami, Florida, who could only conclude that the many white streamers passing their windows were toilet tissue from the untreated sewage of the Miami outfall. The more serious components of megalopolis waste are the toxicants and semi-toxicants that cannot be kept out of sewage—detergents of all kinds, plastics, medicines, pills, pesticides, poisons.

One of the worst polluters of the oceans has been the United States Navy. Any naval task force of any nation can be tracked over the ocean even after many days by the vast quantity of floating trash it leaves in its

Sewer outlet on a beach in
Normandy, France

A Paris river polluted with sudsy detergent

wake. Furthermore, the navies of different nations can be identified by the cultural characteristics of their waste. Raw untreated sewage is dumped into the oceans from ships; the oceans dilute it and eventually convert it to energy or to sediment. But when a navy is in port, the quantity of sewage is serious, especially from ships as large as aircraft carriers. The U.S. Environmental Protection Agency has recently ruled that all United States ships and smaller craft must contain their sewage and that port facilities must provide sewage treatment. It is hoped that other nations will follow this precedent. The initial cost for ships and boats of the United States to provide means to contain their sewage is estimated at over $1.5 billion.

Thermal pollution

Exothermic chemical reactions in discharged effluent, plus the use of estuarial waters for air-conditioning and the cooling of electric power plants, cause thermal pollution. About eighty percent of thermal pollution of water in the United States is from the production of electricity. The significance of thermal pollution in nearshore waters is incompletely known, because it has been overshadowed by the effects of poisons and by oxygen depletion. When and if toxicity and oxygen depletion by phytoplankton blooms are remedied, we may discover the extent to which thermal pollution must be controlled. Problems connected with this kind of thermal pollution and oxygen depletion can not be solved merely by treating sewage. Even when treated, organic and mineral residues still react with the environment to produce heat and increase the temperature of the water. Furthermore, treating and even deactivating sewage only makes it more available to phytoplankton as food. Removal of nutrients and of heat are the only cures.

Although thermal pollution has a number of harmful effects in nearshore waters, perhaps the most serious is the effect on juvenile organisms. In all lifeforms the young especially are adapted to only small changes around a specific optimum temperature. Changes of as little as 3 or 4 degrees are often lethal. Estuaries, such as Chesapeake Bay in Maryland and Virginia, are primary areas for growth and development of the juvenile stages of many temperature-restricted organisms. Raising the temperature of these areas may interrupt the life cycles of these animals and so harm large segments of the food chain. Fishes have temperature sensors and will not swim into thermally lethal areas. Therefore thermal pollution of estuaries might prevent up-

Water sampling in the
Connecticut River to study
the effects of the Yankee
Atomic Plant

stream or other migrations at spawning
time.

Exploitation

Also disturbing to nearshore environments
is the continued search for resources in and
beneath the shallow waters of the continental shelves. For years conservationists
and oyster fishermen have been fighting
with the producers of oyster shell aggregate.
Because some coastal areas have little
sand and gravel, dead and active oyster
reefs have been dredged for aggregate to
be used in concrete and in other construc-
tion-oriented industries. The conservationists
and oyster fishermen maintain that dredging
for oyster shells ruins good oyster grounds,
and muddies the bays until fishing is
reduced if not destroyed. They are partly
right. Bay bottoms spoiled by dredging
for shell aggregate can be rehabilitated,
but the cost of reclamation puts the shell
aggregate producers at a disadvantage compared to the producers of aggregate from
other sources. This raises the cost of aggregate to coastal cities and towns, which
would also have to pay more for construction if crushed limestone or other aggregate
were shipped in to them. In the last analysis, the cost of conservation will be passed
on to the entire community in the form of
more expensive building and highway
construction.

Cleaning oil from the tanker
Torrey Canyon off the
Brittany coast

Dredging for tin, gold, diamonds, and gravel also disturbs bottom communities, but if the dredging is in sandy bottoms reworked by waves, as is true of the diamond dredging operations off the west coast of South Africa, the disturbance is not serious because the waves soon return the beach to its original condition. Obviously, the disturbance of bottom communities in which the life span of individual organisms is short is less serious than in those where the life span is long. The edible quahog—a clam with a maximum life cycle of about fifteen years—is more seriously disturbed by nearshore destruction than the cherrystone clam with a life cycle of three or four years.

Oil leak in Santa Barbara, California

Oil spills

The most widely publicized, but not the most damaging, instances of ocean pollution have resulted from such accidents as those to the oil tanker *Ocean Eagle* off San Juan, Puerto Rico, the *Torrey Canyon* off the coast of Cornwall, England, and the Santa Barbara Channel incident—all oil spills. Dramatic as these and similar events have been, the Santa Barbara Channel incident should not be classed among them. For Santa Barbara Channel beaches have been oily for millions of years. Furthermore, damage seems to have been only tem-

porary. That oil spills have a long history in Santa Barbara is evidenced by the following statement from the ship's log of Captain Vancouver, recorded in 1792.

The surface of the sea, which was perfectly smooth and tranquil, was covered with a thick, slimy substance, which when separated or disturbed by a little agitation, became very luminous, whilst the light breeze, which came principally from the shore, brought with it a strong smell of tar, or some such resinous substance. The next morning the sea had the appearance of dissolved tar floating on the surface, which covered the sea in all directions within the limits of our view. . . .

A common example of water pollution

It may well be that the only way to cure the problem of oil pollution in the Santa Barbara Channel is to draw off enough oil and gas from the continental shelf to take the pressure off the reservoirs and thus prevent leakage. Even this may not completely clean the Santa Barbara beaches because some of the many active seeps are almost certain not to be connected to any producing horizon that can draw off pressure.

Moreover, the Santa Barbara Channel incident has had some good effects. The gulf coast states immediately tightened their rules on offshore pollution. And the extensive publicity made everyone aware not only of oil pollution, but also of other important pollution problems facing the United States. Competent scientists believe, however, that most accumulations of tar on beaches result from natural submarine seeps. Research in progress should enable us, within about five years, to designate the source of tar accumulating in ocean environments. We may then know the ratio of oil pollution by nature to that by man, at least on beaches and the open sea.

Other cases of oil pollution are the accident to the tanker *Yukon* in Cook Inlet, Alaska, and the danger from Hurricane Camille in August 1969 off the Louisiana coast. But damage from storms to ships and oil rigs appears to have been overemphasized. More harmful to the environment are the cleaning of tankers at sea (a common practice), the dumping of oil by world navies, and the dumping of oily bilge by all vessels that sail the seas.

Although single oil spills may produce only local damage, continued dumping may have long-range effects as yet unseen. In any case, oil pollution of beaches is not new, since the United States Bureau of Mines considered it a hazard to public health almost a half-century ago. The cures are difficult but the prevention is not, except for detecting offenders. But, before this is possible, the world public will have to demand the international cooperation, the laws, and the punishment that will prevent further damage.

The Edge of the Sea

Although the open ocean is disfigured by a
certain amount of floating, ugly debris,
it is not in immediate danger of destruction.
This is by no means true of the continental
shelves, and particularly those parts of the
shelves which are partly or totally isolated
from the free circulation of water from the
open sea. The fringes of the ocean *are* in
danger of destruction. Of all marine habitats,
the beaches, bays, lagoons, and estuaries,
the river mouths, tidal channels, and delta-
fronts, are most vulnerable to the ravages of
man, his technologies, and his whims. All
aspects of these areas are endangered—
the life, the bottom, the quality of the water,
the aesthetics, the living space. Those ocean

Sewer outfall that should
be properly recycled

processes which aid in cleansing water of harmful products do not operate fast enough to keep up with the rate of human havoc to these partly isolated areas which receive the brunt of man's carelessness and waste. This problem will grow as population grows, and as the frantic search increases for resources, space, recreation, and escape from the crowd.

Since the understanding of the problems of fringe areas can be greatly improved by comparing a few specific examples, we shall look at five: the bays of southern California, Padre Island in Texas, Chesapeake Bay in Maryland, the Duwamish River Estuary in Washington, and San Francisco Bay. Though their histories vary in detail, they are similar enough in broad outline to be very illuminating.

A familiar picture of man's pollution of his environment

The bays of southern California

Natural bays on recently uplifted coastlines are at first clean almost to the point of being sterile. But as they are more and more used and become crowded by man, they become less desirable residential areas for the wealthy, so that zoning and cleanliness regulations are more and more ignored. The dumping of garbage and other pollutants increases, and the attractiveness of the area is further reduced. Finally, industry moves in, and the area becomes even more polluted. So far, rehabilitation has never been completely satisfactory. Of twenty-three bay environments in southern California, only two have been preserved in their natural state.

Eutrophication

Although poisons, industrial waste, and debris all contribute to the deterioration of bays, the most serious pollutant is the excess nutrients from sewage. At one time San Diego Bay was in such a condition—it had eutrophied. Basically, the term eutrophy refers to good nutrition, but as used in discussions of pollution, it usually signifies the presence of too much nutrient. Too much nutrient results in an all-too-rapid growth of primary producers in the food chain, particularly the algae and other phytoplankton, which sharply deplete the oxygen in the water. Some of the oxygen is used at night by plants that must continue metabolism when they are not engaged in photosynthesis. But the main effect of excess phytoplankton is the depletion of oxygen by the oxidation of dead and decaying phytoplankton in the water. The oxygen level eventually becomes so low that most animals can no longer live in the environment. This process of phytoplankton blooms and oxygen depletion is called eutrophication

San Diego Bay was restored by dumping all sewage out at the edge of the continental shelf, where it is neutralized by a much greater amount of water. The bay can again be used for shelling, fishing, swimming, and other recreational activities.

Beaches

Beaches and lagoons are as vulnerable to waste and degradation as bays. Florida, Oregon, Japan, Australia, and Great Britain are all complaining of the filth left on their seashores by a careless public. As soon as any area (other than a wilderness) is set off as a place for national recreation, it is developed for public enjoyment to justify its purchase with tax monies. But, unfortunately, as paved roads, paths, and other routes of access advance into new areas, the litter of cans, bottles, plastic bags, and other debris keeps abreast of them.

Chesapeake Bay

The Susquehanna River is the largest stream that drains into the Atlantic Ocean from the eastern United States. During the peak of the last ice age it occupied a broad valley extending across the present continental shelf. With the melting of the ice at the end of the last glacial period, the sea level rose, filling much of that old valley. This broad, drowned valley is now known as Chesapeake Bay.

Chesapeake Bay is 195 miles long and up to 30 or more miles wide. Yet it averages just under 30 feet in depth, with maximum depths up to 170 feet in the main channel. The bay's great width and shallow depth have both good and bad effects. The water of this great shallow trough is easily heated, and when enough nutrients pour into it from the land, it is one of the areas of greatest primary production (see page 9) in the United States. At the same time, the water in so shallow a basin circulates slowly and is more easily polluted.

Salinity ranges from 0 at the head of the bay to 30 parts in a thousand at the mouth. Surface waters are fresher, whereas deeper waters are saltier. Food production is primarily from benthic (bottom-dwelling) animals, mainly oysters, soft-shelled crabs, blue crabs, and clams. All these can accumulate toxic metals, and clams and oysters readily transmit hepatitis.

The settlers of Jamestown Colony were the first known Europeans to become acquainted with Chesapeake Bay. Before European settlement, Indians occupied the lands bordering the bay. The bay was their fishing ground and a primary source of their protein. To the colonists, and even to the oystermen of the last century, the oysters in Chesapeake Bay seemed inexhaustible, but oyster wars and overfishing so reduced oyster beds that state intervention was necessary for their preservation.

Although the bay was not damaged in the first 200 or more years after European settlement, as the population of Washington, Baltimore, and surrounding cities has

Nuclear-electric generating station on the Chesapeake Bay

boomed in recent decades, pollution has increased rapidly and the quality of the bay has been steadily deteriorating. Some of the more important pollutants are listed below:

1. Materials dredged up by the deepening of the Baltimore harbor channel cover bottom communities and increase turbidity in the bay.
2. Dredging of the Baltimore and Ohio Canal produces similar effects.
3. Thermal pollution from the Calvert Bluff nuclear-electric plant will be added to the waters of the bay in 1974 when the plant is opened. Accompanying this thermal pollution will be very low-grade, induced radiation pollution.
4. Industrial wastes from Baltimore and other cities are constantly being dumped into the bay.
5. Governors of the states bordering Chesapeake Bay and the Susquehanna River—Pennsylvania, Maryland, and Virginia—have stated that the area will need at least ten 3000 megawatt nuclear-electric power plants during the next two decades.
6. By far the greatest single pollutant is untreated municipal waste dumped into the Susquehanna, the Potomac, other tributary rivers, and directly into the bay. Population has grown, adequate sewage treatment has not been provided, and waste pollution has worsened.

As a result of pollution, 7 percent of oyster bottoms and 13 percent of clam bottoms have been banned to harvesters. In recent decades oysters have been seriously over-fished, and harvests have dropped to one-fourth or one-fifth those of sixty years ago, although Maryland's oyster industry is said to be increasing at the rate of $1.5 million a year. Pollution, as measured by algal blooms, ranges from intense (with bright green water) to moderate to very light.

Maryland has an intensive sewage treatment program, whereas Virginia and Washington, D.C. have not, and there is a lag time of four to ten years between planning for sewage treatment and actual operation. Consequently, immediate improvement of Chesapeake Bay cannot be expected, because sewage treatment plants are being planned in only a few small cities, and the restoration of the bay awaits general treatment of all sewage. Although Chesapeake Bay may be able to hold its own for a few years, solid improvement cannot be expected for at least a decade.

Lake Washington and the Duwamish River Estuary

In contrast to Chesapeake Bay, Lake Washington and the Duwamish River Estuary are an impressive story of how to correct pollution. The Duwamish flows through the southern part of greater or metropolitan Seattle. The Duwamish Estuary is at most about a half mile wide, or about one-sixtieth the width of Chesapeake Bay, and it averages about half the depth of Chesa-

peake Bay, or a little over 15 feet. North of the Duwamish Estuary is Lake Washington, center of a highly desirable area for residential development in the 1950s. The Duwamish Estuary has been receiving wastes, municipal and industrial, since the early 1900s. But it was not until the 1940s that a serious waste problem developed.

Part of the problem of the Duwamish Estuary is concerned with Lake Washington. In the early 1950s the land around Lake Washington became desirable living

Figure 4–6

Greater Seattle sewer trunklines.

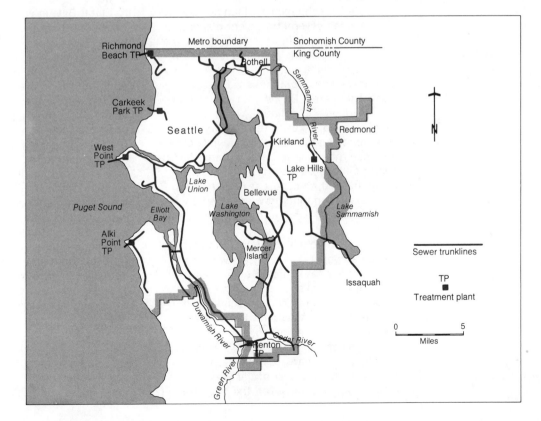

space and was developed rapidly, mostly for residential use. But no sewage treatment facilities were planned, and each new development—village or small city—dumped raw, untreated sewage into the lake. As it began to deteriorate, residents along its shores, and ecologists around the country, became concerned about the quality of lakeside life and the beauty of the lake itself. In 1958 sixteen cities of the area, including Seattle, asked for legislation enabling them to form the Municipality of Metropolitan Seattle to finance a regional sewage and sewage treatment system through public bonds.

The plan for the metropolitan sewage system included several treatment plants and many trunk lines (Figure 4-6). By 1965, seven years after the enabling legislation, the Renton plant went into operation, servicing trunk lines from most of the Lake Washington area. By 1968 Lake Washington, although still polluted, was one of the very few major lakes in the country that was improving in quality. Meanwhile other treatment plants were being built in the metropolitan area. New trunk lines, completed in 1970, transport sewage to treatment plants, and little raw sewage now reaches Lake Washington or the Duwamish River.

In 1963 Municipal Metropolitan Seattle started a cooperative program with the United States Geological Survey (1) to improve the quality of all waters in the greater Seattle area and especially of the Duwamish Estuary, and (2) to ascertain the causes of different types of pollution. Since 1970 much of the Metro's sewage outfall, after primary treatment, enters the Pacific Ocean west of the West Point Treatment Plant. However, the large Renton plant, with a capacity of about 120 million gallons a day, and not yet operating at full capacity, still discharges, and will continue to discharge, treated sewage into the Duwamish River. Although the water is safe, the additional nutrients in the river produce phytoplankton blooms, increase BOD (biological oxygen demand), and reduce DO (dissolved oxygen) (Chapter 5). In other words, eutrophication of the Duwamish River is the result of excess nutrients and the flushing time of the estuary—that is, the time required for water or nutrients to travel from the Renton plant to Elliott Bay. In winter, when the flow of fresh water into the river is high, the nutrients are washed into Elliott Bay in about 20 hours. But in July, August, and September, when the flow of fresh water is low, nutrients spend over 50 hours in the estuary. So, during these three months when temperatures are higher and phytoplankton growth is more rapid, nutrients remain in the shallow estuary longer and become more concentrated. Before 1965 there were only short blooms in the Duwamish Estuary. Since then, additional nutrients from the Renton plant seem to have increased eutrophication and the BOD, and to have lowered DO (Figure 4-7). Phytoplankton blooms are not yet severe enough to prevent salmon migrations, although, as the output of sewage from Renton increases over the next decade, salmon migration could be endangered.

But the story is not yet ended—phosphates and nitrates can be removed by chemical methods during sewage treatment. With a federal grant, Municipal Metropolitan Seattle has installed equipment at Renton to remove these nutrients. If this process is economical

Figure 4–7

Phytoplankton blooms in the Duwamish River Estuary, Washington.

Phytoplankton cells per milliliter × 10⁻³

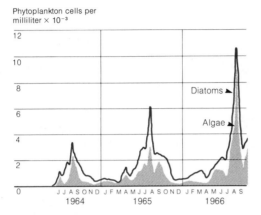

Milligrams per liter

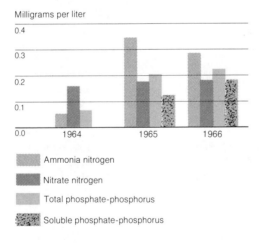

Ammonia nitrogen

Nitrate nitrogen

Total phosphate-phosphorus

Soluble phosphate-phosphorus

it is hoped that the Duwamish Estuary and River can be returned to a nearly pristine condition, and that salmon runs will continue indefinitely.

San Francisco Bay

San Francisco Bay has the same problems of industrial and municipal waste that have plagued Chesapeake Bay and the Duwamish Estuary. It also has the problem of decreased size because of land filling and real estate development, and the pollution created by ships of the United States Navy. Added to all this are the problems brought on by high earthquake risk.

San Francisco Bay is about 55 miles long and from three to five miles wide. It occupies a structural basin in the California Coast Ranges and opens into the Pacific Ocean through Golden Gate at the northwestern end. Golden Gate was the channel of the Sacramento River during low sealevel stages of the Pleistocene. The geologic history of the bay is dominated by the kind of faulting that produces earthquakes. There are two tidal cycles a day in San Francisco Bay, and the maximum total change each day from low to high tide ranges from 10 to 14 feet. The mass of water moved each 24 hours is 1,255,000 acre feet or about 3.5×10^{10} gallons. It is this giant pumping action of the tides that washes pollution and waste into the Pacific Ocean and maintains in the bay a relatively high salinity of 27 to 29 parts per thousand. This salinity, high for bays in nondesert areas, indicates that the fresh water inflow is low, especially from the southern end (Figure 4-8). The salinity rate has been relatively constant over the last several thousand years, and life in and around the bay has adapted to it.

Although Golden Gate is 350 feet deep, 80 percent of the bay is less than 12 feet deep, providing an ideal area of shallow water and tidal flats for primary and secondary marine production.

The area of San Francisco Bay in 1850

Figure 4-8

Landfill and potential land-fill, San Francisco Bay.

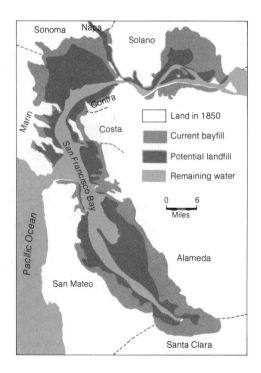

Land in 1850

Current bayfill

Potential landfill

Remaining water

0   6
Miles

gold. All this loosened sediment reached the streams, and by 1970 nearly all that was destined to reach San Francisco Bay had arrived. A total of 16×10^8 million cubic yards of sediment was transported to the bay in the 120 years since the beginning of gold mining. This is perhaps the greatest single removal of sediment on record to result from the activity of man. The bay received almost three times the normal amount of sediment per year for almost 120 years.

Species of wild life common to the bay are salmon, seal, ducks, egrets, pelicans, cormorants, and many others. As natural and artificial land fill and development continues, there is less and less space for marine and shore animals and plants. *Ostrea lurida*, the native oyster, has almost disappeared from the bay area because of pollution. From 1924 to 1970 thirty million tons of oyster shell were dredged from the bay for use in the construction industry. Not only is a food source now gone, but pollution has stopped the renewal of a valuable source of aggregate.

Problems of the future will include (1) the strength of bay muds, (2) municipal and industrial wastes, (3) the San Joaquin drain, and (4) space and aesthetics. Earthquakes are common in the San Francisco area, and one question for the future has to do with the strength of un-consolidated bay muds and fill when earth-quakes occur. In the 1906 earthquake, structures built on unconsolidated sediment were much more severely damaged than structures on firm rock. Industrial and municipal wastes still endanger much of the wild life in the area, and this danger will increase if pollution and waste are not

was 680 square miles, but by 1970 it had been reduced to about 400 square miles. The decrease has resulted from two ac-tivities: (1) the placer gold mining and dredging operations on the Sacramento River and its tributaries in the mountains, and (2) the filling in of the bay in later decades by developers producing more land for building. The bay normally receives about 8 million cubic yards of sediment a year. In the gold rush years, tremendous amounts of rock were dredged loose in the stream channels, and veritable mountains were washed away by the high pressure nozzles of hydraulic mining, spurting water to wash the rock away from the heavier

controlled. Pollution is particularly high when the U.S. Navy is in port, but ships of all kinds and sizes have dumped untreated sewage into the bay in the past.

The San Joaquin drain is a canal 200 miles long. It was built to remove to the Pacific Ocean through San Francisco Bay salt and other agricultural wastes that have accumulated in the San Joaquin Valley of California during one hundred years of farming.

In 1963 many citizens of San Francisco, alarmed over the future, formed the "Save Our Bay Action Committee." This committee has worked with the Nature Conservancy in an effort to acquire bay shore acreage for the benefit of wild life and bay aesthetics. This program is under way but is proceeding slowly because of the high cost of land. Recent laws greatly restrict the process of "making land" by filling in shallow parts of the bay. Still, some of the bay area's special problems are being ignored, such as the earthquake hazard. Building codes do not consider earthquake hazards from unconsolidated sediments (Chapter 8). Many of the problems are being worked on; nevertheless, time is short.

Altering Marine Ecosystems

More grandiose schemes to alter the oceans will affect ecosystems in ways that cannot now be anticipated. The North Florida Canal was intended to decrease the distance for shipping from the Atlantic coast to ports on the Gulf of Mexico. Construction on the canal was stopped to prevent mixing the waters and life of the Gulf of Mexico and the Atlantic Ocean. A sea level canal across Central America has also been proposed.

The general conclusions are that it is feasible, and that without locks not much Pacific water would flow into the Atlantic, although Pacific tides would be up to ten feet higher than Atlantic tides. More recent plans show locks at each end; however, there remains no doubt that many marine animals could swim through, or even spread through during larval stages. The total effect on the ecosystem of allowing elements of Pacific biotas into the Atlantic, or vice versa, cannot now be estimated. On the other hand, several such isthmus destructions occurred during the last fifty million years (during the Cenozoic) without earth-shaking effects on marine fauna and flora. Bioenvironmental research in this area is not difficult. Model environments with controlled input of organisms could be designed to study the effect of Atlantic and Pacific marine biota on each other, that is, to determine the competitive advantages of Pacific or Atlantic species under a variety of controlled conditions.

Summary

Not only are the earth's oceans unique to the solar system, they perform a unique service by cleansing the atmosphere and the rest of the hydrosphere of unwanted waste products through biochemical, hydrologic, and rock cycles. They also provide the atmosphere with desirable trace substances that pass through the interface between water and air by evaporation, bursting bubbles, and breaking waves. Furthermore, the oceans are the primary and long range mediator of climate. In all these respects they improve the environment of man and the other inhabitants of

Water pollution from a
North Carolina mill—before
federal regulation

earth. The ocean is the home of much of man's food, and 90 percent of the oxygen of the atmosphere is produced by its plant life.

Some of these fundamental services of the ocean to life can be upset. It is remotely possible that toxic pollutants from the atmosphere and the land could decrease the growth of oxygen-producing plants.

Bays, lagoons, and estuaries have been polluted to the detriment of life, and since these areas are breeding grounds for even more marine life, pollution of the fringe areas of the ocean can affect the quantity of marine life throughout the ocean.

The fringe areas of the ocean can be improved if people desire it and if different levels of governments cooperate. The open ocean is a different problem. Here national interests interfere with the proper cooperation of governments to prevent pollution, and detection of offenders is difficult. The United Nations must continue to involve itself in settling differences between governments and in applying pressure on governments for the improvement of the oceans.

Selected Readings

Benton, G. S., et al. 1962. Interaction between the atmosphere and the oceans. National Academy of Science-National Research Council, Publ. no. 983.

Bullard, Edward. 1969. The origin of the oceans. Scientific American 221(3):66–75.

Holland, Heinrich D. 1972. The geologic history of sea water—an attempt to solve the problem. Geochimica et Cosmochimica Acta 36:637–651.

Murphy, Robert Cushman. 1962. The oceanic life of the Antarctic. Scientific American 207(3):186–210.

Santos, J. F., and J. D. Stoner. 1972. Physical, chemical, and biological aspects of the Duwamish River Estuary, King County, Washington: 1963–67. U.S. Geol. Survey Water Supply Paper 1873-C.

Stewart, R. W. 1969. The atmosphere and the ocean. Scientific American 221(3):76–86.

Tilson, Seymour. 1966. The ocean. Science and Technology, no. 50, pp. 26–37, Feb.

Welch, Eugene B. 1969. Factors initiating phytoplankton blooms and resulting effects on dissolved oxygen in Duwamish River Estuary, Seattle, Washington. U.S. Geol. Survey Water Supply Paper 1873-A.

Wenk, Edward, Jr. 1969. The physical resources of the ocean. Scientific American 221(3):166–177.

Wurster, Charles F., Jr. 1968. DDT reduces photosynthesis by marine plankton. Science 159:1474–1475.

Yerkes, R. F., H. C. Wagner, and K. A. Yenne. 1969. Petroleum development in the region of the Santa Barbara Channel. U.S. Geol. Survey Prof. Paper 679-B.

Chapter 5

The Water We Drink

Man's two most precious assets—air and water—are the ones that are deteriorating most rapidly. People have always needed potable water. Primitive man had to live near water, and waterless lands could only be used to the extent that water could be carried to them. The lands affected during droughts had to be evacuated until methods of storing and transporting water were developed. As late stone age cultures evolved into agricultures, man became even more dependent on large quantities of water. People in settled communities could not leave their crops to search for water, and the crops themselves required a ready supply. Availability of water thus restricted the spread of early intensive agriculture. The last ice age (see Table 13-1) was followed by a wet period and then by an expansion of deserts. Men were forced to accommodate themselves to this climatic change and did so either by following the rainfall, or by designing sophisticated irrigation systems, such as those in the upper Euphrates Valley, which in their earliest forms are estimated to be 5000 years old.

As the Middle East became more arid after the post-Pleistocene wet spell, early farmers had to find water. As early as 3000 B.C. extensive qanats were developed in the area of the Tigris and Euphrates Rivers. A typical qanat is a slightly sloping tunnel into permeable rock or sediment (Figure 5-1). The tunnel picks up water percolating downward toward the water table, and may even intersect the water table at the upper end of the tunnel. The water is brought through the tunnel to an area which is to be irrigated. A fine qanat provides the water supply for Párras,

Figure 5-1

A typical qanat used in water production.

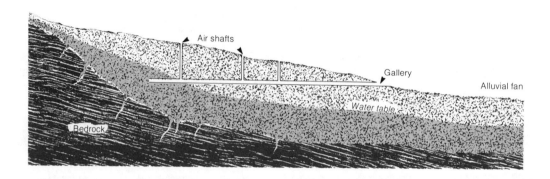

Coahuila, Mexico, where the tunnel extends two kilometers into the mountains through strata that are almost vertical. In Hawaii and the Canary Islands impervious dikes run through fractured and otherwise permeable volcanic rock. Here horizontal or only slightly sloping tunnels penetrate the dikes to tap the underground reservoirs (Figure 5-2).

Nowhere was, or is, a people more dependent on water than in Egypt. Before the first Aswan Dam was built in 1902, the food supply of the entire population depended on irrigation from the Nile River and the waters of its annual flood. In the Old Kingdom, when the civil calendar was established almost 5000 years ago, the annual flood was so important that the calendar makers began the new year with its arrival.

Figure 5-3

The hydrologic cycle.

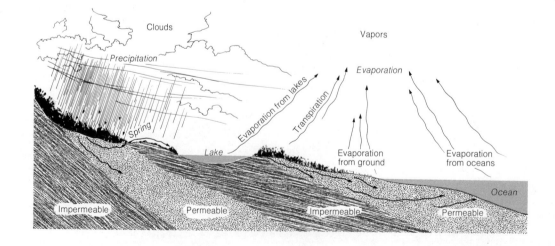

How much water?

The following table shows distribution of the earth's waters:

	Gallons	Percentage
Oceans	1,070,000 × 10¹³	97.2
Ice caps and glaciers	24,000 × 10¹³	2.15
Underground water	7,000 × 10¹³	0.625
Surface water	0.189 × 10¹³	0.017
		99.992

Figure 5-2

A typical gallery that intersects the water table.

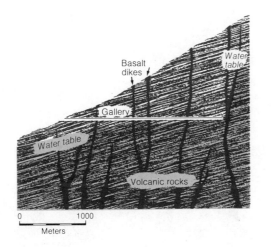

Basalt dikes

Water table

Gallery

Water table

Volcanic rocks

0   1000
Meters

Table 5-1

The larger rivers of the world are those with drainage basins of 1 million square miles or more. From U.S. Geological Survey, "Water of the World," 1968.

River	Square miles
Amazon	2,368,000
Congo	1,243,000
Mississippi	1,240,000
Rio de la Plata	1,198,000
Ob	1,150,000
Nile	1,107,000
Lena	1,000,000

Ocean water is too salty to be used by man. Much of the underground water is unavailable because it is associated with clay or occurs in aquifers that are too small or too deep, and two percent of the earth's water is tied up in the form of ice caps and glaciers. This leaves approximately 0.017 percent of the earth's water available as surface water, plus a few tenths of a percent available to man as ground water.

Surface water

Surface water is produced through the hydrologic cycle (Figure 5-3). It consists of precipitation that collects in streams, ponds, and lakes, and is our major source of potable water. Rivers with drainage basins of one million square miles or more are listed in Table 5-1, and the capacity of some of the world's larger lakes is given in Figure 5-4. Although these represent much of the world's surface water, they supply only a small part of man's needs.

Because of the effects of differences in climate, the volume, flow, and drainage area of rivers may have little in common. For example, the drainage area of the Amazon River is about one and one-half times that of the Mississippi, but its discharge is almost ten times that of the Mississippi, primarily because of much greater average rainfall in the Amazon drainage basin. Moreover, evapotranspiration rates are higher in desert areas, resulting in even greater deficiencies of water per unit area of the earth's surface (Figures 5-5 and 5-6).

About 70 percent of annual precipitation is returned to the atmosphere by evaporation and transpiration (Table 5-2). Of the remaining 30 percent, about 22 percent is not withdrawn for use. (Nearly all of this

Figure 5-4

Water capacity of some of the world's great lakes.

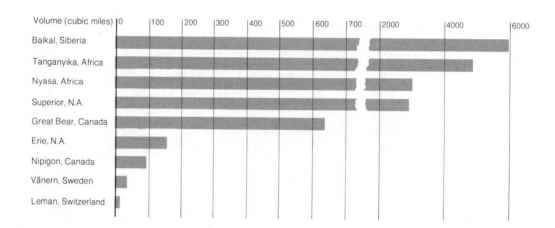

Figure 5-5

Annual precipitation in the United States.

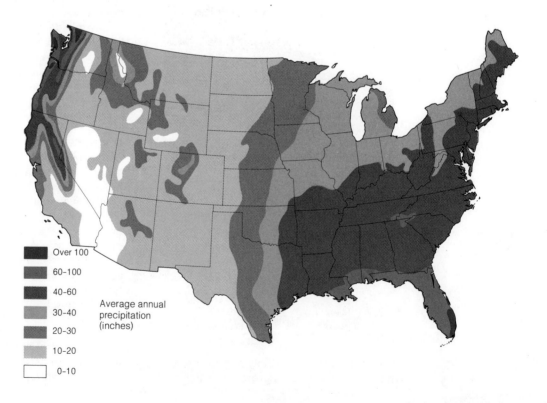

Figure 5–6

Water surpluses and defi-
ciencies, in inches of
rainfall, in the United States.

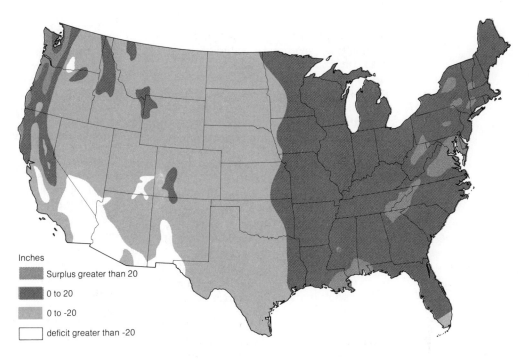

Inches

- Surplus greater than 20
- 0 to 20
- 0 to -20
- deficit greater than -20

Table 5–2

Distribution of precipitated
water in the conterminous
United States.

	10^{12} gals/yr	Percent
Precipitation	1552	100.0
Evapotranspiration	1086.4	70.0
Pasture and crops	357	23.0
Forests and browse	248.4	16.0
Noneconomic	481.1	31.0
Stream flow	465.6	30.0
Not withdrawn	341.4	22.0
Withdrawn and used	124.2	8.0
Irrigation	62.1	3.4
Industry	52.8	3.4
Municipal	9.3	0.6

22 percent is in nondesert areas.) The
remaining water—most of the water of
desert areas—is the 8 percent that is withdrawn and used.

Water chemistry and natural conditions

One of the more important aspects of water
is its chemistry. Water chemistry varies, and
several aspects of the environment affect
the chemistry of the water in an area, just
as several aspects of the chemistry of
water affect the environment and its inhabitants. Geologic formations through or
over which water flows contribute most of
the dissolved minerals that it contains.
The direct processes are those of the hydrologic, biochemical, and geochemical cycles
(Chapters 1 and 7). Rocks contribute dis-

solved solids to water by a natural process. Other natural conditions that affect the chemistry of water are precipitation and evapotranspiration.

Precipitation tends to dilute the mineral content of water. A good example is the Colorado River at Austin, Texas, where in drought years much of the flow of the river comes from underground water passing through limestone formations. As a result, the river's fluorine content may reach 0.7 ppm in those years. In wet years, when a much greater percentage of the precipitation flows rapidly to feeder streams via surface runoff, the fluorine content may drop to less than 0.3 ppm. A second way in which precipitation alters water chemistry is by transferring air pollutants, such as sulfur compounds and nitrous oxides, to the water.

Evaporation removes water to the air, thereby increasing the mineral content of the water that remains. Transpiring plants also transfer water from the surface to the air, leaving the dissolved minerals behind.

Water chemistry and artificial conditions

Artificial factors which affect water chemistry include impoundment, artificial recharge, and disposal of waste material. Impoundment or confinement of water—as in a lake or reservoir—increases the dissolved solid content because evaporation leaves the chemicals in higher concentrations. The effect of impoundment in any area is proportional to the evaporation rate, and is greatest in desert areas. Artificial recharge may add undesirable pol-

lutants and minerals to the underground storage. Disposal of waste matter directly into streams or into underground reservoirs adds the minerals, organic compounds, and toxicants which are present in the pollutants to the water. All these, of course, alter the chemistry of the water.

Significant mineral constituents of water

Chemical elements and compounds—often dissolved in water in varying amounts—affect its taste, odor, medical quality, and industrial usefulness. The major minerals are discussed in Table 5-3. State and federal regulations attempt to maintain continual monitoring of water in order to provide satisfactory quality. Some industries require a higher quality of water than that suitable for human consumption, because certain minerals precipitated from water will clog pipes, valves, and sprays in boilers, sprinklers, and other hydraulic industrial systems and yet have no harmful effects on man. Among chemical substances which affect water quality are silica, aluminum, iron, manganese, calcium, magnesium, sodium, potassium, chlorine, fluorine, sulfate, and nitrates.

Other aspects of water quality

Other important characteristics of water chemistry include hydrogen ion concentration, dissolved oxygen, free carbon dioxide, hardness, turbidity, temperature, and radionuclides—all referred to in Table 5-3—as well as specific conductance, color, and biochemical oxygen demand.

Specific conductance. Specific conductance is the capacity of water to conduct

Water pollution in the
channel, Washington, D.C.

Data from Brian J. Skinner, *Earth Resources* (Englewood Cliffs, N. J.: Prentice-Hall, Inc., 1969).

Table 5-3 Some aspects of water quality as related to water chemistry.

Chemical factor	Normal water content	Effect on biological systems	Industrial effects	Chemistry	Undesirable properties	Desirable properties
Silica	< 30 ppm	usually none	usually none	relatively insoluble in waters of normal pH (6–8.5)	with a low chlorine diet, herbivores may have a reaction to hydrous silicates because of high blood pH	
Aluminum	usually < 0.1 ppm	harmful to eyes in amounts of 0.1 ppm		very insoluble below pH of 9		
Iron	usually < 1.0 ppm	necessary to animal systems; usually not harmful in greater amounts	some industries have a low iron tolerance	more soluble in water of pH < 4.0	stains porcelain and discolors water if more than 1 ppm	
Manganese		necessary to animal systems in trace amounts; toxic at high levels	some industries have a low manganese tolerance	similar to the chemistry of iron	stains porcelain and gives drinking water an objectionable taste	
Calcium	10–100 ppm or more	necessary for biological systems	some industrial processes require low calcium content	serves as a buffer to undesirable toxics; maintains pH above 6		helps to keep soils loose and tillable, aerated, and nitrogenated
Magnesium		necessary in biological systems; cathartic if content is too high	some industrial processes require low magnesium	serves as a buffer to undesirable toxics		helps to keep soils loose and tillable, aerated, and nitrogenated
Sodium		necessary to life, but high concentrations are cathartic and upset K-Na balance in animals		very soluble	packs soils, preventing aeration and nitrogenation	
Potassium		essential to animal systems; cathartic if in excess in presence of sulfate; helps to regulate cardiac rhythm	none	very soluble	packs soils, preventing aeration and nitrogenation	
Chlorine		necessary to biological systems in trace amounts	undesirable in ice, milk, and sugar industries	corrosive	corrosive	

Chemical factor	Normal water content	Effect on biological systems	Industrial effects	Chemistry	Undesirable properties	Desirable properties
Fluorine	0–4 ppm	necessary to biological systems in trace amounts, decreases dental carie, in up to 1.5 ppm as a fluoride			toxic as a gas or in large quantities	
Sulfate		strong cathartic, but traces essential to biological systems	undesirable in ice, milk, and sugar industries	very soluble; usually a measure of pollution in rainwater	forms strong acids	
Nitrate	usually low	carcinogenic and produces blue babies in 10–45 or more ppm	undesirable in fermenting and dyeing industries	reacts readily with organic compounds	promotes excessive plant growth when in excess	
Alkalinity	recommended public water supply range: pH of 5–8.5	no effect except for individual constituents	effect is of the individual constituents	usually produced by hydroxides	irrigation waters should not be too alkaline	buffers drastic pH changes that might be harmful to aquatic life
Dissolved solids (DS)	50–3000 ppm	1000 ppm or more usually undesirable for humans	industry usually requires less than 1000 ppm		more than 2000 ppm undesirable for irrigation	most inland fauna are adapted to about 400 ppm
Hardness			undesirable in industry	calcium and magnesium are the primary constituents	a measure of soap consuming properties	
Hydrogen ion concentration		not important		a measure of the acidity		fishes are usually pH specific
Dissolved oxygen (DO)	4–8 ppm	aquatic faunas require specific amounts; requirements vary with fauna	little effect		measure of pollution; less DO = greater pollution	
Free carbon dioxide		not important	little industrial effect			fishes are usually adapted to specific free carbon dioxide ranges
Turbidity		5 ppm or less recommended for human consumption			esthetically undesirable	little effect on stream life in normal amounts
Temperature		all life is adapted to specific temperature ranges			increased temperature decreases palatability and increases organic growth	warmer irrigation waters increase rate of plant growth
Radioactivity	public water supply standards: gross beta: < 1000 pc/l radium 226: < 3 pc/l strontium 90: < 10 pc/l		harmful to photographic processes	measured in picocuries per liter (pc/l) for each radionuclide	excess radiation is deleterious to biological systems	none

an electric current. Since this depends on the ions of dissolved solids in the water, the specific conductance gives a rapid estimate of the total dissolved solids.

Color. Color is the appearance of water without solids. The primary importance of color is aesthetic, but some colors stain. Both municipal and industrial waters should be free of color.

Biochemical oxygen demand. The biochemical oxygen demand (BOD) is the measure of oxygen required to oxidize organic matter in the water. BOD is not a pollutant, but a general measure of the pollution load in water.

Water Use and Water Needs

Overuse and pollution are endangering our supply of surface water. Since it is not likely that people will be satisfied with less water than that to which they have become accustomed, the available remedies are more extensive recycling and the addition of desalted water to water systems depleted by overuse upstream.

Surface waters are used for human consumption, some agriculture, industry, and recreation. If all these uses increase as expected, water use will rise unless the population decreases or per capita consumption is cut back. This holds not only for the United States but for the rest of the world.

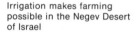

Irrigation makes farming possible in the Negev Desert of Israel

The great Salt Lake in Utah is shrinking

Some surface waters are already being used to their maximum capacity. The volume of water in the Great Lakes seems to be generally decreasing because of overuse, pollution, and the aging cycle of lakes. Enough water should be left in these lakes so they can survive the periodic droughts common to the area, such as the 1967 drought in New York State. A drastic decrease in evaporation due to overconsumption of water in the Great Lakes might also result in less rainfall downwind.

The belief that "rain follows the plow" initiated a dry-farming cycle in the American West which was periodically renewed in the latter nineteenth century, and again in the second and third decades of the twentieth. Today we find the return of the idea that irrigation produces transpiration and evapo-

The Dust Bowl of the United States once received flocks of hopeful farmers

ration, and that this evapotranspiration adds moisture to the air and produces more precipitation downwind from the irrigated area. If this concept is valid, its effect is negligible at best for periodic droughts continue unabated.

Theorists expecting climatic change from irrigation or from particulate pollution forget that the earth's water budget is fairly constant. Increased rainfall at one locality means decreased rainfall somewhere else, though there are two exceptions to this rule: (1) evapotranspiration from irrigation by ground water adds new water to the water cycle, and (2) desalted water, if transported inland for irrigation, would also add new water to the cycle.

Conservation of surface water

It is sometimes claimed that one way to conserve water is to return to the Saturday-night bath, but most bath water, and indeed most sewage, is not wasted; it is recycled and returned to the system to be used again. Some environmental handbooks have suggested putting a brick or two into the tanks of toilets. However, this practice is impractical because sewage systems are designed to accommodate a definite ratio of liquids to solids, and changing the ratio could disrupt the flow of sewage and produce local pollution problems. Besides, water which passes into sewage systems is recycled, and so is not ultimately wasted. Restrictions on the use of water are placed on individual households in times of severe drought, but the greatest users of water are industry and agriculture. It is in these areas that new approaches to water conservation would be likely to produce the most significant results.

Some problems arising from the quantity of water available for use at times of peak load could be solved or at least reduced if it were somehow possible to level off the peaks. For example, one study indicates that there are peaks in water use from 7 a.m. to 9 a.m., at about noon, and from about 6 p.m. to 7 p.m., with a gradual decline from 8 p.m. to 1 a.m., but with minor peaks on the half hour because of TV commercials, and a leveling off after 1 o'clock in the morning. In the South and Southwest and in many cities, peaks also come with the maximum use of air-conditioners and the watering of lawns. Towns and cities should design water and sewage systems which can take care of the maximum capacities instead of the much lower average loads, yet, so far, little progress has been made in leveling off these periods of overloading.

Certain plants, called phreatophytes transpire inordinate amounts of water into the atmosphere to the extent that the federal government pays ranchers and farmers in arid areas to destroy them and thus conserve water. Artificial ponds and lakes give off water by surface evaporation. There are hundreds of thousands of such ponds throughout the middle and western United States, and while the water they give off comes back to earth somewhere by the hydrologic cycle, it comes back elsewhere.

Other conservation measures include curbing pollution and maintaining desirable water chemistry. Maintaining the quality of water is expensive, because it requires sensitive electronic equipment capable of measuring minute amounts of potentially toxic substances. Municipal, state, and federal health departments, and state and

Water purification plant in
southern France

federal water quality boards and commissions, continually monitor the water supply
for all kinds of pollutants and toxicants.
In addition, some good municipal sewage
treatment plants monitor the chemistry of
their product several times daily to insure
optimum water quality downstream.

Sewage treatment and recycling

Sewage may be processed to obtain potable water. Though many people object to

this idea, probably because of inhibitions
about all things scatological, properly
treated sewage yields water purer than
that of most streams. There are many ways
of treating sewage, but the three basic
processes are: (1) mechanical or physical
methods, such as screening out coarse
materials, or flocculating (coagulating) them
by heat, evaporation, or centrifugation;
(2) chemical methods, such as precipitation
by the addition of agents which react with
sewage components to produce insoluble

precipitates; and (3) biological methods, in which sewage components are digested by bacteria or used as food by photosynthesizers, and thus used up and rendered harmless.

Mechanical treatment includes screening, filtration, skimming, and sedimentation. Screening removes coarse particles such as buttons, pebbles, and orange peels. Filtration removes finer particles. Skimming refers to scraping floatables, such as grease, from the surface of the sewage. Sedimentation allows matter heavier than water to sink to the bottom where it is either removed and buried or allowed to accumulate.

Chemical treatment includes coagulation, deodorization, aeration, and disinfection. Coagulation involves the addition of chemicals which precipitate dissolved solids. Sewage may be deodorized by adding specific chemicals which neutralize foul-smelling compounds. Disinfection is the addition of sterilizing chemicals, usually chlorine or ozone, to kill bacteria and viruses. Aeration is the addition of oxygen to reduce the BOD, and in this way neutralize harmful components—usually accomplished by pumping air through the sewage.

Biological treatment consists of allowing organisms to consume and degrade sewage. In septic tanks this is frequently carried out in the environment of the natural soil by soil bacteria, but artificial environments are often provided where organisms are much more concentrated than in natural environments.

Good sewage treatment plants use a combination of mechanical, chemical, and biological methods. In southern states, where cold winter temperatures are not a problem, a very successful combination has been to screen and aerate sewage, and then to deposit it in large ponds where organic activity consumes the nutrients, and the organic debris settles to the bottom. This process can be followed by chlorination, though if the pond is large and residence time is long, disinfection is not always necessary. If water is to be recycled into a municipal system it is usually filtered a second time. Because of the high cost of large eutrophication ponds, the ponding method is used much less than it merits. In Tucson, Arizona, sewage treated in this way is collected in an artificial lake used for such recreation as boating, water skiing, and fishing.

In areas where cold winters prevent the use of natural eutrophication ponds, aeration is usually followed by bacterial digestion kept warm to maintain optimum biological temperatures. (Sometimes the mixture is heated with the methane gas produced by the treatment.) Disinfection follows, and the nutrients are removed chemically.

The combinations just discussed are more advanced than those used by many American cities, which only partially treat their sewage or do not treat it at all. Many cities are being forced by federal legislation to improve their sewage treatment. The manner of treatments is not legislated, but the BOD, bacterial level, and nutrient level of the effluent must meet prescribed standards. Several combinations of mechanical, chemical, and biological treatments can produce effluents which meet present standards.

Desalination plant, Key
West, Florida

Desalination

Desalination is the removal of salt from
water to make it potable. Most people think
of sea water when desalination is mentioned.
But many surface and ground waters contain dissolved solids above the maximum
allowed for human consumption. Since such
waters have fewer dissolved solids than

sea water, it is cheaper to desalinate them.
Desalination presents several problems:
(1) disposal of brine, (2) energy consumption, (3) transportation inland, and (4)
maintenance of the proper balance of trace
elements.

Disposal of Brine.   Normally, saline marine
water contains 30,000 ppm or more of dis-

solved solids. Thus a plant producing 100 million gallons of water per day—enough to supply a medium-sized city—would also produce over 1,000 cubic yards of salt per day. In most methods of desalination, salt is not completely dried, but remains in a mushy liquid form called a brine

In the future some valuable elements may be extracted from brines, but at present most of the dissolved solids are uneconomic and waste. Brine cannot be dumped on land because it will kill plants and pollute soils and water. It cannot be returned to the source because it will increase the salinity of other water later to be desalinated, and thus increase the cost of further treatment. It cannot be dumped in just any bay or lagoon because marine and brackish water organisms are adapted to specific salinites and would be harmed or destroyed by the change. The best—and most expensive—solution is to transport the brine to the open ocean where it can be diluted by ocean currents.

Energy Consumption. Desalination requires great amounts of energy in a time when an energy shortage *is impending.* The use of solar energy might solve this problem (see page 43).

Transportation Inland. Only in a few near-shore desert areas, such as Israel, parts of Arabia, and northwestern Mexico, is it feasible to produce water for irrigation by desalination. Such water must be used locally. The transportation of desalinated marine water inland, even by pipeline, is as yet uneconomic for human consumption, much less for agriculture.

Trace Substances. It is not yet clear whether the different methods of salt removal retain in treated water the proper balance of trace substances for human health. This subject needs more thorough study.

Methods of Desalination. There are several ways to remove dissolved solids from water. In some desert areas of Israel marine water is piped into large, shallow vats with sloping plastic or glass roofs. Solar energy transmitted through the roofs heats the water. The water evaporates and then, free of salt, condenses on the roof and drains into collectors. Another method takes advantage of the fact that most of the dissolved solids are expelled from salt water when it freezes. Freezing water and collecting the ice to remelt will produce potable water. A plant at Roswell, New Mexico removes only moderate amounts of dissolved solids from saline water produced from an underground source. In this operation saline water is forced into a large system where evaporated moisture is drawn off and compressed (Figure 5-7). In another system water is heated to produce steam, which is then condensed in a series of coils and drawn off.

There are many other methods of desalting water. Choice of method depends on the availability and type of energy, the amount of dissolved solids to be removed, the amount of water needed, and the price the user is willing to pay. These factors make desalted water economic in some areas but uneconomic in others.

Figure 5–7

The forced circulation, vapor-compression distillation plant that desalts saline ground water in Roswell, New Mexico.

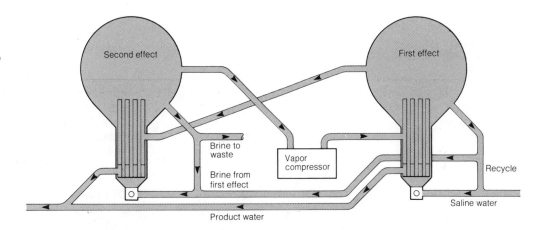

Planning for More Water

The traditional approach to water conservation has been to put it under the jurisdiction of the River Basin Authority. In Germany this Authority has complete control of planning, pollution control, and conservation. In the United States the river basin authorities are structured like those in Germany, but do not have complete or final control. Each such authority has an official organization which controls various aspects of water resources, but most are tightly restricted by their enabling legislation. Even when river basin authorities in the United States have been given some autonomy, they have either failed to act or at best have been lax in the enforcement of standards. Consequently most authorities have relied on cooperation with other governmental levels for success.

The kinds of planning done by river basin authorities include estimates of future needs, of number and kinds of dams, and of volume of water to be reserved for conservation, power, and flood control. Recently some river basin authorities have been given responsibility for monitoring and suggesting ways to control pollution. Upcoming legislation could give river basin authorities and states a much firmer hand in planning and pollution control.

Grandiose water schemes

International Schemes. There are also more grandiose plans afoot. One of these, the Parana River project to dam the Parana River in Brazil, has drawn the ire of Argentina, and objections from both Bolivia and Uruguay. Argentina fears that there will not be enough water left for Buenos Aires and the maintenance of the Plate River Estuary. Similarly, Jordan has objected to Israel's Jordan River project on the grounds that it might cut off Jordan River water

Figure 5–8

India and Pakistan plan to use the same water source (of contention) for irrigation.

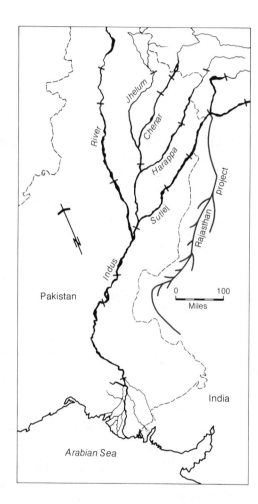

the problems between Jordan and Israel, and Argentina and Brazil, India has had water confrontations with Pakistan. Pakistan is planning and building several hundreds of miles of canals, with many dams, to irrigate the Indus Plain (Figure 5-8) with water from the Indus River and its tributaries, including the Harappa and the Sutlej Rivers. Southeast of the border between Pakistan and India, India is constructing and extending the Rajasthan project to irrigate the Rann of Kutch, a desert area. This plan involves damming the Harappa and Sutlej Rivers upstream from the Pakistan dams. Pakistan is determined that India will not take all the water, including that which "rightfully belongs to Pakistan."

Unforeseen Contingencies. There are also problems of a very different sort. The first phase of the California aqueduct (Figure 5-9) is substantially completed, but already it has been discovered that pelecypods (in California the culprit is a species of *Corbicula*) have cut the transport potential of the canal by as much as 30 percent. It now appears that the canal will have to be shut down several months every three years in order to clean out the *Corbicula*. But no such emergency was anticipated in the original plan, and no water storage was provided for municipalities and other users during the cleaning period. Furthermore, in spite of engineering plans to provide for subsidence and tectonic uplift in the path of the aqueduct, earth movements, including uplift and subsidence of up to one foot since construction, are threatening the life of the canal.

Water projects are being continually sug-

from Jordan. Likewise, any river projects by Jordan that would tend to interfere with the use of water by the Israelis would find strong objection from Israel.

Water produces some of the worst confrontations between nations. In addition to

Figure 5-9

Faults along the California aqueduct.

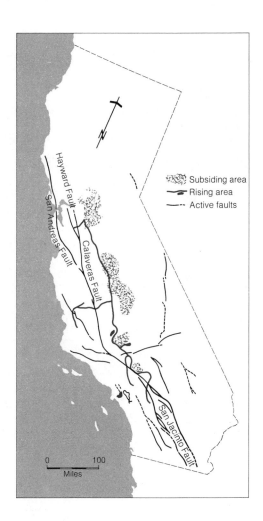

Subsiding area
Rising area
Active faults

0   100
Miles

gested all around the world. Many are needed; some are not. Coordination of water resources and large diversion projects is a goal (most difficult to achieve) of the United Nations Economic and Scientific Organization.

Ground Water

Source and distribution

Approximately 0.625% (about 7000×10^{13} gallons) of the earth's water is underground. This is approximately 37 times the amount of water in all the rivers and lakes of the world. Much of the ground water was trapped within sediments when they were deposited and buried with those sediments. If these sedimentary rocks are marine, the trapped water is salty; if the sedimentary rocks are of alluvial or of fresh water origin, the water is potable or fresh, that is, suitable for human consumption.

Part of the annual precipitation enters the underground supply and becomes ground water. It is potable unless it enters rocks that contain so much salt that the water becomes salty.

Rocks that can contain other mediums, such as water, gas, or liquid hydrocarbons, are porous and are said to have porosity. In clean, mostly uncemented quartz sandstones, porosity may be as much as 30 percent of the rock. In tight, compact claystones, it may be as little as a fraction of one percent. But unless the cavities in porous rocks are interconnected, the liquids or gases cannot pass either through or out of the rocks. If cavities, either minute or large, are interconnected so that their contents can flow through or out, the rock is said to be permeable. Rocks through which liquids cannot flow are said to be impermeable. A large amount of ground water is trapped between the layers of claystones or in other impermeable rocks and cannot be recovered. Other rocks are barely permeable, and

Figure 5-10

Ground water conditions for confined and unconfined aquifers.

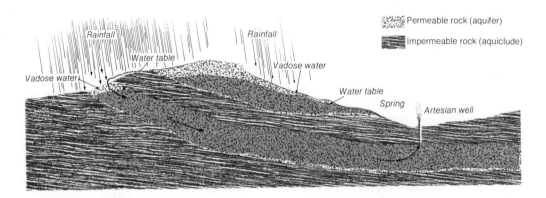

water cannot be obtained from them economically because transmission is too slow.

Precipitation descends through the ground until it meets impermeable rock or joins water already in the ground. A permeable formation can be filled to a certain level, known as the water table which is the top surface of the ground water. Water descending between the surface of the ground and the water table is called vadose water. Where the water table intersects the ground surface there is surface water in the form of a spring, swamp, lake, or stream. A permeable stratum that carries water is known as an aquifer. A confined aquifer is one contained between imper-

meable strata called aquicludes. If a well is drilled to a confined aquifer and water rises in the hole above the aquifer, the well is said to be artesian. If water flows from the well at the ground surface it constitutes a flowing artesian well (Figure 5-10).

Ground water problems

Producible underground water does not occur everywhere. Even where underground water is present it is seldom available in the quantities needed. Essentially there are two kinds of underground aquifers: the sandstone aquifer and the limestone aquifer.

Sandstone aquifers

Even when most permeable, these aquifers carry water very slowly. At one extreme is Carrizo-Wilcox sandstone aquifer in Texas, through which water moves at rates between 5 and 50 feet a year. Water in the Ogallala

aquifer of the high plains of Kansas, Oklahoma, and Texas moves somewhat more rapidly, with a maximum estimated rate about three times that of the Carrizo-Wilcox aquifer, or about 150 feet a year.

Figure 5-11

Decline curves for the Carrizo-Wilcox (sandstone) and the Edwards (limestone) aquifers.

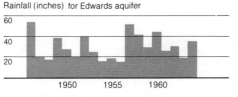

Rainfall (inches) for Edwards aquifer

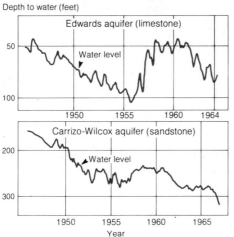

In terms of usable water, the Ogallala aquifer may transmit water at the rate of one mile in 30 or 40 years. Since the water input may be 100 miles away from the area of primary use, natural recharge would take three or four thousand years. Obviously the economy of the high plains cannot wait for this water. Either there must be new water or the economy will revert to the production of animal protein by grazing. The latter solution may be the more desirable in some ways, but a grazing and feeding economy will probably not support the present population of the area.

Since sandstone reservoirs transmit water so slowly, we cannot depend on them to solve immediate water problems. If more water is pumped from an aquifer than annually enters it, the water table is lowered. The declining water level of the Carrizo-Wilcox (Figure 5-11) and Ogallala aquifers demonstrate that withdrawal is greater than inflow.

Limestone aquifers

Limestone aquifers differ from sandstone aquifers, because water laden with carbon dioxide readily dissolves limestone to produce the more soluble bicarbonate [$H_2CO_3 + CaCO_3 = Ca(HCO_3)_2 = Ca^{++} + 2\,HCO_3$]. Large passages and even caves are produced which allow water to flow much more rapidly through this rock than through the sandstone. This is especially true of rocks composed of 97 percent or more calcium carbonate. The more im-

purities—like clay and silica—there are in limestone, the less permeable it is. Thus limestone aquifers are recharged more rapidly than sandstone aquifers. The Edwards aquifer at San Antonio, Texas declined in yield rather regularly and consistently during a long drought for ten years following 1946 (see also Figure 5-11). The drought ended in 1957, one of the wettest years on record, and the aquifer was filled to capacity in that same year.

Ground water is obtained by drilling wells to an aquifer. If the wells do not flow, the

Figure 5-12

Wells obtaining ground water from an aquifer.

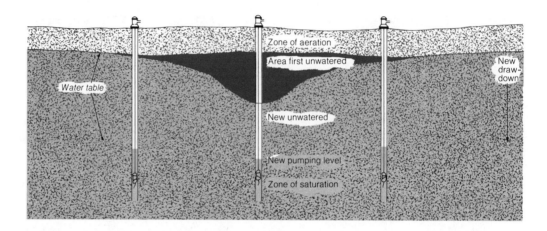

Zone of aeration

Area first unwatered

Water table

New unwatered

New draw-down

New pumping level

Zone of saturation

water must be pumped. If water is withdrawn more rapidly than it will flow naturally through the aquifer, the upper surface of the water cones downward in the vicinity of the well (Figure 5-12). A series of wells close together can produce overlapping cones that lower the water table. This local lowering of the water table is called drawdown. The aquifer fails when either (1) all the water is pumped out and the only water obtainable is the yearly increment of natural recharge, or (2) the water table is lowered until the cost of pumping is equal to or greater than the value of the water obtained.

Better use of ground water can and should be made. Montgomery, Alabama, for example, at the suggestion of geologists of the Alabama Geological Survey, has spread water wells along the length of the aquifer. By spreading withdrawal over a large area, the city obtains enough water for its needs without creating a tremendous central drawdown.

In discussing natural resources in Chapter 2, we said that energy accumulations and ore deposits that took millions of years to form are now being used up in only a few decades. The same thing is true of ground water. Most of the ground water we are using was stored in underground aquifers during the Pleistocene and Pliocene epochs and represents millions of years of accumulation. We should use more of it for human needs, but those areas that continue to exploit it in great quantities for agriculture and industrial cooling will run out in only a few decades.

Water Pollution

There can be no doubt that the rate at which we have been polluting our streams and lakes has been accelerating. The most obvious evidence is increasing ugliness in the form of cloudy and oily water, masses of floating trash and debris, worsening odors, and decreasing palatability of our water.

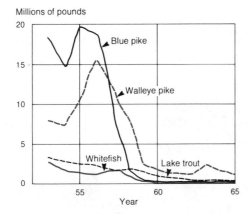

Further evidence is the drastic reduction in the catches of fish from Lakes Erie, Ontario, and Michigan (Figure 5-13). Moreover, the U.S. Department of the Interior estimated that more than 15 million fish in streams alone were killed by pollution in the year 1968. The economic loss due to pollution runs to many hundreds of millions of dollars a year. The loss in food sources has made necessary the rapid shift from one type of food to others, and the damage to the quality of life is great, but still remediable.

There are many sources of pollution to streams and lakes, some more severely damaging than others. But all of those discussed below deserve particular consideration.

Natural aging

Lakes and streams have their own life cycles and natural aging processes, which are accelerated by pollution. Many people do not understand these natural processes and are surprised to learn that their own activities have interfered with them. It is therefore important that we become acquainted with at least the broad outlines of these life cycles.

The aging of a lake is primarily due to the growth of plant life in it and also to sedimentation. Both plant growth and sedimentation are natural processes. If the lake has an outlet, the downcutting of the outlet by erosion is another natural process that assists in the aging of a lake, for it may in time let more water out than comes in. Water plants and nearshore plants continually encroach upon a lake. Decaying plant products trap soil particles, and both accumulate as sediment on the lake bottom. As the lake fills in near the shore, water plants may become so thick that they deplete the water of oxygen, thus making it unfit for most animal life. As nutrients accumulate on the lake bottom, more and more plants grow, the lake becomes shallower and warmer, and the original fish and other animals are replaced by those that prefer the new conditions. Eventually the lake becomes so shallow that plants can gradually take over the last of the water and convert the entire area to a swamp. The original depth of a lake determines the rate of aging. Lake Superior averages 487 feet in depth and has an area of 31,820 square miles. It will therefore take much longer for it to fill and eutrophy than Lake Erie, which averages only 58 feet in depth and has a surface area of 4840 square miles. Nonetheless, Lake Erie is so large that it will not disappear for many thousand years.

The life cycles of streams are less uniform than those of ponds or lakes, and therefore

Pollution in Lake Tahoe

generalization about them is more difficult. The rate of flow of a stream varies from place to place and from time to time, depending on gradient and precipitation. Consequently the environment differs along the length of a river and produces environments very different from those of lakes. The small, colder, clear, rapidly flowing streams at the head of most river drainages, with their steep banks and narrow channels, provide a much different environment from the large, tepid, muddy, sluggish rivers near the sea. Many environments intervene between these two extremes.

Soils, animal life, width of the floodplain, and suitability of the floodplain for agriculture all change along the course of a river. Near the heads of many drainage systems, floodplains are absent or narrow, soils are rocky, and flooding is minimal. Near the mouths of these same rivers, floodplains are wide, soils are more mature, and flooding is common. Rivers differ from lakes primarily because they change in size and gradient from head to mouth and thus provide environmental variety not found on any one lake. Rivers grow old through the downcutting of their channels and widening of their floodplains. Like lakes, rivers can be poisoned and eutrophied with the same accompanying adverse results on man and other organisms. But the aging of rivers may not be accelerated by these processes as much as it is in lakes.

Nutrients

As with oceans (Chapter 4) rivers and lakes suffer phytoplankton blooms that can be produced by many kinds of nutrients.
1. Some degraded industrial wastes become nutrients.
2. Natural turnover in lakes and reservoirs can bring up nutrients, but these are seasonal and last only a few days. Such natural turnovers produce annual algal blooms of short duration.
3. Some minerals and salts, especially if they contain nitrogen, phosphorus, or both, can become nutrients.
4. Some organic pesticides become nutrients when degraded.
5. All treated sewage is rich in phosphorus and nitrogen. Consequently, the normal treatment of sewage does not prevent eutrophication. For this reason it is a good procedure to run treated sewage into ponds where algae can use up much of the nutrient before the effluent is returned to natural waters.
6. The use of soluble fertilizers has increased tremendously in the last few years. Tail-end irrigation water carries large quantities of these fertilizers. East of the Mississippi River, where there is little irrigation, these nutrients can be carried by precipitation runoff directly to the natural water courses, or into the underground reservoirs.

7. Although thermal pollution does not provide nutrients, it speeds up the chemical reactions associated with photosynthesis. Therefore, plankton blooms can be expected to be particularly severe where excessive nutrients are added to thermally polluted waters.
8. Many detergents, soaps, cleaning fluids, and other household compounds, especially those that have phosphates added or depend on enzymes for "activated cleansing," contain large amounts of phosphates that provide nutrients for phytoplankton blooms.

It is suggested that much of the eutrophication of Lake Erie comes not only from sewage, but from the input of soluble fertilizers through runoff from the intensively farmed area of adjacent Ohio. Lake Tahoe, California is also quickly aging from eutrophication, due mostly to sewage from rapidly increasing urban development. Figure 5-14 gives the condition of a number of representative lakes in the United States, where aging is mostly through eutrophication.

Pollution by minerals and salts

Natural Pollution. The following are examples of only a few pollutants: The saline springs flowing into the Colorado River of the Southwest above Lake Mead contribute salt to irrigation water in southern California and northwestern Mexico. Salty irrigation water ruins most soils for cultivation in two to five decades. Similar springs emerging from rocks containing evaporite beds give the name of the Salt Fork to one of the tributaries of the Brazos River; others are responsible for the saltiness of the Red River of Texas and Oklahoma.

Figure 5–14

Relative aging of representative lakes in the United States. (Reproduced with permission from "Dwindling Lakes" by Arthur D. Hasler and Bruce Ingersoll, *Natural History Magazine,* November 1968. Copyright © The American Museum of Natural History, 1968.)

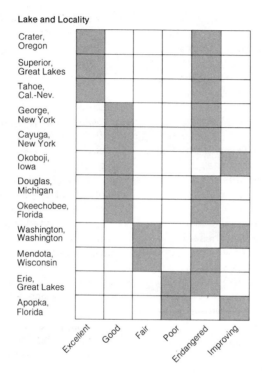

Lake and Locality

Lake and Locality	Excellent	Good	Fair	Poor	Endangered	Improving
Crater, Oregon	▓				▓	
Superior, Great Lakes	▓					
Tahoe, Cal.-Nev.	▓				▓	
George, New York		▓			▓	
Cayuga, New York		▓				
Okoboji, Iowa		▓				▓
Douglas, Michigan		▓				
Okeechobee, Florida		▓				
Washington, Washington			▓			▓
Mendota, Wisconsin			▓		▓	
Erie, Great Lakes					▓	
Apopka, Florida				▓		▓

Artificial.   A source of saline pollution that is not natural is the dumping of oil well brines—produced as an unneeded by-product of petroleum liquids—into streams and sometimes into the underground water system (Figure 5-15). This can be very serious and can quickly pollute the ground water, particularly in limestone areas.

The Rhine River is one of the worst examples of the effects of industrial pollution. In addition to detergents and insecticides, the amount of dissolved mineral matter is phenomenal. It is estimated that 30,000 tons daily, or over ten million tons annually, of waste salt is dumped into the Rhine from potash works in France above Strasbourg. The iron and salt mines of Lorraine, a province of France, contribute dissolved solids to the Meuse River, which shares the Rhine estuary.

In the northeastern United States, salt (NaCl) is used for de-icing highways in the winter months. This salt is killing plants along the highways in some areas. The chlorine content is rising in Lake Erie as a result of the transportation of salt by snow and ice meltwater into streams and lakes. Since chlorine is capable of pulling mercury out of the clays to which it is normally bound in fresh waters, the mercury content of Lake Erie is also increasing because of the use of salt for de-icing.

In areas where fertilizers are used extensively, especially soluble fertilizers, they too may become serious pollutants. Phosphates and nitrates are the fertilizers usually blamed for pollution, although in many streams we do not know which to blame. Both are nutrients that produce eutrophication. Whether a stream or a lake eutrophies is often determined by the limiting nutrient If there is not enough carbon, then the addition of phosphate or nitrate does not increase algal growth. The same is true for potassium. But if there is more than enough of any of these three nutrients, plus necessary trace elements, then the addition of the fourth will produce algal blooms, and, if continued, eutrophication. This condition of limiting and nonlimiting nutrients has made it hard to identify the particular nutrient producing a specific eutrophic event.

Figure 5-15

Contamination of the Ogallala aquifer with brines originally pumped into the Glorietta Formation.

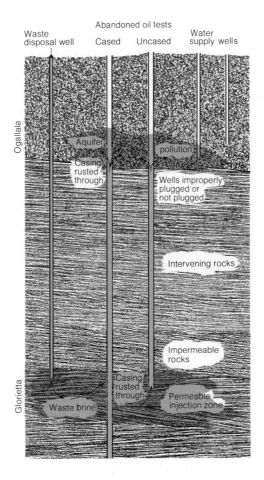

Abandoned oil tests

Waste disposal well
Cased
Uncased
Water supply wells

Ogallala

Aquifer

Casing rusted through

pollution

Wells improperly plugged or not plugged

Intervening rocks

Impermeable rocks

Glorietta

Waste brine

Casing rusted through

Permeable injection zone

Algae

Algae occur in all streams and lakes, and, under natural conditions, are in ecological balance with algal feeders, sunlight and nutrient sources—such as human wastes—which affect their rate of growth. Because of the change in the amount of heat received by sluggish streams and lakes in the spring and fall, the deep and surface waters turn over. This may cause sudden surges of algal growth as nutrient-laden bottom waters come to the top. These natural algal blooms are of short duration, lasting only until the additional nutrient is used up. Artificially added nutrients—from sewage, soluble fertilizers, and some industrial discharges—may feed a stream sufficiently to produce algal blooms of much longer duration, provided that enough carbon dioxide is present. Such algal blooms are usually unexpected because they are not monitored. As a result, municipal purification plants are not prepared for them, and drinking water may develop a bad taste. In addition, algal toxins may be harmful to those few individuals allergic to them.

Industrial pollution

Industry has received a major share of the blame for pollution, and this is correct, to the extent that industries were the first agencies to produce spectacularly observable pollution. Yet, except for toxins (poisons) that certain industries may release, it is now doubtful whether industrial pollution is in any way worse than that created by some of the larger metropolises.

The release of sulfur and copper accounts for the high concentrations of copper sulfate in the streams of many mining areas. Most of the streams in strip mine areas west of the Appalachian Mountains are also high in sulfate, as are abandoned strips. Sulfur, which is usually the result of bacterial oxidation of sulfide to sulfur, and sulfate may produce sulfuric acid. Acid water is neutralized with plain lime or with caustic or hydrated lime. Liming is the primary treatment at the Slippery Rock Acid Mine Drainage Treatment Plant, Pennsylvania.

Some of the worst industrial pollution is produced by the sulfites discharged by paper mills, the cyanides discharged by iron and steel operations, and the chemicals discharged by chemical industries. Accidents are sometimes more damaging than regular or habitual pollution. An accidental dump of ammonia killed thousands of fish in the San Antonio River in Texas in January 1969.

Whereas the major problem in the Hudson and the Potomac Rivers is apparently created by municipal sewage, in the Houston Ship Channel and Galveston Bay, industrial pollution overshadows the effects of untreated sewage from Houston, even though Houston's sewage treatment plants are inadequate. In addition, wherever petroleum is produced there are problems with oil, hydrocarbons, halogens, and oil fractions, as well as the inevitable sulfur.

Pollution can usually be corrected at the source or at a later date, and the cost is high, though correction is much less expensive than curing. If we succeed in bringing pollution under control—as we must in the next two decades—the greatest part of the rise in prices during those decades will result from the cost of correction and cure. Even part of our most recent inflation has resulted from the cost of fighting pollution, and in the spring of 1972, the Environmental Protection Agency estimated that 2.5 percent of our energy in the future would be used in correcting pollution.

Biocides

Serious pollutants are pesticides and herbicides, including defoliants that are now used excessively throughout the world. Water pollution by these biocides is common.

In June 1969 there was a tremendous fish kill in the Rhine River, apparently caused by one or two bags of endosulfan, a chlorinated hydrocarbon, falling from a barge. Endosulfan is banned in Europe, but is manufactured there for export to the United States, particularly for the control of insects in orchards and vineyards.

Several fish kills in the United States have been attributed to DDT. Part of the DDT in the oceans is delivered by streams, although most of it is wind-borne. Studies of the Carmans River marsh on Long Island, New York discovered DDT content as high as 50 ppm in the sediment, where it is probably associated with sedimentary organic debris. The concentration in the overlying water was much less. Carp, catfish, suckers, and other bottom-feeding rough fish, feeding in such areas as the Carmans River marshes, will have fatty tissues with extremely high DDT concentrations that can be passed in ever greater concentrations up the food chain to man. PCB's (polychlorinated biphenyls), also chlorinated hydrocarbon by-products of some industries, are toxic to life, and occur in nearly all inland waters.

Biocides sprayed on crops
are atmospheric pollutants

Figure 5-16

Dissolved solids in the
Great Lakes. (From "The
Aging Great Lakes" by
Charles F. Powers and
Andrew Robertson. Copy-
right © 1966 by Scientific
American, Inc. All rights
reserved.)

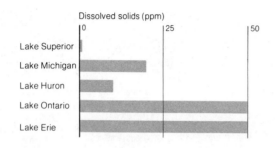

Dissolved solids (ppm)

Lake Superior
Lake Michigan
Lake Huron
Lake Ontario
Lake Erie

Sewage

Most individual consumers of water do not
realize that they are drinking water that
has carried someone else's sewage. Most
water from major streams passes through
municipal sewage systems two or more
times. It was said several centuries ago that
English breweries took their water from the
Thames below London, not above, because
that water gave the beer more body. If

Industrial waste being discharged into a river near
Chicago

municipalities were to recycle their own
sewage knowingly, they would probably
have better water than they are now drinking, because they would take greater care
in purification. Furthermore, in an age when
excellent sewage treatment is available,
surprisingly little sewage is actually treated;
in much of the United States, about fifty
percent of the sewage is unmonitored. The
increased amount of dissolved solids in
Lakes Michigan and Huron, and especially
Lakes Erie and Ontario (Figure 5-16), is a
measure of all the pollution in these lakes,
including sewage.

Thermal pollution

Thermal pollution is the addition of heated
water to natural waters. Most thermal
water has been used as a coolant in some
power plant. Nuclear plants now discharge
up to fifty percent more waste heat than do
fossil fuel plants, though new designs will
reduce their waste heat to comparable
levels. Thermal water is discharged at the
surface and extends downstream for varying distances with gradually diminishing
heat. The Monongahela River, for some
40 miles above the Ohio River, suffers various temperature rises of from 3 to 10 or
12 degrees as a result of municipal power
plants and industrial concentrations in
different cities (Figure 5-17). The main threat
of thermal pollution is to aquatic life, and
it does more damage to the reproduction
of new individuals than to adults. Depending on the species, adults may tolerate
temperature changes up to 20 or 30 degrees,
whereas the eggs and the very young may
tolerate changes of only 3 to 8 degrees.
Higher temperatures may also inhibit spawn-

Figure 5-17

Temperature curve along the Monongahela River, Pennsylvania, illustrating thermal pollution at industrial and population centers. (From "Thermal Pollution and Aquatic Life" by John R. Clark. Copyright © 1969 by Scientific American, Inc. All rights reserved.)

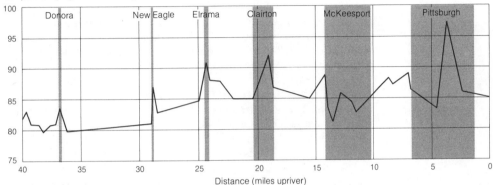

Temperature (degrees Fahrenheit)

Distance (miles upriver)

ing. In addition, fish growing in waters of near maximum temperature seem to have less firm flesh and are less palatable than fish from cooler waters to which they are naturally adapted. Trout and other fish that normally live in white, cold water are less tolerant of high temperatures than are carp, catfish, and other less desirable rough fish. In southern states, where fish are adapted to warmer waters, thermal water seems actually to improve the habitat—for fish at least.

A second undesirable feature of thermal pollution is frequent algal blooms, since higher temperatures increase photosynthetic activity and speed the completion of the life cycle.

Among suggestions for preventing thermal pollution are (1) irrigation with heated waters, (2) construction of cooling ponds with each heat-producing power installation, and (3) the construction of cooling towers. Heated waters could be used for irrigation where waters used for heat dis-

sipation are close to irrigation districts. Greenhouses might also profitably use warmed waters. It has been shown that crop performance increases with the temperature of the water used in irrigation.

Many industries and, in particular, many power facilities, municipal and industrial, use water to remove heat they cannot use. Cooling ponds or cooling towers are constructed by some to dissipate this excess heat. In a tower, water is sprayed or circulated through a forced draft of air to cool it. This water can then be recycled through the system or returned to surface sources. If municipalities can construct cooling ponds or towers for municipal plants, then industry should learn to follow that lead so that the heat will not pollute the natural water courses.

Summary

The United States does not have as much available water per unit area as does the rest of the world, but it uses more water per person. Average water use and projections of water use for the United States are estimated to rise from 320 billion gallons daily in 1960 to around 900 billion gallons daily by the year 2000 (Figure 5-18). This represents a doubling time of about 30 years, and an increase of 300 percent in 40 years. These figures are based on data from the last two decades, which include figures for increases in population, the cooling of power plants, and industrial and residential air conditioning.

Exponential growth of population increases the number of water users. Exponential economic growth further accelerates the increase in water use. These expected increases result in predictions of future water use that may be unreasonable. First, if the increase in population is actually smaller than expected, as seems likely, the exponential growth in number of users

will be less than predicted. Predictions of future use are based on population growth curves more than on anything else. Second, if the air conditioning market becomes saturated, the accelerating demand for power may lessen. Demand for water can also be lessened by increasing the cost to the user. Thus, if actual cost were charged to the user, instead of being paid for by tax monies, there would be greater effort at conservation. Where and when water supply becomes short, preferred use will assume greater importance. Many municipalities already ration water for irrigation of lawns during drought or in late summer. Other preferred use priorities can be designed, and will be necessary in the southwestern United States within a few years.

Let us not underestimate the cost of preventing and curing pollution. In addition to the billions of dollars that federal, state, and local governments, and private founda-

Figure 5-18

Water use in the United States.

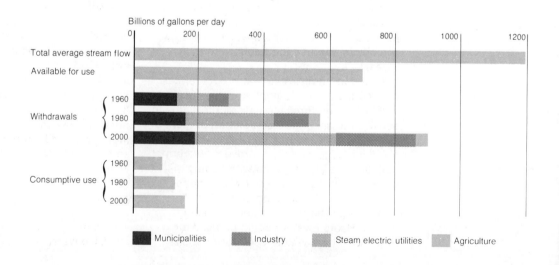

tions, are beginning to pour into the fight against pollution, there will be economic losses to individuals and to communities because of closed industries and loss of income and tax base.

Water pollution is a continuing headache. It is the responsibility of a knowledgeable public to see that its governmental rep-resentatives further the cause of fighting pollution. So many everyday activities pollute water that planning units larger than regional planning commissions may be necessary. River authorities would seem to be in the best position to supervise total planning, because they have control of a single unit of water.

Selected Readings

Avrett, James R. 1966. A compilation of surface water quality data in Alabama. Geol. Surv. Alabama, Div. of Water Resources, circ. 36.

Bergstrom, Robert E., et al. 1968. Ground water resources of the Quaternary of Illinois. In Bergstrom, Robert E., *ed., The Quaternary of Illinois.* Univ. of Ill. College of Agriculture. spec. publ. 14, pp. 157–164.

Bue, Conrad. 1963. Principal lakes of the United States. U.S. Geol. Survey Circ. 476.

Clark, John R. 1969. Thermal pollution and aquatic life. *Scientific American* 220(3):19–27.

Hand, John W. 1970. Planning for disposal of oil shale, chemical and mine wastes. Colorado Geol. Surv. spec. publ. no. 1, pp. 33–37.

Irwin, James R. and *Robert B. Morton.* 1969. Hydrogeologic information on the Glorietta Sandstone and the Ogallala Formation in the Oklahoma Panhandle and adjoining areas as related to underground waste disposal. U.S. Geol. Survey Circ. 630.

Leopold, Luna B. and *Kenneth Davis.* 1966. *Water.* New York: Time Inc.

Odum, W. E., G. M. Woodwell, and *C. F. Wurster.* 1969. DDT residues absorbed from organic detritus by fiddler crabs. *Science* 164:576–577.

Powers, Charles F. and *Andrew Robertson.* 1966. The aging Great Lakes. *Scientific American,* Nov., pp. 94–100.

Scudder, Thayer. 1969. Kariba Dam: The ecological hazards of making a lake. *Natural History,* Feb., pp. 68–72.

Stall, John B. 1966. Man's role in affecting sedimentation of streams and reservoirs. In Bowden, Kenneth L., *ed., Proceedings 2nd ann. Am. Water Resources Cong.,* pp. 79–95.

U.S. Water Resources Council. 1968. *The Nation's Water Resources,* Parts 1–7. Washington, D.C.: U.S. Water Resources Council.

Wulff, H. E. 1968. The Qanats of Iran. *Scientific American* 218(4):94–105.

Chapter 6

The Pros and Cons of Erosion

No more brazen falsehood was ever perpetrated upon a gullible public than the allegation that the dust storms of the 1930's were caused by 'the plow that broke the plains.'
James C. Malin, *The Grassland of North America, Prolegomena to Its History,* 1948

Erosion is a part of ecosystems. As long as there are wind, water, gravity, and relief, soil and rock will be worn away. Without erosion there would be no soils, no trace elements, no minerals in our streams. Without erosion, minerals would not complete their cycles, the rock cycle would not be renewed, and the carbon and nitrogen cycles would have less variety.

Man increases rates of erosion when he removes or decreases vegetation by lumbering, farming, mining, or urbanization. He increases erosion when he increases runoff by the same activities, or by altering the natural courses of streams. Only recently has man come to realize the full extent of the problems he has thus created, and the need to correct them. Wherever his activities have increased erosion, man must reduce it to a natural rate in order to correct or to prevent damage. Wherever his activities have decreased erosion to a level harmful to man and ecosystems, he must change those activities to produce more desirable rates of erosion. As long as man builds, destroys, changes, and rebuilds, the needs for both types of restoration will continue. Erosion is part of the ecosystem, and must be viewed as "too little," "too much," or "just right."

Measuring Erosion

In discussing erosion geologists commonly use the term denudation, which means the lowering of the earth's surface by the removal of soil and rock. Quantitative values of denudation are generally given in tons per square mile or square kilometer, or—in terms of lowering the land surface—usually in inches or centimeters per thousand years. None of these measurements is very accurate, but if one adopts the working idea that a cubic yard, or 27 cubic feet, of soil that is not too wet weighs about a ton, one can eventually reach some figures that are meaningful. Every year about 900 tons of sediment are removed from each square mile of the drainage basin of the Colorado River. Before the construction of Hoover Dam, this sediment was deposited on the delta and in the Gulf of Lower California. As dams were constructed on the Colorado River, much of this sediment was deposited in the reservoirs above the dams. How much sediment is this in terms of acre feet, or in terms of lowering the land surface? An acre of land is 1/640 of a square mile or 43,560 square feet. An acre foot is an acre of material one foot deep, or 43,560 cubic feet; it weighs approximately 1613 tons. Nine hundred tons of solid sediment per square mile is a little more than half an acre foot. Nine hundred tons could be carried by 360 2-1/2 ton trucks or by a railroad train of nine gondolas, which are freight cars of the type used to carry gravel and coal. The Colorado River drainage basin loses 900 tons or a little over one-half acre foot per year per square mile. This is a rate of approximately 1/100 of an inch per year.

Figure 6-1

(a) Cross sections and (b) long profiles of streams. (Redrawn from Henry Gannett, U.S. Geological Survey, 1901.)

(a)

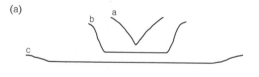

(b)

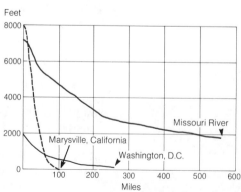

Erosion by Streams

Many topographic features of the earth are shaped or have been modified by water, and on continents most of this shaping is done by streams. Streams create and in turn are shaped by valleys that possess features of environmental importance, both providing and destroying living space, agricultural land, and economically and/or ecologically important areas.

Valley profiles and base levels

The cross-sections or transverse profiles of valleys vary from the sharp, V-shaped valleys found mainly in rugged hilly and mountainous areas, through many gradations to the broad, flat valleys common in more level country of low relief (Figure 6-1). Little agricultural or other human development is possible in sharply V-shaped val-

Sand and wind erosion has provided unique cliff-dwellings for this village in Turkey

leys, whereas the floors of broad, flat valleys may be covered with excellent soils that invite agricultural development—as well as the periodic disasters of recurrent floods. As the floors of more mature valleys broaden and flatten, the valley walls are reduced in height by the lowering of land that is adjacent to the valleys.

Long valley profiles

The profiles of stream channels from head to mouth are long profiles, characteristic of streams (see Figure 6-1). The gradient of a stream is its downward slope, and is measured as fall in feet per mile. Streams have a steeper gradient near the head than near the mouth. Since gravity is responsible

for most of the greater velocity of a stream at higher gradient (the other factor being the depth-width ratio of the channel at any given point), there is usually greater velocity and hence more rapid erosion per unit volume of water at steeper gradients. At any given time a segment of a stream may be aggrading (depositing sediment and raising the level of its channel) or degrading (eroding and removing sediment, and deepening its channel). When it is doing neither, it is said to be at grade. Few streams are at grade or remain so for long, and most that are near grade alternate between aggrading and degrading with the alternation of climatic cycles.

Any stream entering a lake or another stream cannot erode below the bottom of that other lake or stream until the lake is removed or the larger stream deepens its channel. A resistant stratum crossing a stream channel prevents downcutting until it is eroded or cut away. These and similar barriers to the downcutting of a stream valley are called temporary base levels

The permanent base level of erosion is the level to which streams can erode a continent, or a single stream can erode its share of a continent. Because the earth is dynamic, continents are renewed before streams ever reach the permanent base level. Thus the permanent base level is a theoretical level with little practical importance, unless one is concerned with the base level of marine erosion—something quite different—which is the level to which wind-formed waves and currents could erode the land.

The temporary base level of erosion is a concept important to the study of the environment because it can be changed by man, although it cannot always be controlled. For example, every dam constructed by man becomes a temporary base level that results in aggradation above the dam and degradation below the dam. Above the dam the stream can no longer actively erode because it produces a delta in the artificial lake, which in turn causes aggrading of the channel above the delta. Concurrently, because the sediment is deposited in the lake, the stream below the dam is sediment-free; it therefore has a greater capacity to pick up new sediment and therefore to degrade its channel, until it again picks up a normal solid load (Figure 6-2). The high Aswan Dam on the Nile River in Egypt, completed in 1968, has resulted in erosion and downcutting of the channel below the dam.

Thus, altering the gradient or the erosive capacity of a stream by building dams or by other means, changes the stream channel. Degrading rejuvenates tributaries and increases erosion of cultivated fields on the floodplains in broad flat valleys, requiring costly land repair. Conversely, aggrading floodplains can cover low structures with sediment, fill irrigation ditches, and bury crops with mud in the growing season.

Stream channels and meanders

Floodplains and stream channels create other environmental problems besides those due to flooding. Stream channels in flat valleys are not naturally straight, but wind back and forth, broadening the floor of the valley and developing floodplains. If a channel loops back and forth until it is at least twice the length of the valley it occupies it is said to meander, and the curves are called meanders, after the Meandros River in Greece.

The flow of water in a stream is greatest and fastest at the deepest part. This core

Figure 6-2

Changes in a stream gradient resulting from the construction of a reservoir.

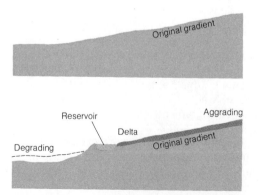

Picturesque example of
water erosion from the
San Juan River, Utah

Figure 6-3

Stream flow and the construction of meanders.

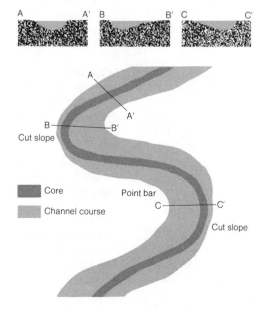

Core

Channel course

Cut slope

Point bar

Cut slope

(Figure 6-3) of fast-flowing water has a momentum of its own, and because it tends to move straight ahead, it will flow along one bank and then the other as the channel curves. Thus, when the stream comes into a curve in the channel, momentum carries the core into the concave bank and away from the point bar or area of deposition on the opposite bank. Thus erosion occurs on the concave bank of the stream and deposition on the convex bank so that meanders are cut ever deeper. And as the momentum causes more erosion on the downstream part of the concave bank, the curves in the channel migrate downstream with time (Figure 6-4). Man sharply encounters the environment when he buys a lot on the concave side of a meandering stream. As his lawn and garden are gradually eaten away by the hungry waters, he complains to the city, county, or state. If he is lucky, the city, county, or state periodically removes sediment deposited on the point bar across the stream back to the owner's lot, and, just as often, the stream again erodes it and deposits it on the next point bar downstream. Sometimes beneficent government grows tired of such continuing complaints. To quiet the grievances of property owners they channelize the stream by straightening it and constructing resistant concrete walls. But channelizing increases flood problems and sets the stage for new environmental crises—as discussed in Chapter 12.

Factors Affecting Rates of Stream Erosion

Three main factors affect the rate of denudation: (1) the amount and distribution of precipitation through the year, largely a response to climate, (2) the resistance of

Figure 6–4

Downstream migration of meanders.

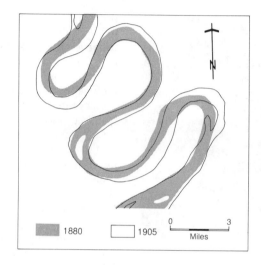

1880   1905

0   3
Miles

Figure 6–5

Effective precipitation causes the removal of sediment.

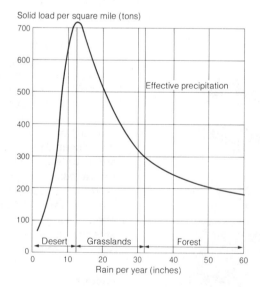

Solid load per square mile (tons)

Effective precipitation

Desert   Grasslands   Forest

Rain per year (inches)

Precipitation

The greatest amount of erosion by water occurs in areas with about 13 inches of precipitation a year, along the border between grasslands and desert (Figure 6-5). Where rainfall is less there is not enough water to remove great quantities of rock. Where there is more rainfall, the vegetative protective cover increases, and less soil is removed. Naturally, erosion also varies with surface. Thus, in areas with the same total precipitation, a red shale, for instance, will produce more solid load than a limestone. Figure 6-6 represents precipitation, evaporation, runoff, and sediment load profiles across the United States along the 40th parallel, which is near the southern boundary of Wyoming. Data for these profiles are given in Figures 6-7, 6-8, and 7-3. The greatest solid sediment load occurs in the areas of least runoff, least precipitation, and greatest evaporation. This is because the yearly precipitation in these areas usually falls in one or two violent storms and there is little vegetative cover to protect the soil. With greater rainfall per year, although there is less solid load, there is more water to dissolve solid minerals from the rocks. The dissolved load of solubles therefore increases with precipitation and runoff.

Figure 6-9 applies these principles to the United States. All the streams of the Southwest have a high solid load, by virtue of low precipitation and little vegetation. Although slopes are steep along the Pacific Coast, increased precipitation produces more vegetation and hence less solid load. The Columbia River Basin, with high precipitation and a low evaporation rate, has

rock and soil to erosion, and (3) the number and abruptness of changes in elevation per unit area.

Figure 6-6

Solid sediment load, precipitation, pan evaporation, and runoff profiles along the fortieth parallel.

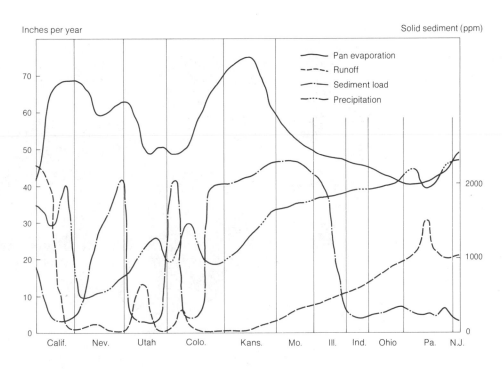

Inches per year

Solid sediment (ppm)

— Pan evaporation

---- Runoff

—·—· Sediment load

—··—·· Precipitation

Calif. Nev. Utah Colo. Kans. Mo. Ill. Ind. Ohio Pa. N.J.

Barren desert soil, Badlands National Monument, South Dakota

excellent vegetative cover and minimal solid load, as do the streams and rivers in the northern and southern Atlantic states and states along the eastern Gulf of Mexico. Streams emptying into much of the western Gulf of Mexico (mostly Texas) would have a higher solid load, but even in those desert areas the hard igneous and limestone rocks do not weather easily.

In contrast to the solid load, the dissolved load is directly related to precipitation. It is higher in the eastern and southeastern United States, and lower in desert areas. This phenomenon is interpreted further in Chapter 7, where it is emphasized that the soils of the eastern and south-

Figure 6-7

Average annual pan evapo-
ration in the United States.

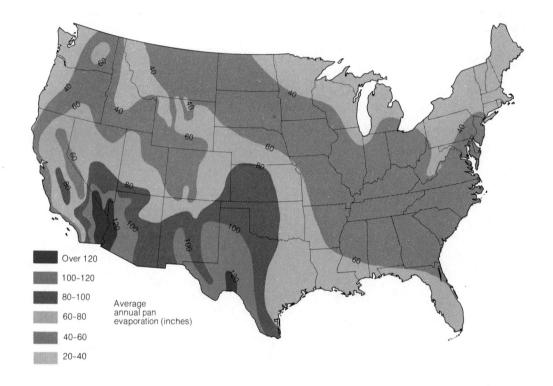

Over 120

100-120

80-100

60-80

40-60

20-40

Average
annual pan
evaporation (inches)

eastern United States are leached of many
desirable minerals.

In summary, there are distinct differences
between the effects of an equitable oceanic
climate, which tends to be gentle and even;
a tropical oceanic climate, often with torrential rainfalls and extensive flooding; and
a continental climate, which is given to
extremes of precipitation and temperature.
In an equitable oceanic climate, precipitation
is distributed evenly throughout the year
and from day to day with no great amount
at any one time. Hence there is good vegetative cover, little erosion, little flooding,

and little solid load. In one such climate,
along the Meuse River in Belgium, houses
are built right at the water level. Also, in
the Champagne region of north-central
France, though the soils are easily eroded,
precipitation is light and consistent, and
there are no heavy storms, so that the solid
load is light. This is not true of the tropical
oceanic climates, where there is great
flooding, and where, if man destroys the
vegetative cover, solid load is great.

On the other hand, it may seldom rain in
a continental climate, but when it does, the
sky falls. Five inches of rain in a few hours,

Figure 6–8

Solid sediment load of streams in the United States.

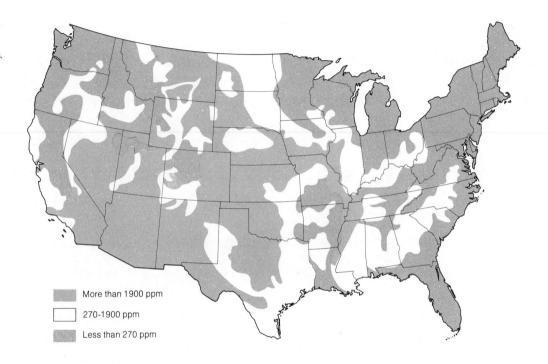

More than 1900 ppm

270-1900 ppm

Less than 270 ppm

Figure 6–9

Relation of dissolved and solid sediment load.

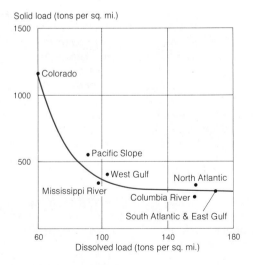

Solid load (tons per sq. mi.)

Colorado

Pacific Slope

West Gulf

Mississippi River

North Atlantic

Columbia River

South Atlantic & East Gulf

Dissolved load (tons per sq. mi.)

or 25 inches in 24 hours, though infrequent, is not rare. When these storms occur in areas where rain is especially effective as a cause of erosion, they flood the countryside and dump tremendous solid loads onto floodplains or into reservoirs. In such a climate, such as along the Rio Grande River in Texas and Mexico, one sees no houses near water level except shanties that will be washed away with the first flood. In the same climate in arid parts of the United States, where soils are of similar composition to those of the French Champagne, the streams carry a tremendous solid load after heavy rains.

Figure 6–10

The confluence of Spring Creek and the middle fork of the Powder River, Wyoming.

Resistance of rock and soil to erosion

Above Kaycee, Wyoming, about two miles east of Barnum, lies the confluence of the Middle Fork of Powder River and Spring Creek. Spring Creek carries much red solid load and is blood red; the Middle Fork of Powder River carries little solid load and is clear. Below the confluence, the waters of the two streams flow side by side without mixing for many tens of feet (Figure 6-10). Above the confluence Spring Creek flows for many miles along the strike of the red sandstones and shales of the Chugwater Group, and easily picks up a heavy solid load. The Middle Fork, however, crosses the Chugwater outcrop on alluvium, and flows over Cambrian orthoquartzite and

Ordovician and Mississippian limestones further upstream. For both streams the climate is identical and the grazing regime comparable. The heavy solid load of Spring Creek is due entirely to the easy erosion of reddish particles from the soil and rock of the Chugwater Group, whereas the Middle Fork is clear because it is hard to produce particles of this size by erosion of ortho-quartzites and limestones. These streams clearly demonstrate that climate and land use are not the only factors responsible for solid load.

Climate and the competency of rock both combine their effects to lower the land surface. The Green River Shale of north-western Colorado is eroded at a rate of 1,100 to 1,200 tons a square mile annually, whereas the accompanying dissolved load may be as low as 60 tons a square mile in this arid region. In less arid areas the solid load decreases as the dissolved load in-creases, reaching a minimum of solid and a maximum of dissolved loads in the south Atlantic Coast states and eastern Gulf states. As precipitation increases, vegeta-tive cover increases and the solid load decreases. At the border of grasslands and desert, an average figure of solid load could be about 700 tons a square mile, but in the South Dakota badlands, where there are steep slopes and soft rocks, the amount might be twice that. In contrast, in some of the limestone terrains of desert areas where the rock is hard, solid load is low. In Trans-Pecos, Texas, the solid load may be as low as 300 tons per square mile or less an-nually, even though the annual precipita-tion is around the 13-inch level that produces maximum solid load.

Relief

Although water flowing down the steeper gradients in the long valley profile (Figure 6-1b) removes more surface rock and soil than water farther downstream on lower gradients, the water may appear clearer and less muddy because most of the sediment is bed load. Bed load, in contrast to suspended sediment, is that part of the load of a stream that consists of fragments too large to be suspended. Instead, they are rolled or saltated along the bottom. In the removal of sediment the gradient over a large area is less important than local, sudden changes in slope. Terrains of soft rock, such as the Green River Formation of the central Rocky Mountain states or the soft rocks of the South Dakota badlands, are particularly vulnerable.

Rejuvenation, an increase in active erosion, has occurred in southeastern Iowa, where easily erodable loess formations are vulnerable. In this part of Iowa the rejuvenation has been laid to agricultural practices. Although the direct connection is difficult to make, it may have been aggravated by the drought of the 1930s. This rejuvenation process is usually explained by the removal of natural vegetation for agricultural uses. For proper growth of most food-producing crops, much of the soil lays bare. This allows greater runoff and more erosion, resulting in a cycle of downcutting. Similar cycles of downcutting occur naturally; they are probably produced in much the same way as those laid to agricultural causes. That is, during extensive drought the vegetation cover is reduced, runoff and sediment load increase, and an erosion cycle starts. With both agricultural and natural rejuvenations, once the terrain is broken and gullies start, erosion will continue even in normal times because an area with local and sudden change of slope has been initiated.

Nature intervenes in some areas to protect bare slopes with the growth of brush. This process can be observed along gullies in southern Oklahoma. It is more likely to occur in semidesert areas because only there does drought occur often enough for a climax cycle, which protects such bare slopes, to have evolved.

Still, the era following the depression of the 1930s has over-indoctrinated us on man's effect on the land. In their book, *The Changing Mile*, Hastings and Turner have shown that in most of the western United States the land appears better covered now than in the last century, a philosophical position long maintained by James C. Malin, Kansas historian, who has thoroughly studied the history of the Great Plains. Hastings and Turner further show that cattle alone have not affected the rate of erosion, but have done so only when accompanied by drought.

Even if relief is great and sudden, vegetation is an effective shield to erosion. The effect of vegetation as a shield on steep slopes can be seen in the vicinity of Orocovis, Puerto Rico, where slopes range up to 45 degrees (Figure 6-11), and on even steeper slopes in some tropical areas.

Stream erosion and urbanization

The erosional history of a small area in Pennsylvania—from the time it was covered

Figure 6-11

Slopes protected by vegetation in the vicinity of Orocovis, Puerto Rico.

Figure 6-12

Erosional history of a small area in Pennsylvania.

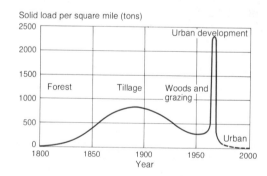

by forest until it underwent urban development (Figure 6-12)—shows the combined effect of vegetation and man on the solid load of streams. When the area was forested and covered with a thick layer of leaf mold, the solid load was very low. As settlers began cultivation, the loss of soil and rock to solid load increased to a maximum of 600 to 700 tons per square mile. As farming became less productive, the

land returned to brush, trees, and pasture, and the solid load decreased to about 250 tons. For a short period before complete development, part of the land lay bare, and the sediment load leaped to 2000 or more tons per square mile. But this lasted only a short time, because the houses, grass, and pavements of the urbanized area reduced the sediment load to that of the original forest cover, or perhaps even less. But the urbanized area will produce more runoff—five to ten times that of the original forested area—with other problems that will be considered in Chapter 12.

Glacial Erosion

During the last four million years large areas in the Arctic and temperate parts of the world were covered by great glaciers, deposits of moving ice. When these occur in mountainous areas, they are called mountain or alpine glaciers. Ice covering great areas of a continent is called an ice sheet. Ice sheets exist today on Greenland and in the Antarctic. Parts of northern North America and Europe were covered by ice sheets several times in the Pleistocene Epoch. The moving ice of both alpine and continental glaciers can pluck loose soil and rock from the surface and push or drag it along, thus eroding the land.

Erosion by glaciers has produced both esthetic and practical effects on the human environment. Certainly the Teton Mountains of the Idaho-Wyoming border, the Canadian Rockies, and the Alps of southern Europe are among nature's greatest beauties. They were all sculptured by mountain glaciers.

Poor landuse planning—
housing development near
Omaha, Nebraska sur-
rounded by erosion.

Drumlins are oval hills with their long dimensions parallel to the direction of the movement of continental ice sheets. The rounded forms of those in eastern Wisconsin and upstate New York add to the beauty of quiet landscapes. Much of the charm of the Finger Lakes district of New York State comes from the many long, narrow lakes whose basins were formed when the continental ice sheet gouged away soft rock that surfaced in long narrow outcrops. This same ice sheet scraped much of the soil from the area, and thus reduced its potential as an agricultural economy.

When glaciers—alpine or continental—melt, they leave a thin deposit of rock and dirt called moraine scattered over the surface of the ground. They also pile up ridges

Glaciers not only leave deposits, but cut out valleys; Lowell Glacier, Alaska

Peters Glacier, Alaska

of material, called terminal moraine, along their distal edges. Such glacial deposits also effect the environment of man. Moraine is not easily tillable; however, lake deposits and other fine-grained deposits beyond the glacier fronts provide good soils when they are weathered. Other good soils are developed on loess (deposits of fine dust many feet thick) deposited when glacial climates resulted in far more wind erosion and deposition than occurs now.

Marine Erosion

Erosion by the sea produces its own environmental problems. Most of these are the result of storms—hurricanes, typhoons, northeasters, and other types spawned by the oceans—all of which cause tremendous property damage. In the long view, perhaps the principal damage done by the sea is through erosion, whether by storms or by less dramatic processes.

Four different oceanic regimes cause the erosion of land by oceans. These are (1) tides, (2) wind-generated waves and currents, (3) storms, and (4) rising sea-level.

Tides

Tides are generated by the gravitational pull of the moon, which causes the waters of the ocean to bulge on the side of the

Sea cliffs carved by marine erosion, Bay of Fundy, Canada

earth nearest the moon and on the side away from the moon. The parts of the earth at 90 degrees to these are at the same time depleted of water mass. Since the earth rotates, each of the two tidal bulges passes around the earth about every 25 hours, producing a rising and falling of the tide about each 12½ hours. The strength of the tide is modified by the sun, enforcing the effect of the moon when aligned with the earth and the moon, detracting when at an angle to the moon-earth axis. The strength of the tide near shore is also modified by the shapes of bays and estuaries and the slopes of beaches.

About every 12½ hours the rising tide runs up onto the shore, into the bays, and up the estuaries, bringing salt water with it. As the tide recedes the outgoing current, pulled down the gradients of beaches or tidal channels by gravity, removes to deeper water anything it has the strength to carry. In this way the currents of the receding tides help to move the continents to the ocean basins, grain by grain, chemical by chemical.

Wind-generated currents and waves

In Chapter 2 we saw that the major ocean currents were generated by the winds and modified by temperature gradients and the west-to-east rotation of the earth. The drifts of these major currents, at the surface of the ocean, may be altered by more local wind effects. When the wind blows onshore, either directly or obliquely, it directs waves toward the shore. If the waves move directly onto the shore, some water is carried up the beach when the waves break. It then returns seaward along the bottom gradient by

Figure 6-13

Behavior of wind-generated waves and currents. (a) Waves wash sediment up the beach. (b) Bottom currents wash sediment down the beach, and longshore drift moves sediment along the beach. (Copyright © 1965 by John Wiley & Sons, Inc.)

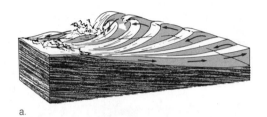

a.

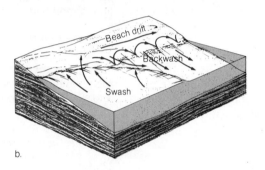

b.

gravity (Figure 6-13). If the waves come into the shore obliquely, part of their energy is redirected parallel to the shore. In this process, particles are moved obliquely toward the beach by incoming waves, but are carried down the gradient at right angles to the beach by outgoing undertow. The total effect is a drift of sediment along the beach. If sea level is rising, sand will also be moved to deeper water by the bottom currents. If sea level is lowering, sand may accumulate on the beach. Silt-laden density currents may also carry sediment out to deeper waters, but these have little direct effect on man's environment.

When man pollutes nearshore environments, tidal and bottom currents aid in flushing the pollutants out to sea, where they are diluted by ocean currents and density currents. Problems arise when the rate of pollution exceeds the rate of flushing.

Changing sea level

Changing sea level is further discussed in Chapter 13. A rising sea aids currents in removing nearshore sediments to deeper waters, and therefore aids in the erosion that damages cultural features created by man. A receding sea provides more land.

Storms

The effect of storms is more dramatic than the effects of other agents of marine erosion, and contributes to most of the damage in the nearshore area. Hurricanes are further discussed in Chapter 12, but some examples of marine erosion follow.

A striking example of the removal of land is occurring in Florida. Florida has been subsiding at a very slow rate since the end of the ice age. Whether its subsidence is caused by tilting that accompanies the rebound of areas that once contained icecaps, or whether it is due to the normal rise in sea level resulting from the melting of icecaps, it has been combatted by the building of groins along Miami Beach and other Florida beaches in an attempt to retain the sand. Groins are barricades extending into the water at right angles to the beach. They trap sand being moved along the beach by longshore currents and hold it at the barricades, and in this way help the beach to maintain itself. In Florida, however, as sand moves along the beach, it is also moved into deeper water. Groins still allow sand to be moved into deeper water in areas of rising sea level, but do not allow replenishment of the sand naturally by longshore currents that move sediment parallel to the beach. As sea level rises—extremely gradually—the ability of

Sand and rock deposition
from storms on Lake
Michigan

bottom currents to move beach sand to
deeper waters is increased. Each year
before the tourist season, hundreds of
truckloads of sand are dumped onto some
of the beaches of Florida. During the year
enough of this new sand is moved along
the shore and also out into deeper water
so that the process has to be repeated the
following year.

Storm erosion by hurricane Alma in June,
1966 (Figure 6-14) illustrates the tremendous erosion of beaches on the Gulf Coast
of Florida that would usually and otherwise
be called stable. During the hurricane
much of the beach and some of the berm

(the narrow higher ridge just shoreward of
the beach) were eroded away. During normal beach action the beach was restored,
but the retreat of the berm may be permanent.

Examples of unusual erosion occur along
the shores of the English Channel. The
Cretaceous rocks of northwestern Europe
contain many silica concretions, usually
black; they are very hard and very resistant
to erosion and are called flint nodules. The
coast at Wissant, on the French side of the
Channel, is continually eroding, though
slowly because of the great height of the
cliffs and the thickness of the rock to be

Figure 6-14

Profile of a beach along the coastal area of northwestern Florida.

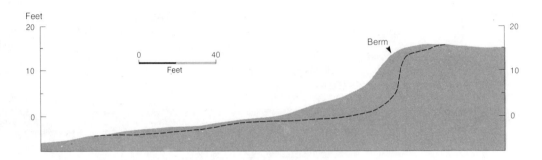

removed. Most of this erosion occurs during the mighty channel storms. The most interesting part of the Wissant coast is the great "wind" row of flint nodules at storm tide level (Figure 6-15). Sediment that gravitates down the cliff is ground to powder in the "ball mill" of flint nodules at the foot of the cliff before it is washed out into the channel. These ball mills or storm rows of flint occur only where Cretaceous chalks

Figure 6-15

Flint nodules along the English Channel near Wissant, France.

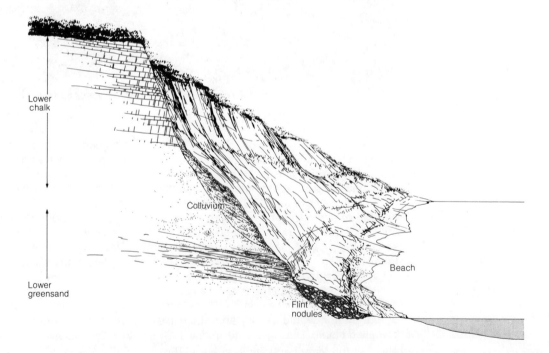

crop out in the channel cliffs and the hard flint nodules roll to the foot of the cliff. There are no flint storm rows, for instance, at Boulogne or below Deauville where the cliffs expose Jurassic rocks, and where valuable land also slides down the steep slopes above the beach where it is under-cut by channel storms.

Destruction of property by removal of the shore can be extensive. If man interferes with natural flow patterns, shore destruction can be increased even more. East of Bolivar, Texas—just before hurricane Carla—a fish pass was cut through the Bolivar Peninsula into East Bay from the Gulf of Mexico. This so changed the circulation of Galveston Bay and East Bay during large storms that the shore of the Bolivar Peninsula was no longer storm adjusted. As a result, during hurricane Carla, water ate into the seaward shore of the Bolivar Peninsula from 100 to 300 yards, carrying out to sea all of the sediment and numerous houses. The tonnage or volume of sediment removed by Carla has not been estimated.

The cutting of passes through barrier bars along the coast for recreation, or to alter salinity, or to enlarge breeding areas for marine life, is more complex than meets the eye. Erosion by hurricanes cannot be prevented, but perhaps we can learn to lessen its effect by studying these storms. Nevertheless, urban development of hurricane-prone areas is poor land-use planning.

Wind Erosion

A final source of erosion is wind, and since some strong wind currents are constant, such as the prevailing westerlies, a certain amount of fine dust from the lithosphere is constantly being moved through the atmosphere. Although it is generally held that such particles have a residence time in the atmosphere of about one year, the amount of airborne dust being circulated and deposited can be considerable. It has been estimated that Europe has received 5 1/2 inches of dust in the past 2000 years. In drought years, north-central Iowa receives one millimeter per year.

That dust storms have been even more serious in the past than at present is indicated by the winnowed Nebraska sand hills and the tremendous deposits of wind-blown silt and clay (called loess) in Iowa and eastern Nebraska especially, but also in other midcontinent states. Furthermore, during times of plentiful food, herbivores immediately out-produce their more slowly reproducing predators, so that grasslands are soon overpopulated. If drought suddenly sets in, grasslands are overgrazed and grasses are eaten into their roots (as they are today with overgrazing). The first prerequisite of dust storms is drought. If overcultivation followed by drought happens to occur during the drying leg of a climatic cycle, wind erosion is even more severe.

In contrast to natural conditions, man's effect in producing wind erosion or deflation by farming is likely to be more pronounced in nondrought years. In western Kansas, western Oklahoma, eastern Colorado, and the Texas Panhandle, continual plowing and summer fallow result in deflation even in good years. Under natural conditions in good years, however, the covering of natural vegetation on the land resists deflation.

Dust storm, Queensland,
Australia

There can be no question that dust storms damage part of the areas they affect, but their effects are not all bad. The highly productive grassland and prairie soils of North America are continually having their important mineral trace elements renewed by the settling of dust that has been earlier removed from the soils of other western lands. In fact, loess deposits and the better soils of the Great Plains are some of the richest agricultural lands in the world, because they have been enriched by the addition of wind-blown dust. Stabilization of areas of wind erosion, if possible, would eventually require new and more expensive fertilizing methods in other areas.

Deposition

The other half of the process of sediment transfer is deposition into lakes and oceans of the sediments that have been washed down streams or blown from arid lands. The accumulation of sediments may be less of an overall problem to man than the loss of ground cover. Nevertheless, sedimentation poses problems. Perhaps one of the important problems relates to the

Figure 6–16

Diagram of a multiple-purpose reservoir.

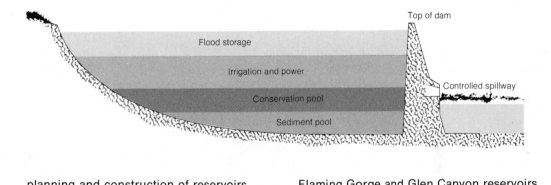

planning and construction of reservoirs (Figure 6-16).

Lake Mead

Today's reservoir is planned with a sediment basin which can accumulate a predetermined amount without interfering with the reservoir's other planned functions. Lake Mead, for example, was planned for a dead storage of 3,223,000 acre feet, or about 5×10^9 tons of sediment. Since the entire drainage area of the lake loses about 900 tons of sediment per square mile annually, it was calculated that the sediment basin built into the lake would last a little over thirty years. Since the construction of Lake Mead, a series of reservoirs, each with its own sediment basin, have been built upstream and on tributaries. These include

Flaming Gorge and Glen Canyon reservoirs on major streams, and many other reservoirs on smaller tributaries. Since these upstream reservoirs receive much of the silt that otherwise would have accumulated in Lake Mead, the life of its siltation basin has been prolonged. Furthermore, much of the sediment that reaches Lake Mead piles up at the head of the lake as a delta and does not enter the dead storage basin. It should be emphasized that the siltation basin of Lake Mead receives a much heavier load of sediment per square mile than is removed from each square mile of the drainage basin. It seems to be true of all such sites that because they occupy smaller areas, rates of maximum accumulation exceed rates of maximum denudation in tons per square mile.

The Mississippi River Basin

The rate of erosion per square mile in the Mississippi River drainage basin is less than that of the Colorado River drainage basin. The Mississippi delta has received

approximately 495×10^6 tons of solid load per year for the last 75 years. The drainage basin of the Mississippi River is about 1.24×10^6 square miles. The denudation

Figure 6-17

(a) Growth of the Mississippi delta in the last one hundred years. (b) The Achafalaya Channel diverts flood waters of the Mississippi River.

The Mississippi Delta in 1874

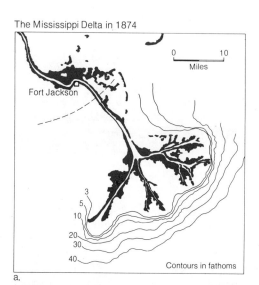

a.

Growth of the Mississippi Delta

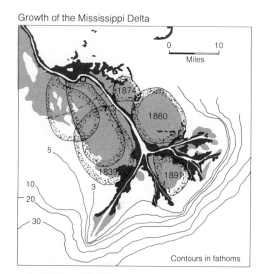

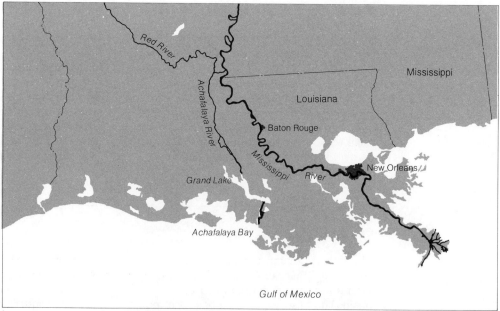

b.

rate for the drainage basin of the Mississippi River averages about 400 tons per square mile, less than half that for the Colorado River Basin. More of the Mississippi River sediments per square mile come from the Missouri River Basin than from the Ohio River Basin. Since there is less rainfall in the western part of the Mississippi drainage basin, the solid loads from western tributaries are much greater than those from eastern tributaries.

Strange things sometimes happen when natural river processes are interrupted. Just as erosion removes land, deposition produces it. Figure 6-17a shows the rate of growth of the Mississippi delta from 1874 to 1940, and the delta has grown at about the same rate for the past several thousand years. As the map shows, the delta area did not fill in evenly, but in segments. The delta area is tectonically subsiding. Deposition and subsidence are usually almost in balance, although the gradual growth of the delta into the Gulf of Mexico is evidence that deposition over thousands of years has

slightly exceeded subsidence. However, delta areas not receiving sediment gradually subside, and the ocean advances at the expense of the land. If the Mississippi River were to be diverted to the Achafalaya Channel (Figure 6-17b), which represents a much shorter distance to the Gulf of Mexico and a more favorable gradient for that section of the river, the present delta shown in Figure 6-17a would gradually subside beneath the waters of the gulf. This would mean that the whole city of New Orleans would gradually sink beneath the waters, and with it much of the Mississippi delta's unique culture. Because dams on the tributaries of the Mississippi River hold back sediment, tectonic subsidence now exceeds deposition. As a result, Louisiana is now losing 17 square miles of land to the Gulf of Mexico each year. As the United States builds more dams on the tributaries of the Mississippi River, the rate of this loss will increase unless methods are found to allow sediment to pass by or through dams.

Loss of Nutrients

Still other changes in the ecosystem result from the retention of sediment in reservoirs. It has long been known that in Atlantic Coast bays and estuaries, oysters attain greater growth in wet years than in drought years. This happens because in wet years more nutrient minerals are washed into the bay by surface waters than in dry years. Not only do soluble minerals provide nutrients for organisms, but sediments, particularly clays, carry adsorbed organic

material and minerals that are released into salt water. Similarly, the tremendous decreases since 1968 in Mediterranean fish catches by Egyptian fisheries has been attributed to the retention of sediments and the nutrients they contain behind the high Aswan Dam. As would be expected, a rejuvenated Nile whose waters bear little sediment is also degrading its channel below the dam.

One need not go to bays and estuaries, however, to discover that deficiencies in nutrients result from alteration of stream

Figure 6–18

Land and water reclamation since 1900 in the Netherlands.

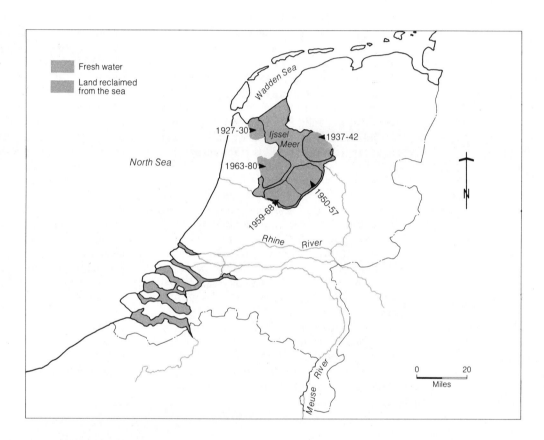

processes. Much of the world depends on the cultivation of floodplains for the production of its food. Under natural conditions, floodplains are periodically covered and a new layer of mud and nutrients is added to the soil. Modern flood control retains the sediments and a large part of the nutrients in reservoirs, so that downstream soils are never fully restored. It is expected that great amounts of chemical fertilizers will be needed downstream on the Nile River bottoms now that the sedi-

ment is being retained behind the Aswan Dam.

Boundaries Between Land and Sea

Changes in the boundaries between land and sea are occurring steadily all over the world. A major change by deposition is the filling of the Persian Gulf, an area about 130 miles long and 75 miles wide, which has been going on since about 3000 B.C. The North Sea has long been a problem to the Netherlands. The Dutch people have taken advantage of Rhine River sediments and the dredging of shallow shoreline deposits along the North Sea to fill in swamps and estuaries in order to create more land (Figure 6-18). This is one way of producing new farmland to support an increasing population, but it has also destroyed most of the Zeider Zee (Ijssel Meer), the largest single spawning ground for North Atlantic fishes.

The combination of changing climate and deposition also alters land forms. At some time in the not-too-distant past the Caspian Sea was much larger than it is now—before precipitation on the mountains along the borders between Iran-Afghanistan and Siberia decreased the supply of water to the Aral and Caspian Seas. Since 1930 the Caspian Sea has decreased by about one-half (Figure 6-19), as a result of decreased rainfall and—according to some writers— the retention of water in upstream reservoirs for irrigation.

Flood Deposition

The examples just considered are the effects of deposition over thousands or hundreds of thousands of years. There are also records of very rapid deposition in geologically short periods of time. As much as one foot a day of sedimentation has been measured on the Mississippi delta, and rates as high as 19 feet a year have been recorded at the mouth of the Fraser River in British Columbia. Nowhere else have such rates been measured, which must mean that the areas of most rapid sedimentation are less numerous than the areas of most rapid erosion, even though total erosion must equal total deposition. This, in turn, means that the total areas of deposition, less the areas of most rapid deposition, accumulate sediment at a considerably slower rate than an equivalent area of erosion loses sediment.

The highest rates of deposition occur during floods. Depositions from really great floods are known both in recent times and in the geologic past. A flood—perhaps the one mentioned in Genesis—covered a

Figure 6-19

The shrinking of the Caspian Sea since 1930.

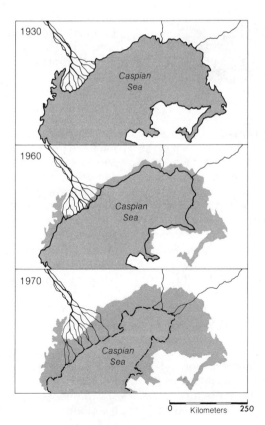

1930

Caspian Sea

1960

Caspian Sea

1970

Caspian Sea

0   Kilometers   250

Figure 6-20

Exhumed sequoia tree from the Eocene, Powder River Basin, Wyoming.

the trees remained upright. This is a volcanic area, where as much as one hundred feet of volcanic debris can accumulate over a large area in a geologically short time. In the Powder River Basin, about 20 miles east of Buffalo, Wyoming, on the old Dry Creek Road, there is a buried forest of Eocene sequoias. Some of these trees, still standing upright, were buried by many feet of water-laid sediments during the Eocene. If these trees grew new sets of near-surface roots after the deposition of each load of sediment, as modern sequoias do, many feet of sediment were laid down on these Eocene floodplains after the trees died but before they fell, since no near-surface roots have been exposed (Figure 6-20).

Problems of Erosion and Deposition

It is now becoming apparent that the problems created by sedimentation are not to be solved easily. Geologists, of course—even James Hutton[1]—would have predicted this. The only way at present to slow down the silting of a reservoir is to build a second reservoir upstream to pick up most of the sediment, then another above the second, and another above that, *ad infinitum*. The Army Corps of Engineers has successfully followed this method by constructing many dams on a single watershed. To counter this predicament, engineers are experimenting with techniques to remove sludge from the sediment pools. Such experiments include the use of dredges, suction pumps, and siphons to draw sediments out of the reservoirs. But it is difficult to locate an environmentally justifiable storage site for the sludge.

large part of the lower Tigris and Euphrates valleys sometime before 2000 B.C., filling houses in many of the cities. There was a tremendous flood on the Snake River in Idaho in the late Pleistocene, which can be recognized by the thick deposits it left in its wake. G. C. Swallow, an early U.S. geologist, records sediment from six to eight feet thick in the Kansas and Missouri River bottoms, resulting solely from the flood of 1844. In November 1966, a flood on the Arno River at Florence, Italy, more severe than any recorded in the history of that river, deposited mud and debris to depths of eight feet in the streets of the city.

Although there are many known floods and other catastrophes by deposition, rates of deposition are difficult to estimate. In Yellowstone National Park, entire forests (now petrified) were once buried while

[1]Eighteenth century geologist who expressed the concept of erosion, sedimentation, and deposition.

If we cannot wholly prevent erosion, there are areas where its effects can be lessened. And indeed some ecologists would even ask whether the silting should be lessened. As we succeed in preventing floods, we shall force ourselves into using more sophisticated fertilizers within a few years. Whenever men spend money to solve one problem imposed on them by nature, they seem always to upset some other natural system.

Other ecologists would maintain that more than normal sediment in streams reduces the penetration of solar energy into the aqueous environment and so reduces the entire food cycle. They would further maintain that we should prevent erosion to maintain the balance of nature. But these ecologists tend to ignore the natural system as it was before the development of land by man, when the balance of nature was maintained in spite of extensive erosion.

Estimates of the total cost of erosion control and damage by sediments in the United States are substantial (Table 6-1), yet they may be too low.

There are many engineering techniques that can lessen harmful erosion, and they in turn decrease sedimentation downstream. The following simple techniques are common knowledge to engineers and geologists, though the general public seems largely unaware of them.

1. Real estate developers should not denude an area until they are ready to develop. This is a normal procedure, but high interest rates starting in 1967 cut down on home construction, so that some areas were left sitting for periods much longer than had originally been intended by the developers.

2. Rip-rapping (covering with closely spaced stones) steep slopes of artificial and natural terraces or cut benches is cheaper than continual repair. Prospective buyers should make sure before buying that such areas are rip-rapped, or insured against erosion in other ways because after purchase the cost of repair becomes the responsibility of the purchaser.

3. Revegetation of areas naturally or artificially devegetated should proceed immediately.

4. Engineers are now experimenting with suction devices and siphons to remove deposits from sediment basins and thus increase the useful life of reservoirs. The primary problem with that is to find suitable disposal sites for the waste sediment.

5. In many places groins are used to prevent beach erosion, yet as often as not eddies develop between groins that result

Table 6-1

Annual damage by sediments in the United States. Modified from John B. Stall, "Man's role in affecting sedimentation of streams and reservoirs," p. 82, Table 2. In Kenneth L. Bowden, ed., Proceedings of the 2nd Annual American Water Resources Association, 1966.

Annual damage	Cost (millions of dollars)
Damage to floodplain soils by covering them with sediments that are not immediately fertile	50
Storage space destroyed in reservoirs	50
The cost of removing sediment from drainage ditches that have been filled	18
The cost of dredging sediment from navigable waterways	83
The cost of removing sediment from irrigation canals	16
The cost of removing sediment from public waters and waterways	14
Sediment damage accompanying floods	20
Other damages (e.g., to highways, fisheries, etc.)	11
Total	262

in more rapid removal of beach sand to deep water instead of movement of sand along the shore. Short sea walls have proved beneficial in preventing shore erosion, but they are not useful enough to justify the cost of sea walls hundreds of miles long. The effect of long sea walls on the ecosystem is not known, and can probably be determined only by actual experiment.

In short, whereas the effects of erosion and deposition are extensive they are as yet far from completely understood, especially by decision-makers in sensitive positions and by the general public. More and more is being learned about erosion and deposition all the time, including the techniques and devices that have or have not proved successful in controlling them. For the layman, the best approach to any erosion problem is to consult a capable geologist or geological engineer.

Erosion resulting from strip mining in Kentucky

Selected Readings

Campbell, C. A., et al. 1967. Applicability of the carbon-dating method of analysis to soil humus studies. *Soil Science* 104:217–224.

Engel, A. E. J. 1969. Time and the earth. *American Scientist* 57:458–483.

Fisk, H. N., et al. 1954. Sedimentary framework of the modern Mississippi Delta. *Jour. Sed. Petrol.* 24:76–99.

Frazier, David E. 1967. Recent deltaic deposits of the Mississippi River: Their development and chronology. Gulf Coast Assoc. Geol. Socs. 17:287–311.

Hastings, James Rodney, and Raymond M. Turner. 1965. The changing mile: An ecological study of vegetation change with time in the lower mile of an arid and semiarid region. Tucson: Univ. of Arizona Press.

Judson, Sheldon. 1968. Erosion of the land—or what's happening to our continents. *American Scientist* 56:356–374.

Malin, James C. 1948. *The grassland of North America: Prolegomena to its history*. Ann Arbor: Edwards Bros., and Lawrence, Kans.: James C. Malin.

Moss, R. P. 1968. Soils, slopes and surfaces in tropical Africa. In Moss, R. P., ed., *The soil resources of Tropical Africa*. Cambridge: The Univ. Press, pp. 29–60.

Scruton, P. C. 1960. Delta building and the deltaic sequence. In Shepard, Francis P., Fred B. Phleger, and Theerd H. Van Andel, eds., *Recent Sediments, Northwest Gulf of Mexico*. Tulsa: Am. Assoc. Petroleum Geologists, pp. 82–102.

Smith, W. O., C. P. Vetter, and G. B. Cummings, et al. 1960. Comprehensive survey of sedimentation in Lake Mead, 1948–49. U.S. Geol. Survey Prof. Paper 295.

Stall, John B. 1966. Man's role in affecting sedimentation of streams and reservoirs. In Bowden, Kenneth L., ed., Proceedings 2nd Ann. Am. Water Resources Assoc., pp. 79–95.

U.S. Water Resources Council, 1968. The nation's water resources; Parts 1–7. Washington, D. C.: United States Water Resources Council.

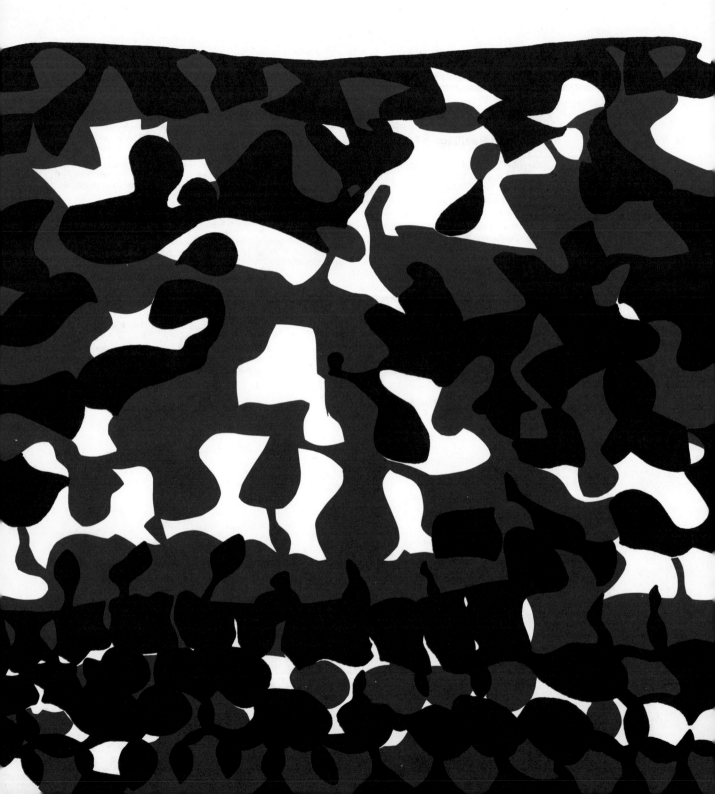

Chapter 7
Soil, Sustenance, and Pollution

When we ourselves die, the soil that has nourished us over a life-time claims us back into its natural rotation, regardless . . . of whether we have cared or not about the facts and rules of its existence and formation.
Kubiena, 1970

Soils are the weathered parts of the lithosphere in which water, mineral nutrients, and organic nutrients form a remarkable combination that nurtures all biogeochemical cycles by transferring new mineral nutrients from the rock cycle of the lithosphere to the ecosystems. Many undeveloped nations are deficient in protein not only because of overpopulation, but also because of poor soils. Only a part of the world's tillable soil is now used for food production (Table 7-1), and much of this is deficient in nutrients that produce proteins or trace substances that maintain or improve health. Persons who argue that the world can support a human population of many billions more overlook two important facts: (1) most of the unused tillable soils in the world are deficient in mineral plant foods at the proper depth, or are desert soils deficient in water and humus; and (2) fertilizers are not only an exhaustible resource, but when used in amounts large enough to produce a rapid increase in food supply also become pollutants in their own right. The conservation of soil for future generations requires care not only to prevent erosion, but also to prevent the overcultivation that causes protein deficiency, shortage of trace minerals, and hypersalinity, and to preserve the ability of the soil to take in oxygen and nitrogen.

When we clean the air, water, or ourselves, we bury the residue. These pollutants may also deteriorate the soil if little care is exercised in siting and burial.

Factors in Soil Formation

Soils are of many kinds and compositions, and range from a few inches to many feet

Figure 7-1

Soil profile of an accreted soil (chernozem) compared to the soil profile of a leached soil (podzol).

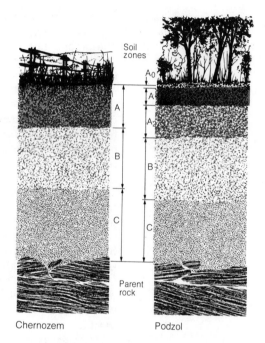

Chernozem   Podzol

Table 7-1

Percent of tillable soil under cultivation. From Sterling B. Hendricks, "Food from the Land." In Preston E. Cloud, ed., *Resources and Man: A Study and Recommendations* by the Committee on Resources and Man of the Division of Earth Sciences, National Academy of Sciences-National Research Council, with the cooperation of the Division of Biology and Agriculture. San Francisco: W. H. Freeman and Company. Copyright © 1969. Table 4-3, p. 68.

Tillable Soil

Continent	Acres cultivated per person	Percent of tillable land cultivated
Australasia	2.9	2
U.S.S.R.	2.4	64
North America	2.3	51
Africa	1.3	22
South America	1.0	11
Europe	0.9	88
Asia	0.7	83

zon is that of primary mineral accumulation in soils that collect minerals, though in some soils the B horizon is also subject to leaching (Figure 7-2). The C horizon is composed of weathered parent rock. Some soils without differentiated horizons, termed auto-ingesting soils, are homogenized by the effects of alternate wetting and swelling, drying and shrinking. Such soils have only one level above the C horizon. Transported soils, such as floodplain alluvium and sand dunes, may have only one horizon.

Soils are formed over long periods by complicated processes involving climate, the weathering of parent rock, the biological communities that live in and help to condition soil, and by topography (particularly slope), chemical activity, mechanical activity, and time.

Climate

Many early soil experts, particularly the Russian pioneers in the study of soils a century ago, believed climate to be the only important factor in the development of soils. Although still considered important, climate is now thought to be only one of several factors in soil formation. The two aspects of climate that are of greatest importance are rainfall and temperature. Optimum rainfall is around 30 inches a year. This amount of precipitation grows enough vegetation to produce a good humus content in the soil profile, but does not wash and leach away humus and mineral nutrients. Less rainfall supports insufficient vegetation to produce enough humus, and further decreases may cause soils to become too saline. As rainfall increases above the optimum, more and more leaching of both

in depth. The richest mature soils contain three horizons, which together make up the soil profile (Figure 7-1). Some soils do not have all three horizons, and some have none. The A, or uppermost, horizon contains humus from decaying plants, and is subject to the greatest loss of mineral and organic nutrients by leaching. The next, or B, hori-

Figure 7-2

Relationship of climate and soil.

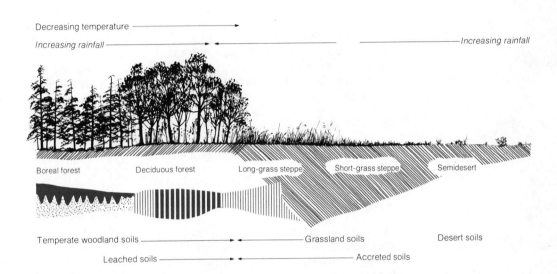

humic and mineral nutrients occurs, and soils are consequently not as rich.

Temperature is less important than rainfall, but biologic and chemical processes are accelerated at higher temperatures. This means that a given rainfall of over 40 inches will probably produce a better soil in a temperate climate than in a tropical climate, because chemical processes of leaching are enhanced in the tropical climate.

Parent rock

Different types of parent rock produce weathered particles of different size and chemical composition. Quartz sandstones produce sandy soils, and limestone areas receiving 60 inches of precipitation a year or more produce *terra rosas* (Table 7-2). Formations high in a particular trace mineral produce soils high in that mineral, ex-

Table 7-2

Soil horizons and their properties.

Horizon	Chemical transfer status	Description
A		Top of A zone, if organic material only, is called A₀.
	Zone of eluviation or removal (leaching) of minerals	Zone of maximum leaching; upper part may be dark-colored because of organic content; remainder is usually light-colored because of leaching.
B	Zone of illuviation or addition (accretion) of minerals	Zone of maximum accretion or accumulation of mineral materials, either leached from the A zone, or brought by rising waters from the C zone.
C	Little transfer except in desert soils	Weathered parent material.
R	None	Unweathered bedrock.

cept when it is soluble and leaches under a particular rainfall regime. Conversely, formations deficient in one or more trace substances produce soils deficient in those substances. Only under peculiar conditions of climate or ground water is a soil completely unrepresentative of the parent rock. Yet the influence of parent rock diminishes with time, since the stable products of weathering are broadly similar regardless of time.

Biology

The biota (a collective term for all life in a particular environment) of a soil is extremely important to its properties. Worms and insect larvae loosen the soil and turn it over, eat organic material, and evacuate it in more degraded form. Bacteria and fungi feed on organic matter, including the usually indigestible cellulose fiber of plants, and further degrade it to aminoproteins and other organic building blocks which both plants and animals convert to their own uses. If we destroy the biotas of soils, as we may be doing with biocides, we destroy one of the steps in the important protein cycle. Organisms also aid in the aeration of soil by keeping it loose, so that air can circulate slowly through it and supply oxygen and nitrogen for important soil, plant, and bacterial processes.

Topography

Topography, primarily slope, is such an important modifier of soils that the U.S. Soil Conservation Service uses different names for soils on steeper and shallower slopes, even if the soils are formed on the same parent rock under the same climatic regime. Many soil modifiers produce different effects with different slopes. An example is creep (page 271). Soil and subsoil creep more rapidly on steeper slopes, thickening the soil profile as the slope gentles toward its base, and thinning it upslope. The soil on top of a hill in an area of active creep may be very thin, particularly in the horizon in which creep is most active. Montmorillonitic soils and soils of other shrink-swell clay minerals creep more rapidly than loams. Groundwater in soil moves downslope. Consequently, lower slope soils or valley soils may contain more water and undergo more chemical alteration and leaching (Figure 7-3).

Furthermore, in the northern hemisphere soils on north-facing slopes receive less direct sunlight and dry less rapidly than soils on south-facing slopes. (The converse is true in the southern hemisphere.) With time, these variations may create chemical differences that produce different soil types. It has been suggested that the low evaporation rate of north-facing slopes accounts for their being timbered in the eastern grassland plains of the United States.

Chemical activity

Major chemical reactions in soils add nitrogen or oxygen, produce carbon compounds, and add water to or remove it from compounds. Water in soil provides hydrogen (H^+) and hydroxyl (OH^-) ions that are available for reaction with other elements and compounds, affording greater mobility to necessary chemicals. A dry soil provides

Figure 7-3

North and south hill slopes and soil formation.

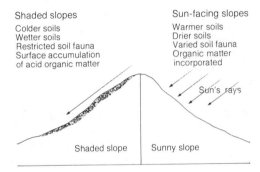

Shaded slopes
Colder soils
Wetter soils
Restricted soil fauna
Surface accumulation
of acid organic matter

Sun-facing slopes
Warmer soils
Drier soils
Varied soil fauna
Organic matter
incorporated

Sun's rays

Shaded slope   Sunny slope

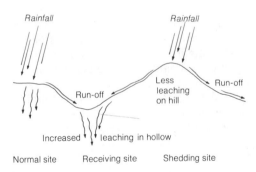

Rainfall   Rainfall

Less
leaching
on hill
Run-off   Run-off

Increased   leaching in hollow

Normal site   Receiving site   Shedding site

Figure 7-4

Plants exchange hydrogen ions through their roots and consequently alter the soil.

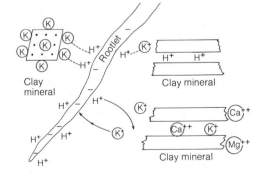

leach necessary minerals to depths beyond the reach of the roots of desirable plants.

Usually the effects of soil factors occur in combination. Chemistry affects the biota, and vice versa. Large amounts of humus produced by plants increase the amount of humic acids, which react with soil particles to mature a soil. Acid soils have lower microbial content than do alkaline or neutral soils. Thus, microbial populations vary in kind and number from soil to soil and produce different rates of cellulose degradation and soil maturation.

Carbonation is the addition of carbon dioxide to chemical compounds. In many compounds the bicarbonate is more soluble than the single carbonate. Carbon dioxide, then, makes some minerals more readily available to plants. Oxidation and reduction reactions (respectively, the addition and removal of oxygen) occur throughout the soil. In essence, the procuring of food by plants involves the exchange of hydrogen ions or hydroxyl ions for other elements (in ionic form) (Figure 7-4).

Mechanical activity

A number of mechanical activities aid in the modification of soils. Some minerals expand and contract with temperature changes. Expansion produced by increased temperature loosens soil, but such changes penetrate only a few centimeters at most because rock is a poor conductor of heat.

Freezing expands porespace in soil because the water in the spaces expands when it freezes, and the soil may remain looser after it thaws. In soil already well-aerated, expansion will be minimal because air already in the porespace will contract

a growing plant little mobile mineral matter. On the other hand, too much water may

as the water expands to form ice.

Many soils contain clay minerals that expand with water. These soils tend to remain loose in areas of alternate drying and wetting, especially if calcium is present to form minute nodules around which the clay minerals are distributed. Under other conditions, similar clay soils pack to hardpan, especially those high in sodium or potassium and low in calcium and magnesium.

Time

Time is important in soil formation. Most soils are older than was formerly believed. Some tropical soils are said to date from the Miocene, or even the Cretaceous, and it is almost certain that many of these soils continued leaching through most of post-Miocene time, or for 12 million years or more. Autoingesting calcimorphs of the Austin, Texas area, on the other hand, have developed a profile approximately three feet deep on landslides that occurred in some wetter period during the last 10,000 years. There are three- to four-foot soils on 15,000-year-old glacial deposits of the Lower Wisconsin, whereas 25-foot soils on Illinoisian glacial deposits are several times as old. We need to know much more about the rates at which soils are formed on different kinds of parent rock under different climatic regimes. Such knowledge would help us to understand how rapidly we are using up our soil. Obviously, on a long-term basis, soil formation (including artificial processes) must equal soil depletion if man is to produce food for the many generations yet to come. At some

point, the ecosystem must be maintained in a steady state.

Soil Classification and Distribution

There are many different classifications of soils, each of which has usually been designed for a specific purpose. The agronomist, largely interested in potential food production, proposes a different classification than does the engineer, who is interested in shear strength or particle size.

About 1900 the U.S. Department of Agriculture adopted, and over the years modified, a classification of soils developed by Russian soil scientists in the latter part of the last century. Although the U.S. Soil Conservation Service no longer uses this particular classification, it is adopted here because it is simple and internally consistent, and because it has been used by most of the nations of the world. Based on the development of profiles and horizons, the system classifies soils in a hierarchy of three major categories, called orders, suborders, and great soil groups (Table 7-3). Many soils, if reasonably mature, differ according to the climate in which they matured; these constitute the zonal order. Others, such as bog soils and saline soils, are related to local conditions of drainage and make up the intrazonal order. A third group, young soils without profile development, fall within the azonal order. Among the latter are recent sanddune formations and alluvial soils on floodplains. Most classifications take into account such important phenomena as climate, temperature, rainfall, nature of parent rock, topog-

Table 7–3

Major categories of soil classification.

Order of Zonal Soils

Suborder	Great soil group	Description
Soils of the cold zone	Tundra	Thin, rocky, but may be high in humus and nutrients.
Desert soils	Reddish desert soils	Thin, reddish or light brown soils with little or no A horizon and with calcareous horizons; frequently saline; hot, arid climate.
	Sierozems (light gray desert soils)	Thin gray soils with little or no A horizon and with calcareous horizons; cool or temperate arid climates.
Grassland soils	Prairie soils	Dark brown soils with some leaching in the A horizon, calcareous B horizon.
	Chernozems	Dark brown to nearly black soils with deep humus profiles and A and B horizons accreting calcium minerals.
	Chestnut soils	Dark brown soils with moderately deep humus layer and accreting A and B profiles.
	Reddish chestnut and brown soils	Brown soils with thinner humus profile and highly accreting A and B horizons.
Temperate forest soils	Podzols	Light-colored soils with leached A zone and thin dark humus zone (A_0); B zone may be thin.
	Gray podzolic soils	Gray soils with leached A zone and thin dark humus zone (A_0); B zone may be thin.
	Brown padzolic soils	Brown soils with leached A zone and thin dark humus zone (A_0); B zone may be thin.
	Red-yellow podzolic soils	Red and yellow soils with leached A zone and thin dark humus zone (A_0) or leached humus; B zone may be thin.
Latosols	Lateritic latosols	Red and reddish soils leached of soluble minerals and of alumina and silica, retaining iron oxides and hydroxides.
	Bauxitic latosols	Red and reddish soils leached of soluble minerals and of iron and silica, retaining aluminum oxides.
	Siliceous latosols	Gray and reddish soils leached of soluble minerals and of aluminum and iron, retaining or accreting silica.
Mountain soils		Rocky, usually thin, extremely variable, spotty.

Order of Intrazonal Soils

Saline & alkaline soils of arid & nearshore regions	All of those saline soils produced by peculiar & local environments.	Soils of various colors accreting soluble salts in all horizons, and particularly the A horizon.
Hydromorphic soils	Various kinds of bog and marsh soils formed in peculiar and local groundwater conditions.	Mostly dark peaty or mucky soils, poorly drained.

Suborder	Great soil group	Description
Calcimorphic soils	Soils in which a limestone parent rock overrides the leaching effect of climate.	Dark or light soils with calcareous A and B horizons; some have been termed false chernozems.
Terra rosa soils	Usually developed on limestone parent rock.	Reddish, thin, lightly leached.
Autoingesting soils	Usually developed on mont-morillonitic claystone parent rock.	Shrinkage during dry periods allows cracking. Humus and surface materials fall into the cracks. Through time, the soil profile is homogenized and distinct horizons do not develop.

Order of Azonal Soils

Transported soils	Usually alluvial; sand dunes and sand fields also.	Horizons are not developed; original depositional features still preserved.
Regosols and lithosols		Rocky soils or soils made up of fragments of unaltered parent rock or unaltered fragments of parent rock.

Figure 7-5

Podolization in relation to the water table: (a) grassland soils and (b) podzolic areas.

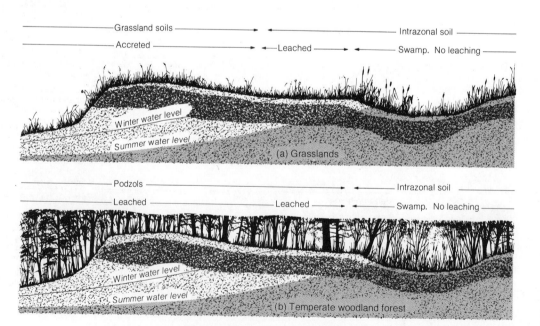

raphy, capillarity, precipitation, and ground-water in relation to leaching (Figure 7-5).

Zonal soils

The zonal soils are those that are related to climate, a phenomenon that can best be observed by comparing precipitation and soil maps (Figures 5-5 and 7-6).

The soils of the alpine or cold zone constitute the great soil group known as the tundra soils. Tundra soils may be high in humus and nutrients, but little is known of their properties except that they are very

thin. Desert soils are light red, reddish brown, or light gray. They are not leached in the A horizon, if it is present, though it may be thin or absent due to lack of humus. The B horizon is rich in salts and available minerals, and may even be saline. Desert soils are fertile and productive but require irrigation and nitrogen because of the absence of water and humus. With irrigation, evapotranspiration transforms them to intrazonal saline soils that gradually become less and less productive.

The grassland soils include many important soil groups, including the prairie

Figure 7–6

Soil map of the United States.

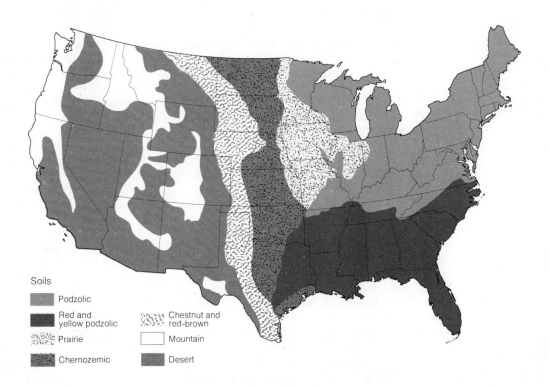

Soils

- Podzolic
- Red and yellow podzolic
- Prairie
- Chernozemic
- Chestnut and red-brown
- Mountain
- Desert

soils, chernozems, chestnut soils, and reddish chestnut soils. All these have a B horizon that accumulates minerals useful to plants. The A horizon may undergo some leaching, as in prairie soils, but there is usually not enough precipitation to cause extensive leaching of nutrients required by shallow-rooted plants.

Soils of the temperate forest suborder, in contrast to the grassland soils, suffer extensive leaching in both A and B horizons, usually enough to make the lower part of the A zone and even part of the B zone a pale or light-gray color (see Figure 7-5). This leaching removes valuable mineral nutrients, and after several years these soils become unproductive for shallow-rooted plants, though some deep-rooted plants can extract necessary mineral nutrients from the C horizon. The soils of the temperate forest suborder include the podzols. Gray podzolic soils, primarily associated with north temperate deciduous and conifer forest areas with more than 25 inches of rainfall, are gradually replaced in the south temperate areas by red and yellow podzolic soils under deciduous and conifer forests with 40 to 50 inches of rainfall.

Another suborder, latosols or "soils of the tropical woodland," are not now restricted to the tropical woodland, although most may have formed under that condition millions of years ago. These soils are now relict if they survive in the midst of changed surroundings. Latosols include a variety of red ferruginous soils, including laterites, bauxites, and siliceous soils. Latosols carry the leaching of nutrients several steps farther than do podzols. The typical horizons of other zonal soils are usually lost or only faintly observable because leaching of humus and nutrients leaves approximately the same minerals at all levels of the soil profile. Latosols, due to leaching of both humus and mineral nutrients, are the most unproductive of the zonal soils.

Mountain soils vary in thickness, quality, and other characteristics. Although important to the individual landowner, they are not significant in world food production.

The most productive soils for the shallow-rooted plants that provide most of man's food—cereal grains, potatoes, sorghums, and the like—are those of the grasslands suborder. Of these, the best are the chernozems, with a thick humus zone in the A horizon, accreted mineral nutrients in the B horizon, and 25 to 40 inches of rainfall annually. Some montmorillonitic, auto-ingesting, and calcimorphic soils (see Table 7-3), like those that produce the blacklands of Texas and Mississippi, are also good. Prairie soils and the red and brown chestnut soils also have a substantial humus content in the A horizon and good mineral nutrients in the A and/or B horizons.

Soils of the grasslands suborder constitute the breadbaskets of North America and eastern Europe. Large parts of Siberian U.S.S.R., a small area in Argentina, and other minor areas of the world also have chernozem soils. That is, nearly all the developed nations have, or are associated with an economic block that has, extensive grasslands soils.

Fertilization of grassland soils is at first very simple. Nitrates and phosphates are simply poured on, and the natural minerals will last for many years. But eventually more and more trace elements must be added,

Egyptian farming on poor desert soil

though the dust storms that result from alternating wet and dry cycles add many trace minerals to U.S. grassland soils, bringing trace elements from the bedrock and desert soils farther west.

Soils of the temperate forests, the podzols and podzolic soils, are less productive than grassland soils because more nutrients are leached out of them. In pioneer days in the American East, the burning of forests to clear farms provided nutrients that lasted for a few years. As the soil deteriorated due to farming, settlers abandoned their farms and moved farther west to fresher lands with unused soils. Podzols and podzolic soils are harder to fertilize than grassland soils because they are more rapidly depleted of many trace elements and minor mineral nutrients.

As mean average temperature increases,

Thailand's wet climate produces better soil for crops

the potential for chemical activity increases. And as precipitation increases, the pathways for transmitting chemical activities increase. Thus the most leaching occurs in the tropical rain forests. In fact, leaching is the primary characteristic of latosols. In laterites and related ferruginous latosols, most minerals, except iron and related metals, are leached. Some ferruginous latosols retain enough nickel to be considered nickel ores when the iron content is too low for development of iron. In bauxites and related aluminous latosols, most minerals except alumina are leached. Thus even the more insoluble minerals are leached from latosols. The soil profile may be only two or three feet deep, as in some parts of Australia, or up to 80 or 100 feet deep, as in other parts of Australia and parts of Rhodesia. Some latosols are forming now, while others are as old as the Miocene or even the Cretaceous. Such extremely old and leached soils lower the

agronomic value of an area. Because their mineral nutrients and humus are minimal, the latosols are the poorest of the soil suborders.

Fertilizers produced by developed nations are appropriate for their own soils, but not for latosols. Not only do latosols need the nitrates, phosphates, and potash that other soils need; they also need most of the thirty or more nonmetallic trace elements (Table 7-4) necessary to sustain healthy plant and animal life. Furthermore, latosol fertilizers must have a pH of 7 or greater, so that nickel, cobalt, cadmium, zinc, silver, and other retained metals are not drawn up into the plants in amounts dangerous to animal systems. Common phosphate fertilizers contain many of the trace elements needed by latosols, but are so acid that they increase the mobility of these metals, especially in the absence of calcium.

The desert soils are used extensively wherever water is available. They have little humus but are high in mineral nutrients. In desert areas evapotranspiration is so great that minerals are not carried away by groundwater, but are deposited by water evaporating from soil and transpiring from plants. Unless good organic fertilizers are used, the food products from desert soils may lack essential proteins. Many organic food faddists who eat only "natural" foods may be subjecting themselves to the protein deficiencies of plants grown (especially in desert soils) on nonfertilized farms. Nevertheless, the main problem in desert-soil farming is lack of water. Since these soils are located in arid lands, they must be irrigated. And since potential evapotranspiration far exceeds precipitation plus irrigation, little water runs off, either on the sur-

Table 7-4

Some common trace substances used as fertilizer supplements. From Z. Kalix, "Utilization of Fertilizer Raw Materials in Australia." In *Economic Commission for Asia and the Far East*. Mineral Resources Development Series, No. 32, 1968, p. 38.

Element	Compound used
Copper	sulphate
	oxide
	carbonate ore
Zinc	sulphate
	oxide
	carbonate ore
	(refined)
Cobalt	sulphate
	oxide
Molybdenum	sodium molybdate
	oxide
	trioxide
	calcium molybdate
Manganese	sulphate
Magnesium	sulphate
	oxide
Iron	sulphate
Boron	borax

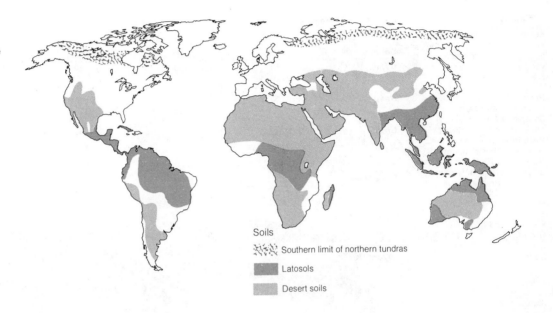

Figure 7-7

Distribution of desert and latosol suborders. (Compare with Figure 7-8.)

Soils

Southern limit of northern tundras

Latosols

Desert soils

face or underground. The soils of such irrigated deserts eventually become so saline that normal human food will not grow in them (see page 125).

Soil Distribution

Only the zonal soils are distributed widely enough to be significant in food production on a national and continental scale. Because of the importance of soils in any national economy, or in any program to feed increasing numbers of people, the kinds of soils on different continents play a part in determining the nature of national development. For example, lands presumably formed from the ancient continent of Gondwana (see Appendix B) generally have poor soils, largely because of the climatic

and geologic history of Gondwana during the Tertiary migration of the Gondwana continents to their present sites. Most of the undeveloped nations are on these continents—South America, Africa, India, Australia, and Antarctica. Except for areas too small to be significant, the soils of the Gondwana continents are chiefly latosols and desert soils (Figure 7-7). With their increasing populations, naturally poor soils, and shortages of energy (see Chapter 2), the undeveloped nations face almost insurmountable problems.

Soil and the Food Problem

A diet can be deficient in either calories or protein (Table 7-5; Figures 7-8 and 7-9). Although there is a general relationship

Table 7–5

Estimated calories of food and grams of protein in the average daily diets for various nations. From Sterling B. Hendricks, "Food from the Land." In Preston E. Cloud, ed., *Resources and Man: A Study and Recommendations* by the Committee on Resources and Man of the Division of Earth Sciences, National Academy of Sciences-National Research Council, with the cooperation of the Division of Biology and Agriculture. San Francisco: W. H. Freeman and Company. Copyright © 1969. Table 4-1, p. 66. (Source: United Nations, 1966)

Caloric and Protein Consumption

Country	Prewar calories	1951/53 calories	1957/59 calories	1963/64 calories	Protein, grams
U.S.A.	3280	3130	3110	3110	92
France	2880	2840	2940	3070	100
Sweden	3120	3020	2930	2980	84
Spain		2490	2590	2850	78
Brazil	2190	2380	2590	2850	69
Japan	2020	1930	2170	2280	73
India	1950	1750	1910	1990	50
Philippines		1690	1760	1990	46

Figure 7–8

The geography of hunger. Population figures estimated for 1975 in millions of persons. (From *Population, Resources, Environment: Issues in Human Ecology*, 2nd Ed., by Paul R. Ehrlich and Anne H. Ehrlich. W. H. Freeman and Company. Copyright © 1972.)

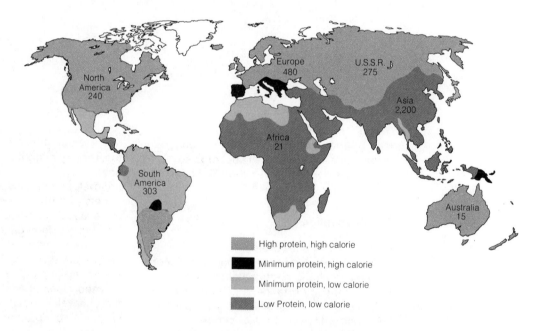

between the intake of protein and calories in the diets of different countries, it is not exact, and people of undeveloped nations have a relatively lower protein intake per calorie than do people of developed nations.

Furthermore, Oceania is the only undeveloped area that enjoyed a significant per capita food increase in the decade 1956–1966 (Table 7-6). Eastern Europe and the U.S.S.R., by developing areas of good soil,

Figure 7-9

Per capita protein consumption of selected nations. (From *Population, Resources, Environment: Issues in Human Ecology*, 2nd ed., by Paul R. Ehrlich and Anne H. Ehrlich. W. H. Freeman and Company. Copyright © 1972.)

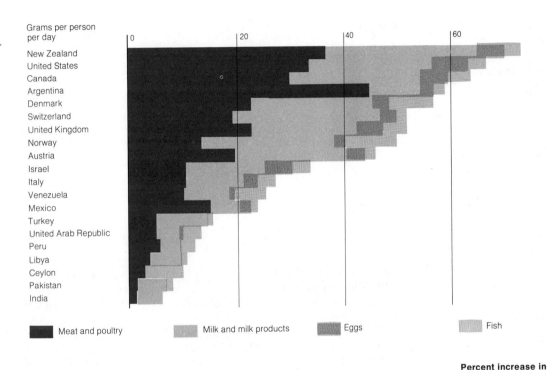

Grams per person per day

New Zealand
United States
Canada
Argentina
Denmark
Switzerland
United Kingdom
Norway
Austria
Israel
Italy
Venezuela
Mexico
Turkey
United Arab Republic
Peru
Libya
Ceylon
Pakistan
India

■ Meat and poultry   Milk and milk products   Eggs   Fish

Table 7-6

World per capita production for 1958 and 1968 with the base 100, an average for 1948–1952. Population increase for 1960–1970 is also given. Data from Food and Agricultural Organization, 1970; *Production Yearbook: 1969*, vol. 23, part 1, United Nations, Rome; and "Special Topic: Population Trends," *Demographic Yearbook: 1971*, 22nd ed., United Nations, New York.

Region	1958	1968	Percent increase in food production	Percent increase in population 1960–1970
World, excluding China	106	114	+7.5	+24.3
Africa	98	99	+1.0	+27.4
Northern America	101	107	+5.9	+14.2
Latin America	105	101	−3.8	+37.5
Asia, Near East	107	110	+2. }	+24.5
Asia, Far East	104	110	+5.8 }	
Eastern Europe and U.S.S.R.	122	148	+21.3 }	+10.3
Western Europe	106	129	+21.7 }	
Oceania	107	129	+20.5	+29.1

accomplished a remarkable increase in protein intake per capita in the same decade. Little change occurred in other parts of the world, despite great effort and expenditure. Latin America and the Far East seemed to improve slightly in the late 1950s,

Table 7–7

Population estimates 1960–2000. Data from *The World Population Situation in 1970.* New York: The United Nations, Population Studies, No. 49. High, medium, and low estimates were made. Those given here are intermediate.

Population Estimates

Region	1960	1970	1980	1990	2000
East Asia	785	930	1005	1265	1424
South Asia	865	1126	1486	1912	2354
Europe	425	462	497	533	568
Soviet Union	214	243	271	302	330
Africa	270	344	457	616	818
Northern America	199	228	261	299	333
Latin America	213	283	377	500	652
Oceania	15.8	19.4	24.0	29.6	35.2
World total	2986	3632	4457	5438	6494
Undeveloped regions	2010	2542	3247	4102	5040
Developed regions	976	1090	1210	1336	1454

but lost that gain in the 1960s because of increasing populations. Table 7-7 projects the future population of the world, and of some undeveloped areas. In Latin America food production will have to double over the next 30 years even to maintain the marginal starvation diet now available.

The protein cycle

The protein cycle in a natural ecosystem does not involve much new protein, and parallels the nitrogen cycle (see Figure 1-15), since nitrogen is of primary importance in protein. Phosphorus is also important in nearly all protein-manufacturing enzymes, and many trace elements are important catalysts in enzyme function. As animals and plants die, their remains accumulate in the humus profile of the A horizon of the soil. Here they are reworked by soil organisms, degraded by soil bacteria and fungi, and converted to usable form for a new generation of plants. Ex-

cept for that lost through leaching, the protein of the entire community is constantly recycled. Therefore, the only new protein introduced into a stable system is that produced by plants to replace the loss from leaching. In agricultural areas where natural fertilizers (manure, plowed-under crops, and so on) are not used, and where seeds that concentrate protein are harvested and removed, the ecosystems degenerate. The natural humus of the soil is used up, and greater quantities of nitrogen, phosphorus, and potash fertilizers, as well as trace substances, become necessary to provide sources from which plants may manufacture new protein to replace that removed or lost.

Protein in the soil

Trace elements have an incompletely understood affinity for organic materials. As humus in a soil is depleted, its trace mineral content drops, even though nitrate, phos-

Spraying sulfur dust (a biocide) on grapes near Fresno, California

ticides and herbicides on the soil microbiota. If the soil microbiota is seriously depleted, the protein content of the crop can be lowered. Short-term tests on soil bacteria seem to indicate that they recur in optimal numbers soon after the application of biocides. Whether this occurs through selection of more resistant forms is not yet entirely clear. Nor is it known whether the selected or mutated forms are as efficient at degrading cellulose and other humus material, or at building more soluble organic materials, as the previous forms. Long-term studies of this problem are in progress. The entire protein cycle is so complex that adequate answers should not be expected soon. It seems possible, however, that certain combinations of fertilizers and biocides could give an increase in bushels per acre but an actual decrease in protein per pound.

Various chlorinated hydrocarbons remain in the soil for different periods of time (Figure 7-10). The persistence of these pesticides in soil means that either (1) all soil microbiota are destroyed, or (2) soil microbiotas become tolerant and regain their former populations. It has been shown that treatment of boll weevils with certain pesticides destroys the boll weevil predator, but that the weevils themselves become tolerant of the pesticide (in this case, azodrin) and even more numerous! Such phenomena are not rare, and lead one to believe that more natural approaches to insect control may produce more satisfactory results over many decades.

Protein depletion resulting from overuse of biocides and herbicides, plus the depletion of humus and trace substances, produce soils that can best be called

phate, and potash fertilizer are added. Moreover, trace substances that are bonded to organic chemicals are more readily used by plants than those in inorganic form.

Over a short period of time, the use of a combination of pesticides and soluble fertilizers appears to increase crop return. The effect of such use over many decades is not so clear, especially if herbicides and defoliants are applied to the crop to destroy weeds. Considering the role of bacteria and fungi in the protein cycle, there is some question about the long-run effect of pes-

Figure 7-10

Persistence of insecticides in soils.

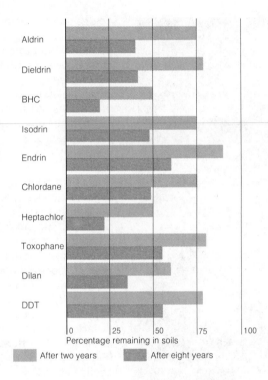

Aldrin

Dieldrin

BHC

Isodrin

Endrin

Chlordane

Heptachlor

Toxophane

Dilan

DDT

Percentage remaining in soils

After two years   After eight years

Figure 7-11

Population trees for India and the United Kingdom, undeveloped and developed nations, respectively.

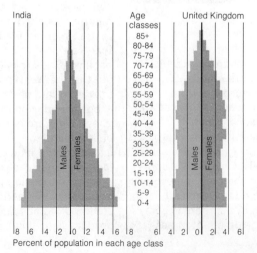

India

Age classes

United Kingdom

Males Females

Males Females

Percent of population in each age class

agrosols, because they resemble nothing in nature. We know so little as yet about soil processes that we are not taking proper care of our soils; they are deteriorating in spite of our best efforts.

Pollution of the Lithosphere

Whatever bothers the conscience of man, he hides. The puritanical attitudes of most Americans toward all things scatalogical may have produced better health in this country than in most parts of the world: human waste was to be hidden, and the best place to hide things is in the ground. The same has been true of solid waste, particularly if it is odorous. In less fastidious undeveloped countries, the "best" disposal site may be the railroad cut just outside of town, or even the nearest street. Until population and the volume of wastes outgrew simple remedies, it was usually true that "out of sight" meant improved sanitation because waste was buried "in the ground." This is no longer always true, even in the United States, and in the undeveloped countries the problem is more acute. Modern medicine and sanitation have been highly effective against infant mortality resulting from filth, and are largely responsible for recent population increases. This is partly demonstrated by the larger proportion of young people in the populations of undeveloped than of developed nations today (Figure 7-11). And with more people the problem grows more acute.

Solid waste

It has been estimated that by 1980 the United States will produce enough solid waste each year to fill the Panama Canal,

A neglected attempt at controlling environmental pollution

or over 700,000 tons a day.

There are three chief ways of handling solid wastes, each requiring a different technology: (1) burial, (2) a new approach to the container problem, and (3) recycling. Most engineers seem to think that with the use of sanitary landfill techniques, proper cover, and an impervious sill, nature reclaims garbage dumps rapidly. The time involved is thought to range from 5 to 20 years. On the other hand, archaeologist Aurel Stein tells of excavating a 1700-year-old refuse dump in the eastern Takla Makan Desert in Sinkiang, China. He describes its odor as almost unbearable, and states that after years of experience excavating garbage dumps he could identify a 1000-year-old refuse pile as Tibetan from its odor alone, long before finding corroborating antiquarian evidence.

The space required for solid waste could

Poor method of solid-waste disposal—an open dump

Recycled garbage marketed in the United States as a soil conditioner

perhaps help bring about a more favorable balance of trade.

Geology enters into the solid waste problem when waste is buried. Burial sites need to be selected carefully, not only to adhere to aesthetic values but also to maintain health. Most city administrations suffer from the delusion that once solid waste is buried it is no longer a problem.

Surface Disposal.   In an arid environment with a deep water table, there is little danger of pollution (Figure 7-12a). If the water table is in permeable rock, and is near or touches sanitary landfill, water downslope can be polluted (Figure 7-12b, 7-12d). The best solution is to bury solid waste in impermeable rock (7-12c). Excavation costs usually require that the impermeable rock be claystone; excavation in other impermeable rocks is usually much more expensive. Approximately 92 percent of the landfills in Texas are in old gravel pits in permeable rocks, from which the leachates can drain into surface waters. The situation in other states does not greatly differ. In many areas the great danger of using permeable rocks for sanitary landfill is drainage into the underground water system. This problem has been studied by the Illinois Geological Survey, especially in areas where hundreds of communities depend on potable water from shallow, permeable aquifers in surficial glacial deposits. These deposits are particularly vulnerable to pollution, and in Illinois the only solution seems to be the use of rare, impermeable rocks for disposal sites. Disposal of waste, whether sewage or garbage, in limestone is particularly unsatisfactory, because channels are more open in limestone than in other

be reduced if manufacturers ceased making disposable containers—though it would still be necessary to convince people to return the reusable substitutes. One way of insuring returns would be to require a much more substantial deposit on returnable containers. We might then dispense with most of the 20 million tons of plastic manufactured each year, most of which is non-degradable and builds up continuously in the environment.

Recycling is long overdue. In a few years we may find ourselves short of many of the substances we are burying with solid wastes, including tin, chromium, nickel, and manganese—commodities for which we depend on trade with other countries. By recycling eight million cars each year, we could appreciably reduce our imports of some of these metals and of foreign steel—and

Figure 7-12

Examples of different hydrologic conditions at sanitary landfill sites.

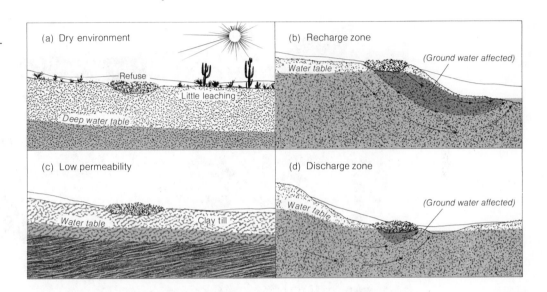

(a) Dry environment

Refuse

Little leaching

Deep water table

(b) Recharge zone

(Ground water affected)

Water table

(c) Low permeability

Water table

Clay till

(d) Discharge zone

(Ground water affected)

Water table

rock, and migration of leachates many times more rapid.

Subsurface Disposal.   In areas without impermeable formations, other methods of waste disposal must be sought. Underground or subsurface disposal has been suggested, but it has limitations. For example, the impermeable volcanic and granitic terrains of the northwest United States—Washington, Oregon, and northern California—prevent the disposal of liquid wastes at depth. New England, underlain by granite and metamorphic rocks, is likewise unfavorable for subsurface disposal (Figure 7-13).

If liquid waste is disposed of below the surface, improperly plugged water wells and oil tests, or deep water wells, will have to be located and monitored to insure that there is no leakage. This problem is the

same as that of brine disposal (see Figure 5-15). Cities that eventually employ subsurface disposal methods must be prepared to budget more money than they now do for waste disposal. In addition, each will need two or three geologists or subsurface hydrologists to advise them on proper monitoring and disposal.

Composting.   Some ecologists have urged every citizen to bury all of his soft garbage in his own backyard or compost pit, and journalists have passed this recommendation on to the public (e.g., *Denver Post,* June 26, 1970, p. 27). However, the individual compost pile as a private sanitary fill is nothing less than ridiculous. If widely adopted, this practice could cause an entire community to smell like the old Chicago stockyards within a few months. Composting must be undertaken with a great deal of caution;

Figure 7–13

Subsurface waste-disposal prospects for the United States.

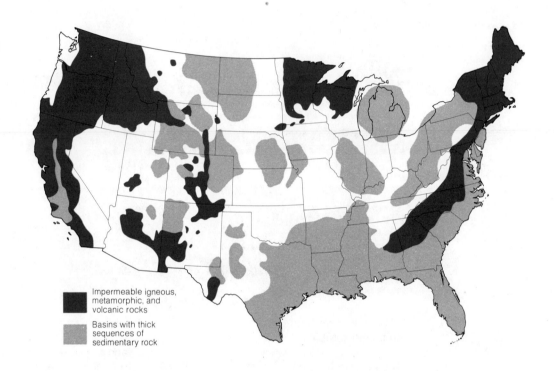

Impermeable igneous, metamorphic, and volcanic rocks

Basins with thick sequences of sedimentary rock

Figure 7–14

The relationship of composting and drainage conditions.

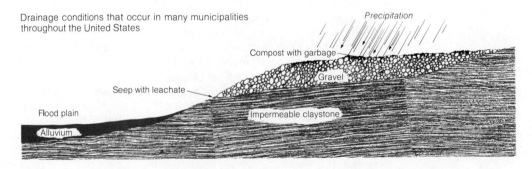

Drainage conditions that occur in many municipalities throughout the United States

Precipitation

Compost with garbage

Gravel

Seep with leachate

Flood plain

Impermeable claystone

Alluvium

not all soft garbage can be buried at any time anywhere.

Figure 7-14 illustrates a geological situation that is replicated at thousands of places in the United States. A gravel deposit rests on impermeable clay. The gravel is permeable and transmits water so that small springs or seeps occur at the ground surface contact of gravel on impermeable claystone. If the occupants of houses on the gravel indiscriminately buried soft garbage, it would leach and produce leachate springs. In addition to various poisons, it has been shown that such leachates can also carry viruses. Permeable limestone overlying impermeable rock creates a comparable situation.

Sewage

Sewage disposal also becomes a geological problem when sewage enters the groundwater system or is disposed of in the subsurface. (see Chapter 5). Wastes from septic tanks may escape through permeable rocks and pollute groundwater, or may reach an outcrop and pollute surface waters. The groundwater system can be monitored and protected by passing proper ordinances to prevent pollution. Such ordinances restrict septic tank fields to the proper kind of rock and the proper density for that rock.

Sewage, if disposed of in the subsurface, requires the same precautions as do liquified garbage or industrial waste: (1) disposal must be well below fresh water aquifers, and (2) monitoring to assure that the waste is being contained at the desired level must be constant. If pollution is discovered in a municipal aquifer, it is too late. Pollution must be detected before it reaches the site to be protected. Since in most areas this is impossible, the most satisfactory alternative is a backup sewage disposal system or a backup water supply. For most cities the expense of two such systems argues strongly against extensive underground storage of sewage wastes.

Toxic Waste—A Special Problem

The term biocide refers to chemical agents, usually artificial, that kill animals or plants; it is probably a better term than pesticide, since most such agents are not selective for animal pests only. Defoliants are special chemicals used to destroy the leaves of plants. In Viet Nam defoliants were used so ground activity could be observed from the air. Defoliants were originally developed to aid in crop production. Cotton, for instance, can be picked much cleaner mechanically if the leaves are first removed by a chemical defoliant. Arsenic trioxide is one of the original cotton defoliants. Since common pesticides, such as dieldrin and DDT, will also kill plants if used often, and since some defoliants interfere with the hormonal development of plants and animals, biocide seems to be a more accurate term for them also. It should not be forgotten that human beings subjected to excess polychlorinated biphenyl (a chlorinated hydrocarbon) have been known to suffer physical deterioration.

Biocides break down and disappear from the environment at different rates. Some persist only a short time, whereas hard biocides persist in the soil for as long as many years (see Figure 7-10).

Biocides and fertilizers may now be necessary to feed a hungry world, but if they do not destroy the world, they may at least produce conditions that make starvation and malnutrition the fate of future generations.

Metallics

We are adding some metals to soils at rates sufficient to constitute a health hazard.

Excesses of metals have always been perilous to mammals, and the danger posed by a few metals has been recognized for many years. Cadmium, in a concentration of 5 ppm in some latosols in Malaysia, is a definite health hazard. Cadmium plating solution, dumped into sewage, caused massive fish kills in the sewage eutrophication ponds at Austin, Texas, in the spring of 1970.

Recent examination of pheasants in Wyoming revealed a mercury content of 0.03 ppm, a level considered unsafe for continuous human consumption. This mercury comes from the soil. To preserve seed grains from rodents, most are treated with mercuric salts; as a result, mercury builds up in the soil, and growing plants concentrate it and pass it on to animals.

Many metals are relatively insoluble in normal environments, but are made more mobile by the addition of phosphate fertilizers, which lower the pH of the soil (see Chapter 8). A few metals create serious problems that will become more severe if municipal wastes and garbage are recycled into the soil, as in experimental waste disposal in Chicago.

Radioactive isotopes

Not all atoms of an element have the same number of neutrons; hence they do not all have the same atomic weight. Those that have the same number of neutrons (and the same atomic weight) constitute an isotope. Many elements have more than one isotope, and many isotopes are radioactive, that is, they emit radiation. Most radioactive isotopes are products of radiation waste. A number of radioactive isotopes are introduced into the soil cycle, and may become geological problems. In Montana, for example, iodine-131, falling on grazing lands from the atmosphere after nuclear explosions, has contaminated milk to unsafe levels for as long as 30 days; iodine-131 has such a short halflife (8 days) that there is insufficient time to cycle it through the soil. By the time it is deposited in the soil, converted to soluble salts, and picked up and utilized by growing plants, the radioactive residue is at safe levels.

Strontium-90 has also, but not often, reached high if not lethal, levels. Strontium-90 has a halflife of 28 years, long enough to cycle through the soil. Strontium is chemically related to calcium, and tends to concentrate with calcium in the bones; the increased incidence of leukemia throughout the world since 1945 has been attributed to greater amounts of strontium-90 in those bones where bone marrow produces red blood cells.

Another isotope thought to be harmful is cesium-137, which has been studied in Australia, New Zealand, and Alaska. Cesium-137 has a halflife of 30 years, and builds up in the A horizon of soils because of its affinity for humus and clay minerals. Once absorbed on clay or joined to an organic compound, it can be released steadily over a period of several years (Figure 7-15). Cesium-137 in Alaska is the product of Chinese nuclear tests, and in New Zealand and Australia it is the product

Figure 7-15

Pathways of Cesium-137
fallout.

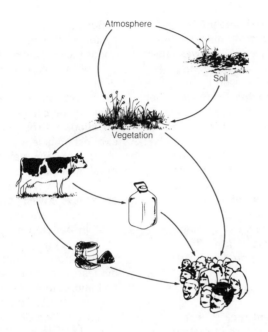

potassium, and a few other elements have
one or more radioisotopes with halflives
long enough to accumulate in the soil.

Radiation from the Plowshare program
(see Chapter 2) has not endangered man;
we must assure that it does not.

Secondary radionuclides (isotopes escaping from nuclear reactors) are now building
up in some waters. Over 100 radioisotopes
have been identified in the Columbia River
below the Hanford nuclear site. Though
such radioisotopes do not pose individual
hazards, the gradual accumulation in the
biosphere and lithosphere of those with a
halflife of nine months or more may eventually create local health problems.

At present radiation may be less of a
hazard than other forms of pollution such
as, say, destruction of phytoplankton by
chlorinated hydrocarbons. Because we are
afraid of radiation, we take pains to avoid
it. Thus we take comfort in the knowledge
that four networks alone have over 100
radiation-monitoring stations in the United
States. And we are encouraged that long-
lived fission products are decreasing in
the environment, and will continue to do so
unless we inaugurate another ambitious
program of nuclear detonations.

of French nuclear tests. Adverse effects
are suspected, but have not been documented. Some Eskimos are carrying doses
of cesium-137 greater than are considered
safe.

Zinc, cesium, manganese, ruthenium,

Chemical and biological warfare (CBW) agents

If we are correctly informed, the worst
CBW agents have been banned in the
United States. But less potent agents are
still in use, and if man insists on fighting
it may be that such weapons as tear gas
and pepper gas are more humane than

guns, clubs, and knives. Chemical warfare agents include defoliants, nerve gases,
and other chemical toxicants. But classifications change. White phosphorus was once
considered a chemical warfare agent; it is
now classified as a general-use weapon.

The Dugway incident in Utah illustrates
the dangers of CBW. In 1968-1969, the
accidental misapplication of nerve gas

from a test killed approximately 2,500 sheep up to a distance of 40 miles downwind. That no people were killed was sheer good luck. Such chemical agents may be buried, if great caution is used in selecting impermeable rock sites that will prevent them from making contact with the biosphere, hydrosphere, or atmosphere for a long period of time. Such burial sites should be continuously monitored. For long-term biological agents like anthrax, which are likely to produce tough spores or cysts, decontamination is quite difficult. Again, the best means of disposal is neutralization or degradation, but this must be planned at the time of manufacture, not after the stored agents have become dangerous.

Summary

Soils evolve from parent rock as a result of precipitation, temperature, the effects of living creatures, topography, chemistry, and time. Soils profoundly affect cultures, because good and poor soils are distributed unevenly around the world.

Soils play an intricate part in the ecosystem as the interface connecting the lithosphere to the biosphere, atmosphere, and part of the hydrosphere. Hence the misuse of soils can seriously upset the cycles of the ecosphere. Burying waste is one way to upset these cycles, and speeding up food production through artificial practices that exhaust soils is another. Indeed, the wearing out of soils may be the most serious of all man's abuses of the environment.

The only ways we know to restore soils are nature's ways, and we must practice them. For example, we must return more protein-building humus to the soil. When we do not know the natural cures, we must not prevent nature from curing itself by destroying her instruments, such as soil microbiota, trace substances, and soil structures. Future generations may be better off if we do not try to produce as much food as possible per acre. The sustenance of future generations still depends on conserving good soils.

Selected Readings

Bal, L. 1970. Morphological investigation in two moder-humus profiles and the role of the soil fauna in their genesis. Geoderma 4(1):15–36.

Bergstrom, Robert E. 1968. Disposal of wastes: Scientific administrative considerations. Illinois State Geological Survey Environmental Geology Notes, no. 20.

Comar, C. L. 1966. Fallout from nuclear tests. U.S. Atomic Energy Commission Division of Technical Information, Understanding the Atom series.

Edwards, Clive A. 1969. Soil pollutants and soil animals. Scientific American 220(4):88–92, 97–99.

Hendricks, Sterling B. 1969. Food from the land. In Resources and Man, ed. Preston E. Cloud, pp. 65–85. San Francisco: W. H. Freeman.

Hight, Clyde W., Jr. 1967. How a corn grower increased his yield from 90 to 200 bushels in five years. American Society of Agronomy special publication no. 9, pp. 87–92.

Jackson, Richard M., and *Frank Raw*. 1966. *Life in the soil*. London: William Clowes & Sons.

Kovda, V. A., et al. 1966. Microelements in the soils of the U.S.S.R. Translated by N. Kaner. Washington: U.S. Department of Agriculture and National Science Foundation.

Moss, R. P. 1968. Soils, slopes and surfaces in tropical Africa. In *The Soil Resources of Tropical Africa*, ed. R. P. Moss, pp. 29–60. Cambridge: Cambridge University Press.

Nash, Ralph C., and *Edwin A. Woolson*. 1967. Persistence of chlorinated hydrocarbon insecticides in soils. *Science* 157:924–927.

Shea, Kevin P. 1968. Cotton and chemicals. *Scientist and Citizen*, November: 209–219.

Woodwell, George M. 1963. The ecological effects of radiation. *Scientific American* 208(6):2–11.

——, 1967. Toxic substances and ecological cycles. *Scientific American* 216:24–31.

Woodwell, George M., et al. 1971. DDT in the biosphere; where does it go? *Science* 174:1101–1107.

The human body, of course, is no less complex, responding with equal variability in accordance with the genetic, cultural, experiential and physiological make-up of the individual.
Cotzias, 1968

Many human diseases have geographic associations. Aristotle pointed out that miasma and ague (malaria and similar ailments) were most common in low swampy areas, often near water. He considered it more healthful to live on higher ground away from swamps and water. The geographic associations of particular diseases have often been described but seldom explained. Circulatory diseases are common in developed countries and respiratory diseases in undeveloped ones. Cancer is more prevalent in developed countries, and digestive diseases in the tropics. Infectious diseases are not common in developed countries because medical standards are high.

Diseases are generally associated with geographic areas for one of four reasons: culture, inheritance, climate, or a deficiency or excess of some trace element in the ecosystem.

Examples of Geographically Associated Diseases

A disease caused by a cultural practice is cancer of the colon among the Bantu of central Africa. Once thought to be hereditary, the incidence of colonic cancer in different Bantu tribes has recently been shown to vary with the amount of cyanide extracted from the major source of carbohydrate in their diet. Cyanide is both toxic and carcinogenic (cancer-causing); among the Bantu, the more cyanide, the more cancer: a clear case of a direct relationship between culture and disease. Again, the high incidence of stomach cancer among the Japanese is culturally induced. The Japanese prefer their rice polished and powdered;

Figure 8-1

The distribution of Burkett's tumor in Africa. The shaded area represents areas below 5000 feet elevation and with more than 20 inches of rainfall; the dots show sites of Burkett's tumor.

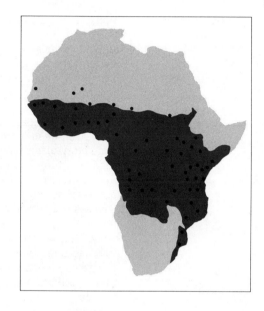

Figure 8-2

Fluoridation and dental caries in 12- to 14-year-old children.

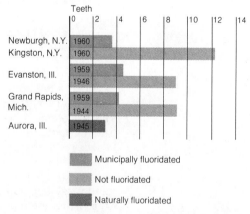

herited disability manifested in old age is related to bone density. American blacks have much denser bones than do American whites. This explains why elderly black women in this country suffer fewer hip fractures than do elderly white women. The absence or low incidence of hip fracture is associated geographically with predominantly black populations.

Two of the worst climatically controlled diseases are malaria and schistomiasis (snail fever), both primarily tropical. The distribution of both is controlled by the climatic adaptation of the parasites and nonhuman hosts. Another climatically controlled disease is Burkitt's tumor, a debilitating but seldom lethal cancerous disease of the lymph system. Burkitt's tumor is associated with three climatic factors, rainfall, elevation, and temperature. Figure 8-1 shows that part of Africa which receives more than 20 inches of rainfall per year, lying below 5000 feet in elevation, and with temperature above 60° Fahrenheit. The area of incidence of Burkitt's tumor is represented by the dots and is almost identical to it.

In geomedicine we are mainly interested in the fourth category previously mentioned, those diseases caused by deficiencies or excesses of trace substances. Common examples are the high incidence of dental caries (decay) in areas that lack fluorine (Figure 8-2) and of goiter in areas deficient in natural iodine (Figure 8-3).

asbestos, a known carcinogen, is a primary impurity in the powder which is produced from the mineral talc.

Many diseases are inherited, as can be seen in any textbook on genetics. One in-

Relating Trace Substances to Disease

It is not easy to prove that particular diseases are attributable to deficiencies or

Figure 8-3

Geographic relationship of goiter and iodine deficiency in the United States.

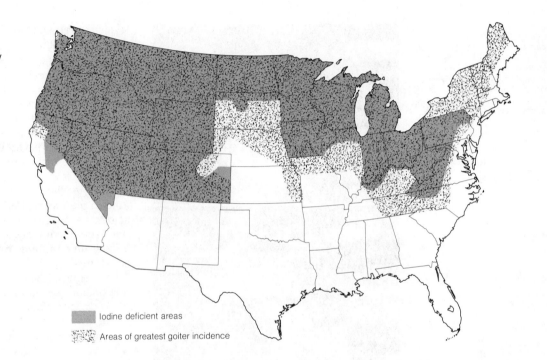

Iodine deficient areas

Areas of greatest goiter incidence

excesses of trace substances, for their symptoms may be very subtle. Deficiencies are particularly hard to prove because of the minute quantities—parts per million or billion or, for tin, even parts per trillion—with which the researcher must work. Only since 1968 have analytical methods become accurate enough to measure these minute amounts. That people are not used as experimental animals further impedes progress in this area, since information about the effects of trace substances on rats, chicks, and other experimental animals must usually be extrapolated to human subjects, adjusting for body weight, fluid content, and diet. Individual trace substances behave differently with different diets or in

the presence of differing amounts of other trace substances. For example, a small amount of selenium in the diet offsets the toxic effects of a little mercury or cadmium, though greater amounts of selenium may be toxic.

Procedures

The first step in research with trace substances is to examine the geochemical background, which involves analyzing many samples of water, soil, and air. If one does not have one's own analytical equipment, each neutron activation analysis can cost $150 or more. Are trace substances entering plants? Plants of different species,

The upper, healthy, white mouse was kept under conventional conditions on a purified diet. The lower white mouse, exhibiting trace element deficiency, was maintained for 20 days on the same diet in a trace-element-sterile environment.

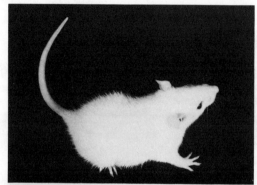

levels must be closely correlated with pathologies—through epidemiological studies involving the consultation of local doctors—hospital records, and the records of individual patients—information not usually available to the average researcher. In the United States and Great Britain, this kind of information is now being assembled in computer banks so that the incidence of any disease may be correlated with trace substance levels and localities.

Persons who have lived most of their lives in old mining districts have long been said to be "peculiar," but their behavior has not been adequately described. It could arise from continual loneliness, from trace substance toxicity, or from other causes. To determine whether trace substances are responsible would require the cooperation of (1) geochemists to analyze soils, plants, blood, and water; (2) organic chemists to identify the organic molecules combined with the trace substances; (3) doctors to describe the physical effects on the inhabitants; (4) psychologists to catalogue their behavior; and (5) computer analysts to program and analyze the data for the significance of various correlations. It might then be possible to reach a decision—provided that the data and correlations are all straightforward.

Because adequate analytical equipment has only been available a few years, it is not surprising that there are great gaps in our knowledge about the effects of trace substances on health.

The Geology of Trace Elements

Much has been written about the geochemistry of trace minerals, but little about

including vegetables from private gardens, must be collected and burned, and their ashes analyzed. The animal body has different barriers to different trace substances, and one substance may offset another. Blood samples from humans must be analyzed, and the findings from plants, blood, water, and soil compared statistically to determine the significant high and low levels of trace substances. Even with such a program, which may represent a five- to ten-year commitment, levels of toxicity or deficiency cannot always be determined. Such

their relation to environmental health. Maps from a reconnaissance survey of trace substances in soils and other surficial deposits in the United States appeared in 1971. Though not detailed enough for epidemiological studies, these maps point up important occurrences and processes. When trace substances are present or absent by virtue of parent rock, pollution, or soil and hydrologic processes such as leaching and accretion or deposition, they can be considered geologic problems.

Parent rock

In localities where there has been no leaching or accretion, surficial deposits formed in place will have the same trace elements as parent rock. In the mountains of Oregon and Washington, rocks high in usually insoluble metallics—such as aluminum, cobalt, gallium, iron, and manganese—produce soils high in those elements. Conversely, limestone terrains in the southeastern United States, low in aluminum, copper, gallium, iron, manganese, and zinc, produce soils low in these metals. This does not mean that such areas are dangerous to health. Only unusual concentrations or deficiencies are dangerous to the environment. Mining areas in British Columbia and Wales produce soils, particularly on mine tailings, that contain dangerous amounts of toxic metals because special geological

Uranium ore mine in the Colorado Plateau

processes have enriched these rocks to ore status. Thus different processes produce rocks with differing amounts of trace substances which, in turn, can be altered by other processes.

Pollution

In 1968 it was pointed out that lead carried in the air is 1000 parts pollution to one part natural lead, and that lead in water is nine parts pollution to one part natural lead. When lead is extracted from the atmosphere by the hydrologic cycle it becomes a geological problem. Plants can be analyzed for toxic metals and used as indicators of atmospheric pollution. Although we know

that lead and mercury are being added to the soil more rapidly than at any time in the geologic past, this effect on geochemistry is not yet completely understood. If mine tailings are pollutants, many families are raising home gardens on soils heavily polluted with toxic metals.

Leaching

Leaching removes soluble ions from soils, producing deficiencies and concentrating the substances that remain to produce some excesses. The soils of most of the southeastern United States, for instance, are short of barium, calcium, magnesium, potassium, sodium, and strontium. Maps

Figure 8–4

Abundance of calcium in soils and other surficial materials of the United States.

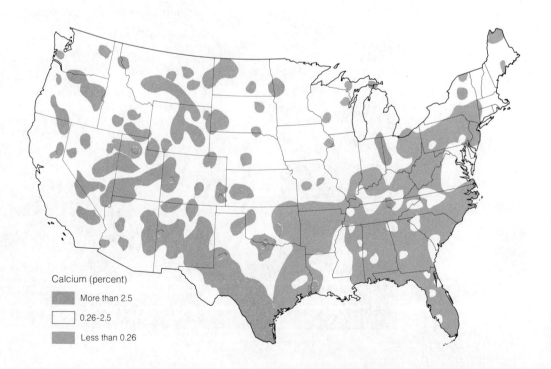

Calcium (percent)

More than 2.5

0.26–2.5

Less than 0.26

for these elements all have the same basic appearance (Figure 8-4). The substances that remain may include concentrated toxics (see the discussion of laterites in Chapter 7). These are usually sesquioxides of iron and iron-related elements.

Accretion

Accretion, or the addition of substances to soil, is usually beneficial if not too extreme. The higher pH of accreted soils, usually because of added calcium, produces a harder water. Hard water, even in the soil, forms carbonates and hydroxides of toxic metals (except nickel), which are usually insoluble and therefore not taken up as rapidly by plants. Though moderate accretion is beneficial, soils may become saline if more soluble products are precipitated from evaporating water under climatic conditions inducing excessive accretion. Some cereal-rich grasses growing on boron-rich soils have evolved mechanisms as yet unknown to reduce the uptake of boron. But most cereals are usually intolerant of boron in water at more than one ppm. Some plants adapt to soils with peculiar amounts of trace elements, whereas others fail to do so.

Soils with moderate accretions of calcium can cause widespread infection of cattle with anthrax disease, caused by *Bacillus anthracis. B. anthracis* forms spores that are not viable on acid soils of pH less than 6, but can survive on calcimorphic soils, if not too heavily leached, and on soils with accreted calcium.

As an example of excess accretion, that area of Iran around the southern shore of the Caspian Sea has an exceptionally high incidence of cancer of the esophagus. The exact cause is unknown, but incidence rises with an increase in the salinity of the soil in this desert area where soils are naturally saline.

Depositional processes

The lack of certain trace substances in limestones, and therefore in soils derived from limestones, occurs because those substances were not deposited along with the calcium carbonate. At the point at which calcium carbonate becomes saturated in water and precipitates, the trace substances are still well below saturation levels, because there is so much less of each of them in solution.

The sandy soils of the Nebraska sandhills are almost totally lacking in copper, cobalt (Figure 8-5), gallium, iron, lead, manganese, nickel, phosphorus, titanium, vanadium, zinc, and magnesium, because these substances occur in either smaller or heavier grains than quartz, and did not move with the quartz grains that produced these ancient sanddunes. Conversely, the alluvial soils of the South Platte River and lower Platte River drainages are high in content of cobalt, copper, molybdenum, gallium, lead, manganese, and nickel because the headwaters of the South Platte drain the highly mineralized areas of the Colorado Rockies and deposit these metals in the river alluvium.

Clay deposits, marine or nonmarine, may contain relatively high percentages of certain trace elements because many ions are adsorbed to the clay minerals. The high content of lead and mercury in some claystones and coals is the result of the affinity of these metallic toxics for clay minerals and organic molecules.

Figure 8-5

Distribution of cobalt in the
United States.

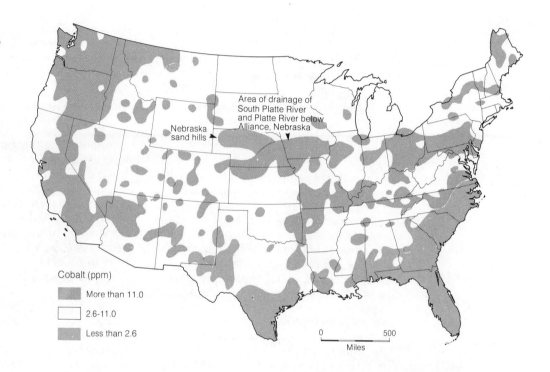

Cobalt (ppm)

More than 11.0

2.6-11.0

Less than 2.6

Figure 8-6

Effects of trace substances
on plants and animals.

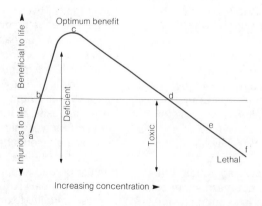

Tolerance in Biologic Systems

Concern about pollution has brought about a fear of pollutants of all types and concentrations, even minute. Although we know that there are potentially dangerous substances in the air, water, and soil, we seldom know the extent of the danger posed by any given concentration. So we make educated guesses. Thus the federally imposed limit of 5 ppb of mercury in food or water is an educated guess, and will probably be changed when more exact knowledge is available. The federally recommended limit of 45 ppm of nitrate in drinking water is an educated guess that will probably be lowered. Various plants have thresholds below which there is no danger; different species of plants have evolved different thresholds. Animals, including man, have come to utilize many trace substances and to function optimally in their presence. As is true of other animals, man almost certainly has thresholds below which trace substances are tolerated and pose no danger. The task of environmental health today is to determine these thresholds and to identify the geographic areas where they may be exceeded, whether naturally or by pollution. But evolution did not make all men equal in this respect. Some will be more tolerant of one toxin than another. The problem also has genetic ramifications, and though 100,000 people have inherited a satisfactory threshold, this does not assure us about the 100,001st. Theoretically, if we lower the concentration of a biologically required potential toxin so that no one will be injured by it, the amount available to the individual who inherits an excessive need will probably be insufficient. An average figure may be optimal for nearly all, but not for the individual with an inherited intolerance or greater than average need.

Health Effects

Deficiency problems

Some deficiencies create structural pathologies. For example, it takes a certain amount of calcium to produce bone of required density and strength. If not enough calcium is available, bone development will be in-

complete. A deficiency or enzyme inhibition that blocks calcium metabolism affects bone structure. Still other deficiencies affect catalytic processes. Enzymes are responsible for most molecular and cellular functions. An enzyme is an organic catalyst that governs the rate and direction of certain metabolic reactions; a catalyst activates or accelerates a chemical reaction. A negative catalyst is called an inhibitor For example, chromium is necessary to catalyze the enzyme responsible for the metabolism of the sugar glucose.

Toxicity problems

The effects of toxicity are usually the reverse of the effects of deficiency, although symptoms may be similar. Some toxicities result from replacement and occur when necessary substances, whether or not they are deficient in the environment, are replaced by toxic or deleterious substances—e.g., when barium, strontium, or cadmium harmfully replace calcium in bone. Toxicity also occurs through neutralization, as when a trace substance combines with protein to make it nonfunctional. Thus nitrate combines with the active substances in red blood cells, rendering them incapable of carrying oxygen. Some trace substances are enzyme inhibitors—that is, they prevent enzymes from performing their normal function. Mercury, lead, and silver usually inhibit the action of particular enzymes, and impede the proper functioning of part of the animal or plant system.

It should be emphasized that many trace substances are required in minute amounts, but are toxic in larger—though still minute

Table 8-1

Trace elements known to exist in biological systems (excluding radioactive isotopes).

Trace Elements

Aluminum (Al) 1, 4	Dysprosium (Dy)	Manganese (Mn) 1, 4	Silver (Ag) 1, 3, 4
Antimony (Sb)	Erbium (Er)	Mercury (Hg) 3, 4	Sodium (Na) 1, 4
Arsenic (As) 4	Fluorine (F) 1, 4	Molybdenum (Mo) 1, 4	Strontium (Sr) 1, 4
Barium (Ba)	Gadolinium (Gd)	Nickel (Ni) 1, 3	Sulfur (S) 1
Beryllium (Be) 4	Galium (Ga)	Niobium (Nb)	Tellurium (Te) 4
Bismuth (Bi)	Gold (Au) 2	Nitrogen (N) 1, 4	Terbium (Tb)
Boron (B) 1, 4	Holmium (Ho)	Oxygen (O) 1	Thalium (Tl) 1, 4
Bromine (Br) 4	Hydrogen (H) 1	Phosphorus (P) 1	Thulium (Tm)
Cadmium (Cd) 3, 4	Iodine (I) 1, 4	Platinum (Pt) 1	Tin (Sn) 1, 4
Calcium (Ca) 1	Iron (Fe) 1, 4	Potassium (K) 1	Titanium (Ti)
Carbon (C) 1	Lanthanum (La)	Rubidium (Rb) 1	Tungsten (W) 1
Cesium (Cs) 3	Lead (Pb) 3, 4	Samarium (Sm)	Vanadium (V) 1
Chlorine (Cl) 1, 4	Lithium (Li) 1, 3	Scandium (Sc) 2	Ytterbium (Yb)
Chromium (Cr) 1, 4	Lutetium (Lu)	Selenium (Se) 1, 3	Zinc (Zn) 1, 3, 4
Cobalt (Co) 1, 4	Magnesium (Mg) 1	Silicon (Si) 1, 4	Zirconium (Zr)
Copper (Cu) 1, 4			

1: necessary for life processes in one or more plants or animals
2: not necessary for life processes
3: an enzyme inhibitor
4: toxic
Unmarked: significance unknown

—amounts. Figure 8-6 illustrates the deficiency at minute amounts, the rise to beneficial amounts, and the toxicity with excess amounts. The entire range from deficiency to toxicity may be less than 10 ppm.

Categories of Trace Substances Related to Health

Over 60 elements are known to exist in biological systems. Of these, over 30 are necessary, and others are added to this list annually (Table 8-1). Most of the needed elements are required by both plants and animals, but a few required by plants do not seem to be necessary to animals and vice versa. Some trace substances, like mercury, are of no known biological benefit and are thought to be always toxic. Other trace substances, such as tin and platinum, are not known to occur in biological systems in sufficient amounts to be toxic, and hence can be listed as only beneficial. A third category is trace substances required in minute amounts, but toxic in larger amounts; examples are zinc and selenium. A fourth category is substances that have no apparent effect, except to indicate the amount of that substance in the environment that is breathed or eaten; an example is gold.

Effects of Trace Substance Imbalance

Organisms, particularly plants, differ in their content of trace substances. Deciduous trees growing in the same soil with conifers will have a much higher potash content, whether the potash content of the soil is high or low. In soils high in selenium, woody aster and greasewood will carry much more of the element than most other neighboring plants. Where there is considerable copper in the soil, certain plants will contain far more of it than others. Similarly, some plants will contain more uranium. These differences do not always affect the plants in question, but animals that eat those plants may become poisoned or suffer other symptoms. In general, animals seem much less tolerant of toxic imbalances than plants, which have a closer association with and dependence on soil, and may be better adapted to specific trace substance suites.

Common soils contain only small amounts of most elements. Plants and animals have evolved to use the minute amounts of trace substances existing in soils. Since evolution of most organisms did not occur where soils contained gross amounts of trace substances, most organisms did not develop tolerances of more than trace amounts. This process accounts for the general acceptance and use by organisms of minute amounts of many elements and for their intolerance of larger amounts. Iron, for instance, is more abundant than other metallic elements. As a result, organisms have a greater tolerance for iron than for other metallic elements. On the other hand, iron is usually much less soluble than the other abundant elements that buffer iron

chemically, and as a consequence organisms are much more tolerant of these other soluble elements—for example, calcium and magnesium—than they are of iron.

In the discussion of the ecosystem (geobiocoenose) cycles, Chapter 1, the importance to all life of nitrogen, phosphorus, carbon, hydrogen, oxygen, and other elements was pointed out. It was also observed that there is a similar ecosystem cycle for each of the 60 or so trace elements found in organic systems. We are now interested in some of the special characteristics of these trace elements. Many are just as necessary to life as are those elements that make up the bulk of organic systems.

Calcium

Almost everyone in our culture knows that calcium is necessary for healthy bones and teeth. But it has only recently been found that calcium is necessary to transport metabolic waste products through cell walls in animal systems. Further, calcium and magnesium buffer soils so that many of the metallic elements discussed below are less active and don't appear in toxic amounts in plants consumed by animals. Calcium deficiency in the environment produces Urov disease, which is particularly prevalent in certain isolated areas of southeast Asia. Urov disease is the incomplete calcification of bones, with enlarged joints that become stiffened, sore, and sometimes even gelatinous.

Lithium

Lithium, especially lithium carbonate, has been used in the orthochemical treatment of mental illness, particularly depression. There is a significantly higher incidence of mental illness in areas of Texas deficient in lithium than in areas with sufficient lithium.

Arsenic

Arsenic is an insidious poison that has received too little attention. It has been estimated that during the past two centuries in England and northern France dozens of people were unjustly executed for murder with arsenic. When autopsies on elderly people showed arsenic in sufficient quantities to produce death, the normal course of the law was to look for a poisoner; relatives were usually suspected, and often someone was convicted and hanged. Many houses were decorated with wallpaper, and arsenic was put in the paste to control rats, mice, and other vermin. In the damp climates of these countries, arsenic gradually sublimates into the air; as a result, individuals who lived in wallpapered houses for long periods of time, particularly elderly women who did not often go outside, accumulated excess quantities of arsenic through the air they breathed. They were not intentionally poisoned. In the Waiotap Valley of New Zealand, arsenic is not measured in parts per million. Soils there contain up to 1 percent arsenic, which enters the water and kills livestock.

Fortunately, arsenic forms relatively insoluble products in most soils, and does not readily enter the environment of animals. But the addition of excess acid phosphate fertilizers to such soils can mobilize the arsenic, making it more available to plants and hence to animals. It certainly seems un-wise to use arsenic trioxide as a defoliant in cotton fields fertilized with acid phosphate and on which animal feeds are raised as a second crop.

Radon

Another substance dangerous to health is radon, a gas that collects in uranium mines and produces lung cancer. The incidence of lung cancer is far greater among uranium miners than in the general population. Tailings from uranium mines have been used as construction bases for houses in many parts of the Four Corners area (the area where Utah, Colorado, Arizona, and New Mexico meet). Escaping radon can enter houses built on tailings from uranium mines and subject the inhabitants to a higher risk of cancer. In Grand Junction, Colorado, the air in one elementary school built on uranium mine tailings exceeded the federal limits for radon content in the air of uranium mines. Radon is not a widespread environmental hazard, but is specific to areas in which radioactive minerals are mined.

Zinc and lead

Zinc and lead are trace elements with which many humans come into daily contact. The low tolerance of many plants to zinc was pointed out two decades ago. Zinc is absolutely necessary for plant growth, but in quantities of more than 0.07 ppm it reduces growth in some plants. It has now been determined that the normal tolerance of plants to zinc ranges from at least 0.02 ppm to not more than 0.2 ppm. As an enzyme catalyst, zinc promotes cell growth. In the absence of zinc there is no growth, and with insufficient zinc maturation is either slow or incomplete.

Alcohol needs zinc to be metabolized, and zinc salts have been useful in treating cirrhosis of the liver. Zinc is a catalyst to many enzymes, and it is probable that much of the population suffers from some zinc deficiency.

Zinc promotes the healing of wounds, and relatively large quantities are transferred from storage sites in the body to the healing area during wound recovery; zinc sulphate has been used to promote healing. But if the concentration is too great, zinc is toxic, at least to plants. Zinc phosphate is a very strong rodenticide that has been used in the Netherlands and Germany, where many wild animals have been killed by its use.

Lead, unlike zinc, with which it is nearly always associated, is not known to be necessary to biological systems, and is extremely toxic. Some historians have argued that one reason for the decline of imperial Rome was the fondness of the aristocracy for wine that had been mulled in lead containers, forming lead acetate in the wine. The resulting lead poisoning produced sterility before the reproductive age, and the aristocracy died off.

It had long been known, but unexplained, that there was a greater incidence of cancer in the Devonian outcrop area of England than in outcrop areas of other geologic systems. It has been shown that the Devonian rocks contain from 100 to 1,000 ppm of lead, far more than other rocks in the area; and lead is known to be carcinogenic.

Lead is also an enzyme inhibitor and prevents proper maturation, particularly in young children. There may be as many as 200,000 to 300,000 mentally deficient children in the United States as a result of lead poisoning (see Chapter 3).

The Greenland glaciers show the gradual increase in the lead content of the atmosphere as it was picked up by evaporation and then fell with snow. The increase of lead in the atmosphere since the Industrial Revolution, and particularly since the advent of the automobile, has been enormous. The lead content in plants near highways has increased. And whereas 2 percent of ingested mineral lead enters the human biological system, 40 percent of airborne inhaled lead—mostly ethylated—enters the system. Organic lead is about ten times as toxic as mineral lead; so, ion for ion, leaded gasoline exhaust is much more dangerous than lead paint.

Copper

Copper is also essential to biological systems but toxic in excess. The range for normal plant growth is very narrow, from 0.01 to 0.05 ppm. In animal systems copper is a necessary catalyst for the enzyme cytochrome oxidase, which is required for cell respiration. Hence, a deficiency may result in ataxia, a loss of muscular coordination, and in anemia.

Copper sulfate is a poison first used as a fungicide in 1807. Bordeaux mixture, containing copper sulfate, was put on the market in 1885, and copper sulfate has been used as an herbicide since 1895. Paris green, containing both copper and arsenic,

Automobile exhaust adding toxic lead to the atmosphere, Los Angeles Freeway

Figure 8-7

Minor amounts of molybdenum in the diet decrease the uptake of excess copper in the liver.

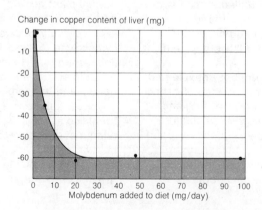

Change in copper content of liver (mg)

Molybdenum added to diet (mg/day)

is moderately toxic to vertebrates and bees if sprayed in wet weather, and extremely dangerous to fish.

The distribution of copper in soils in the United States suggests that there is generally enough, and that in many areas there would be too much if it were not buffered by calcium, magnesium, or other substances in the soil to prevent excess uptake by plants. In addition, molybdenum in biological systems seems to help the body rid itself of excess copper (Figure 8-7). Unfortunately, the exact range of tolerance for copper in animals, including man, has not been determined.

Silver

Silver is an enzyme-inhibitor, but does not seem to be readily available to biological systems. In some areas enough silver iodide has been used in seeding clouds for rain to deposit large amounts of silver iodide on the leaves of plants (Table 8-2). If such plants were eaten by man or other animals before being thoroughly washed, silver poisoning could result. Cloud seeding also rapidly increases the amount of silver in the general environment. Whether cloud seeding adds harmful amounts of silver to the ecosystem is still being debated.

Mercury

Some years ago children near Los Alamos, New Mexico, were poisoned by eating pork from hogs fed seed grain coated with mercury salts to poison fungi and rodents (see page 188). Until then, mercury poisoning had been most common in Japan, where it was known as minamata disease, and there have been sporadic outbreaks since 1953. Mercury poisoning affects the central nervous system, impairs speech, and produces spastic, ataxic failure of mobility and other signs of mental deterioration. In Japan at the same time, 61 cats that had eaten enfeebled fishes washed in by the tides also contracted the disease, and most died.

Mercury pollution has apparently existed for many years, but only since 1968 has it been possible to measure mercury in quantities small enough to allow for monitoring. It is possible to contour mercury content of surface soils in coal-burning areas. As one approaches a coal-fired industrial area, the surface soils increase in mercury content. The mercury content of Illinois coals ranges from 100 to 500 ppb. Mercury attaches to organic molecules very easily, either as methylmercuric chloride (CH_3HgCl) or as methylated mercury (CH_3HgCH_3), and since organic mercury is soluble it remains in water. In this way, through geologic time, mercury has been deposited with organic compounds and is particularly abundant in richly organic shales or coals. Before the Japanese studies, it had been assumed that the mercury discharged into streams sank to the bottom, became part of the bottom sediment, and posed no further danger to man. Only since the Japanese studies has it become apparent that some organic mercury remains in the water, is available to all biological systems that consume water, and is even transferred to lakes and oceans, where marine organisms may develop a high mercury content. When eaten, mercury builds up in biological systems, so that top carnivores contain more mercury than do organisms at lower levels of the food cycle.

Geological evidence indicates that mercury can never be completely removed from the environment. Because the environment in

Table 8-2

Possibility of local silver pollution from silver iodide seeding to produce rainfall. Modified from Charles F. Cooper and William C. Jolly, "Ecological effects of silver iodide and other weather modification agents: A Review." *Water Resources Research* 6(1):89, Table 2, 1970. Copyright by American Geophysical Union.

	Concentration (10^{-12} g/ml)
Seeded storms	10–1760
	20–200
	1–700
	10–4500 (snow)
Unseeded storms	0–20
	0–20 (snow)
440 lakes in northern Maine	10–3500 (mean 94)
Major North American rivers	0–940 (median 90)

many places naturally contains almost as much mercury as is safe, it is essential that most mercury pollution be prohibited. The figure of 5 ppb, the amount of mercury considered safe for human consumption, is so low that it is presumed safe under all conditions. It may be several times lower than necessary. Yet one should not criticize excessive caution, for decisions about tolerances must often be no more than educated guesses. It is safer to assume a figure obviously below safe tolerance than to take a chance.

It has recently been shown, using specimens in museums, that marine animals contained as much mercury a century ago as they do now. It has also been demonstrated that the mercury content of terrestrial birds and animals has steadily increased since the beginning of the Industrial Revolution. Studies of the Greenland icesheet show a steadily increasing mercury content in precipitation over the last century. It appears that some of the original assumptions about mercury were correct: much mercury, but not all, is adsorbed on clay minerals in fresh water, where it remains unavailable to animal systems. As soon as the clays enter a high chlorine environment, including the alimentary tracts of animals, the mercury is removed from the clays to become mercuric chloride. Salt used to de-ice winter roads is taking mercury from clays and releasing it into Lake Erie, giving the lake a much higher mercury content than is normal in fresh waters. Likewise, when mercury-laden clays reach estuaries or other more saline environments, the mercury is removed from the clays and becomes available to marine life. Filter feeders, animals that filter organic food from mud and water, should

have relatively high mercury contents because they ingest mud into the more acid environment of the alimentary tract. Like DDT and PCBs, mercury will decrease the thickness of the eggshells of some birds.

Chromium

Chromium is an essential catalyst to a number of body enzymes; it catalyzes the enzyme that metabolizes glucose. Hence chromium deficiency simulates all the symptoms of diabetes. In the fall of 1970, and again in the summer of 1971, it was reported that pills prescribed for mild diabetes were not curing or controlling the disease. It is possible that some of the patients were not true diabetics, but were suffering from chromium deficiency and needed a different medicine. Organic chromium appears to be 90 to 99 times as effective in treating chromium deficiency as is mineral chromium. The form of the organic chromium is an ethylate ($C_2H_5CrC_2H_5$). Ethylated chromium does not seem to be manufactured by either plants or animals; its place in the ecosystem is not yet understood. It can be manufactured synthetically by man and used for treatment.

Chromium salts mixed with arsenic and copper are used in wood preservatives outdoors (e.g., on telephone poles) but do not seem to enter the ecosystem in significant amounts. Workers in chromium industries have a higher incidence of cancer than the general population, because the sexvalent compounds of chromium are more toxic than the bivalent, and the ion of greater valence is thought to be carcinogenic. Low-order chromium pollution causes rhinitis, bronchitis, and emphysema.

Figure 8-8

Increased toxic metal content of urine in hypertensive patients. The lower bars represent cadmium in urine of hypertensive patients; the top bars represent normal controls.

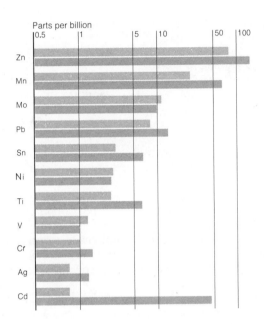

Cadmium

Cadmium is a factor in hypertension (Figure 8-8). Hypertensive patients have higher than normal amounts of most metals (except for vanadium and molybdenum) in their urine, but cadmium is by far the highest. Most other metals have now been eliminated as causative, because patients with a high content of those metals and low amounts of cadmium are not hypertensive.

We are indebted to Japanese studies of "itai-itai" (or "ouch-ouch") disease for demonstrating the more serious side of cadmium poisoning. A zinc, lead, and cadmium mine releases wastes into the upper reaches of the Jintsu River, from which irrigation water is drawn for rice paddies many miles downstream. The rice concen-

trates the cadmium, and some plots show abnormal development because of the heavy metal concentration. Sufferers of itai-itai have severe osteomalacia (deterioration and softening of the bone), and suffer intense pain in the bones and joints. Osteomalacia is a disease of calcium and vitamin D deficiency. It is probably significant that the Jintsu River area is characterized by calcium deficiency and by prolonged overcast skies, causing reduced vitamin D. Cadmium toxicity can also lead to pulmonary emphysema, proteinuria (the breakdown and elimination of proteins through urine), and cancer of the prostate. Significant cadmium in cigarettes is now thought to contribute to emphysema in many smokers. Cadmium consumption in large doses is not usually instantly lethal, but it can result in vomiting and severe gastritis.

Sources of cadmium are waste waters from mines and smelters, and the manufacture, transportation, or use of lead and zinc. Cadmium sinks with sediments and is gradually released to the waters. Organic cadmium has not been studied. Cadmium carbonates and chlorides are used as fungicides in the United States but are outlawed in England, where the general public is more aware of cadmium poisoning. Cadmium products appear to be as useful for fungicides as do mercury products. Cadmium enters the roots of plants, and edible roots grown on or near mine and smelter tailings are especially dangerous. In some vegetables, such as lettuce, the toxic metallics are concentrated in the tops instead of the roots.

Cadmium is 67th in abundance of the 92 common elements on the earth. But it is 21st in prevalence in adult males of western Europe and the United States, and 9th in the kidneys of these males. Since cadmium is closely related to mercury, much more research into the symptoms of associated diseases is needed.

Cobalt, nickel, molybdenum, tungsten, and vanadium

A number of metallic elements are enzyme catalysts in trace amounts, but are toxic in higher concentrations. Cobalt is one. Nickel is necessary to growth, at least in chickens, but interferes with bone structure if the concentration is too high. Vanadium and tungsten seem to catalyze nitrogen fixation in some plants. Molybdenum, like selenium, is a necessary catalyst in the reproductive and growth cycles of animals, but vanadium will perform some of the functions of molybdenum. The ratio of male to female births in both humans and sheep decreases on soils deficient in molybdenum. The range of tolerance to molybdenum is very narrow, from 0.01 to 0.05 ppm.

Manganese and iron

Manganese is also an enzyme catalyst, toxic in high concentrations; the range of tolerance is from 0.1 to 0.5 ppm for normal plant growth. Maneb is a manganese-bearing dithiocarbamate fungicide with a low toxicity for mammals. Manganese cyclopentadienyltricarbonyl (MCT) has been suggested as a substitute for ethylene lead in gasoline, but it has experimentally produced functional disorders in nervous systems and kidneys.

Iron, of course, is essential to all plant and animal growth and functioning. The tolerance of plants for iron varies from 0.5 to 5.0 ppm. Since iron is common in many old soils, the buffering effects of calcium and magnesium salts are very important. Most humans can eliminate excess iron, but there is one mutation that blocks elimination and can lead to iron toxicity.

Iodine and fluorine

The relationship of iodine deficiency to goiter has been mentioned (see Figure 8-3). The incidence of breast cancer in women is higher in iodine-deficient areas (Figure 8-9), but the nature of cause and effect has not been established.

Fluorine has long been used to prevent tooth decay in children, and the U.S. Geological Survey has published a map of the fluoride content of ground water in the conterminous United States. Fluorine is a prime example of the problem posed in Figure 8-6. We do not know the minimum requirement to produce good teeth (point *b* in Figure 8-6), but it probably lies somewhere between 0.8 and 1.2 ppm, and there is some indication that the optimum is around 1.5 ppm (point *c*). Somewhere between 1.7 and 4.0 ppm, brown spotting of teeth occurs (point *d*). The deleterious

Figure 8-9

Geographic distribution of
iodine deficiency compared
to incidence of breast cancer
in the United States.

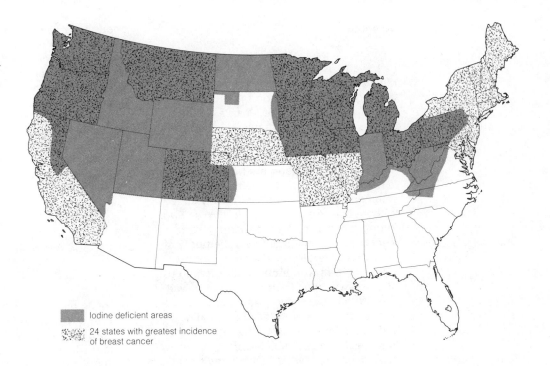

Iodine deficient areas

24 states with greatest incidence
of breast cancer

effect producing brown spotting varies with
the amount of zinc and tin in the individual's
environment. Lethal amounts of fluoride for
humans are not known, but point *f* for fish
is 6 ppm. These observations simply under-
score the amount of research yet to be
done even in this long-studied field.

Increased incidence of osteoporosis (decreased strength of bones because leaching
of calcium increases their porosity) in areas
of North Dakota with fluoride-deficient
waters has been reported. The function of
fluoride in both teeth and bones is to hold
the calcium in place. Perhaps more important

is the evidence that hardening of the aorta
can be prevented by taking calcium fluoride
pills. This information correlates fairly well
with data that show the death rate from
cardiovascular diseases to be lower in
areas with hard water, which usually contains more fluoride. There is also usually
more calcium, lithium, and magnesium in
hard water, and deficiencies in these substances have been implicated in high
incidence of cardiovascular troubles. It
has also been demonstrated that low
amounts of fluoride impair fertility in mice,
and perhaps other mammals as well.

Selenium

Selenium in excess has long been known to be toxic to livestock. Early in this century, ranchers in Wyoming learned that they could not graze sheep on the Frontier, Niobrara, or parts of the Mesaverde formations in the spring, particularly in wet years. The sheep lost muscular control and many of them died.

Soils in New Zealand are notoriously poor in trace elements, and sulfate fertilizers of commercial grade were periodically added to them before and during World War Two. During the war the New Zealand government became concerned with the purity of the sulfur in fertilizers, but could not procure fertilizers with a pure sulfur fraction. After the war such fertilizers became available. Within a very short time the production of lambs began to drop drastically, and sheep began to lose muscular control. In a crash program, soil experts from many countries were invited to study the trace elements. There was no selenium in New Zealand soils. It was found that in fertilizers with impure sulfur fractions very small amounts of selenium replaced some of the sulfur. But with a pure sulfur fraction there was no longer any selenium, and the livestock began to suffer a selenium deficiency. Selenium is a catalyst to the enzymes that operate with vitamin E, which is necessary to the production of viable offspring. Oddly enough, some of the symptoms of selenium deficiency are similar to the symptoms of selenium poisoning of sheep in Wyoming.

Organic selenium is more toxic than its inorganic counterpart because it enters biologic systems more easily. In addition to causing motor ataxia, selenium can damage the alimentary canal, teeth, skin, and blood-forming organs. It is usually acquired not from water, but from food. Selenium toxicity is widespread and has been reported in the United States, Canada, Mexico, England, Wales, Ireland, Israel, Colombia, Argentina, Venezuela, Zaire, Malagasay, Nigeria, Kenya, Japan, India, the Republic of South Africa, and Australia.

Boron

Boron is necessary for the satisfactory growth of plants and animals. The tolerance of boron in plants of normal growth is from 0.1 to 1.0 ppm, a very narrow range. There is normally no danger of excess except in sierozems and other saline desert soils. (Boron pollution as a result of borax in soaps made from degradable steroids was discussed in Chapter 5). It is now necessary to monitor streams from which irrigation water is drawn, since rice and most irrigated cereal grains will not grow properly in water containing more than 1 ppm of boron.

Nitrate

Although nitrogen is necessary for all life, nitrate (NO_3^- ion) can be poisonous: 45 ppm or more nitrate, the equivalent of about 10 ppm nitrogen, can produce cyanotic or "blue" babies. Furthermore, nitrates reduce to nitrites, which can react with amines to

Figure 8–10

Increase in the size of animal feedlots in Texas.

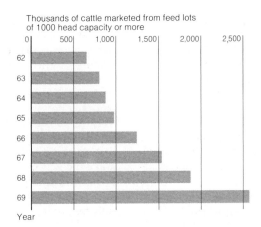

Thousands of cattle marketed from feed lots of 1000 head capacity or more

Year

produce various substances that can cause cancer, stillbirths, deformities, and perhaps even harmful mutations. Excess nitrates in the environment may stem from untreated animal waste, untreated sewage, water flowing through rocks that carry a large amount of nitrate minerals, and extensive use of nitrate fertilizers. Tail-end irrigation waters of the high plains contain excess nitrate.

Nitrates have changed the entire economy of the area around Ballinger, Texas. This district has long been cattle-raising country, but in the last ten years hundreds of cows have died from nitrate poisoning. Since the area was never farmed, the source cannot be fertilizers. Moreover, the nitrate waters do not seem to be related to any particular geologic horizon, so that the source can hardly be nitrate minerals. This leaves only one possibility: the waste produced by the generations of cattle and buffalo that have occupied the area.

In the past two decades, the number of stockyards in use has steadily decreased, but the size of those that remain has increased more than proportionately because increased human population has created greater demand for animal protein (Figure 8-10). Naturally, more animals produce more waste. One cow produces as much waste as do 16 human beings. Therefore a 2500-head stockyard or feedlot produces as much waste as a city of 40,000 people and a 10,000-head feedlot produces as much as a city of 160,000. We usually treat waste from cities, but not from stockyards or feedlots; some states, however, do have nitrate programs. Alabama, through the State Geological Survey, aids farmers in placing pig sties in locations that minimize chance of animal wastes escaping into surface waters used for human consumption. Other states will have to provide such services if population continues to increase and the nitrate problem becomes more severe.

Summary

Trace substances can create problems in the environment, and adversely affect the animals and plants on which man depends, for one of two reasons: they are too scarce or they are too plentiful. Often the beneficial effects of trace elements operate within an extremely narrow range, and the symptoms of excess are sometimes bafflingly similar to the symptoms of deficiency. Trace substances create problems because they (1) are or are not present in parent rock, (2) are added to soils and water by pollution, (3) are drained away by leaching, (4) are present or absent through deposition, or (5) are added to soil by natural accretion.

Because of the uneven distribution of trace substances, all persons, and especially food faddists, should remember that the broader their selection of food, the less likely they are to be poisoned or to suffer a deficiency, and the greater their chances of achieving a balanced diet.

There is still a dearth of data on trace substances, and most local problems cannot yet be solved. The first step in any remedial program is research into the geochemistry of trace substances in an area, followed by descriptions of behavioral and physical symptoms and causes of diseases. When this data has been gathered, epidemiological and remedial studies can proceed. Not only people who live in areas deficient in trace substances, but the entire population should be informed about the foods that do or do not contain optimum trace substances; this will ultimately require a broad educational program. Geomedicine is a growing field, and epidemiologists are only beginning their fascinating search.

Selected Readings

Andersson, A., and Nilsson, K. O. 1972. Enrichment of trace elements from sewage sludge fertilizer in soils and plants. Ambio 1:176–179.

Clarke, V. de V. 1972. Some aspects of the epidemiology of Bilharziasis in Rhodesia. Rhodesia Scientific News 6:312–317.

Dinman, Bertram D. 1972. "Non-concept" of "no-threshold": Chemicals in the environment. Science 175:495–497.

Dubos, René, and Pines, Maya. 1965. Health and disease. New York: Time Incorporated.

Epstein, Samuel S. 1970. Control of chemical pollutants. Nature 228:816–819.

Fleischer, Michael. 1962. Fluoride content of ground water in the conterminous United States. U.S. Geological Survey Miscellaneous Geological Investigation map I–387.

Friberg, Lars; Piscator, Magnus; and Nordberg, Gunnar. 1971. Cadmium in the environment. Cleveland, Ohio: CRC Press.

Glen, A. I. M.; Bradbury, M. W. B.; and Wilson, Janet. 1972. Stimulation of the sodium pump in the red blood cell by lithium and potassium. Nature 239:399–401.

Hemphill, Delbert D., ed. 1968–1973. Trace substances in environmental health, vols. I–VI. Columbia: University of Missouri.

Kline, Nathan S. 1970. Depression: its diagnosis and treatment. Lithium: the history of its use in psychiatry. New York: Brunner-Mazel.

Koval'skii, V. V. and Petrunina, N. S. 1965. Geochemical ecology and evolutionary changes in plants. In Problems in geochemistry, ed. N. I. Khitarov, pp. 613–627. Akad Nauk, U.S.S.R.

Kovda, V. A.; Yakushevskaya, I. V.; and Tyrykanov, A. N. 1966. Microelements in the soils of the U.S.S.R., trans. N. Kaner. Washington: U.S. Department of Agriculture.

Lyster, William Ronald. 1972. The sex ratios of human and sheep births in areas of high mineralization. International Journal of Environmental Studies 2: 309–316.

Moore, N. 1969. Heavy metal pesticides. Swedish National Scientific Research Council Ecological Research Committee, bulletin no. 5, pp. 36–42.

Nilsson, R. 1969. Cadmium effects; Swedish National Scientific Research Council Ecological Research Committee, bulletin no. 5, pp. 56–63.

Oldfield, J. E. 1972. Selenium deficiency in soils and its effect on animal health. Geological Society of America Bulletin 83:173–180.

Patterson, Clair C., and Salvia, Joseph D. 1968. Lead

in the modern environment; how much is natural? *Environment* 10(3):66–79.

Sauchelli, Vincent. 1969. *Trace elements in agriculture.* New York: Van Nostrand Reinhold.

Shaklette, Hansford T., et al. 1970. Geochemical environments and cardiovascular mortality rates in Georgia. U.S. Geological Survey Professional Paper 574-C.

———. 1971. Elemental composition of surficial materials in the conterminous United States; U.S. Geological Survey Professional Paper 574-D.

Schroeder, Henry A. 1971. *Pollution, profits and progress.* Brattleboro, Vt.: Stephen Greene Press.

Warren, H. V.; Delavault, R. E.; and Fletcher, K. W. 1971. Metal pollution—a growing problem in industrial and urban areas. *Canadian Mining and Metallurgical Bulletin,* July.

Chapter 9

The Earth Shakes

The bay was filling again; not slyly, as it had emptied, but in a great rushing wave, climbing the shores.

Mary Renault, *The Bull from the Sea*

Few people have personally defied an approaching earthquake and tsunami, as did Theseus in Mary Renault's *The Bull from the Sea*. Yet man has been at the mercy of a shaking earth ever since he migrated into areas of periodic earthquakes. Then as now, he could neither predict nor prevent. But he did wonder why and how, and except for crediting one or more divinities with the earth's behavior, the answers evaded him until the last century.

Proponents of plate tectonics theorize that great plates of the earth's crust move around the globe, accounting for continental drift (page 440). Whatever the cause of continental drift, energy tends to be released along certain identifiable fracture trends in the earth. The opposite sides of these fracture zones, e.g., San Andreas fault in California (Figure 9-1), move in opposite directions. A fault is a fracture along which there has been earth movement, but where the rocks impinge against each other. The rock along a fault does not move easily, because rough surfaces on one side of the fault impinge against rough surfaces on the other side. When the steady pressure along both sides of the fault creates forces stronger than the friction that prevents movement, the two sides of the fault part suddenly and with great force (Figure 9-2). Some of this stored energy is released as seismic (earthquake) waves, and the crust of the earth vibrates as these waves pass through it.

In Figure 9-2, AB is the original line of survey. As stress builds up in the rocks because A is moving north and B is moving south, A'B' is a second position of the line of survey. When the stress exceeds the friction on the fault plane, there is sudden movement along the fault, and A"X' and

Figure 9–1

Active faults in California.

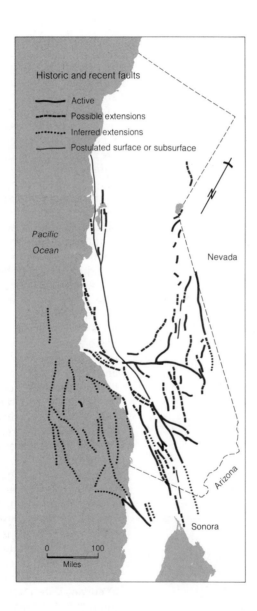

Historic and recent faults

—————— Active

------------ Possible extensions

·············· Inferred extensions

—————— Postulated surface or subsurface

Pacific
Ocean

Nevada

Arizona

Sonora

0   100
Miles

B″X″ become the new position of the original survey line. But the final resting place is not achieved until AX has reached A″Y′ and snapped back to A″X′, and BX has reached B″Y″ and snapped back to B″X″.

Describing Earthquakes

The magnitude of an earthquake is the total amount of energy released. The magnitude is not measured directly, but is expressed on an arbitrary scale independent of the place of observation. At the present time the magnitude of an earthquake is most often expressed in terms of the Richter scale (Table 9-1). In contrast, the intensity of an earthquake is the amount of shaking of the earth's surface at any given locality, and may vary with the degree of consolidation of the geological materials at that locality.

Though faulting from an earthquake may extend for tens of miles, the first waves recorded by a seismograph, an instrument for measuring the different parameters of earthquakes, behave as though they originated at a point beneath the surface of the ground. This point, called the focus, may be located from a few kilometers to 600 kilometers below the surface. The epicenter is the point on the surface directly above the focus.

Movement on faults can be lateral (horizontal), vertical, or both. In an earthquake the ground motion is the back-and-forth and/or up-and-down movement of any given spot of ground. Like an ocean wave, a seismic wave moves laterally, but a house on the ground bobs up and down like a cork on rippled water, and is twisted as the groundswell passes beneath it. Acceler-

Figure 9–2

Elastic rebound theory of earthquakes. AB is the original line of survey.

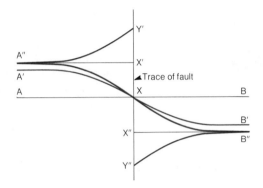

Death and Destruction

Between 50,000 and 70,000 people died in the Peruvian earthquake of 1970. The great earthquake at Ardabil, Iran, in 893 or 894 A.D. is thought to have killed at least 100,000 people. To the earthquake specialist, or seismologist, there are not only too many people on earth, but too many in the wrong places. As population continues to expand and urbanization to accelerate, it becomes possible that a single earthquake could be responsible for a million deaths! This figure was nearly reached by the Shensi, China, earthquake of 1556, responsible for an estimated 830,000 deaths.

The relatively large number of deaths in earthquakes of magnitude 7 in the Iranian Plateau is attributable to the design of buildings and villages there. The buildings are constructed of cobbles or adobe, and the mortar is weak and crumbles under the rolling motion of the groundswell. Consequently, houses collapse on their occupants, and many who escape are killed in

ation, the change in the rate of movement of any one spot on the ground per unit of time (Figure 9-3), is usually measured in centimeters or inches per second; with earthquakes of great magnitude, it is expressed in meters or feet per second. Acceleration is described as vertical and horizontal, since the structural design of buildings is different for these two components.

Table 9–1

Energy equivalents of earthquakes compared to the Richter scale of earthquake magnitude. From *California Geology* 22:72, 1969.

Earthquake magnitude	TNT equivalent	Example	Earthquake magnitude	TNT equivalent	Example
1.0	6 ounces		6.0	6,270 tons	
1.5	2 pounds		6.3	15,800 tons	Long Beach, 1933
2.0	13 pounds		6.5	31,550 tons	
2.5	63 pounds		7.0	199,000 tons	
3.0	397 pounds		7.1	250,000 tons	El Centro, California, 1940
3.5	1,000 pounds		7.5	1,000,000 tons	
4.0	6 tons		7.7	1,990,000 tons	Kern County, California, 1952
4.5	32 tons		8.0	6,270,000 tons	
5.0	199 tons		8.2	12,550,000 tons	San Francisco, 1906
5.3	500 tons	San Francisco, 1957	8.5	31,550,000 tons	Anchorage, Alaska, 1964
5.5	1,000 tons		9.0	199,999,000 tons	

Figure 9-3

An accelerogram from the Pacoima Dam during the San Fernando earthquake of 1971, showing ground velocity and displacement.

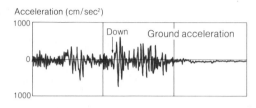

Acceleration (cm/sec²)

Ground acceleration

Down

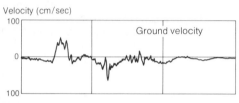

Velocity (cm/sec)

Ground velocity

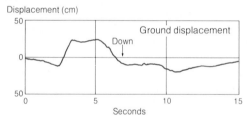

Displacement (cm)

Ground displacement

Down

Seconds

Ruins from a Sicilian earthquake

the narrow streets. Wooden houses would decrease the death toll by 90 percent, but wood is scarce in Iran, and wooden dwellings would not easily shut out the chill winters of the high plateau.

Ideal behavior during an earthquake is to remain calm and stand in a doorway, where the lintel, or the keystone in an arched doorway, will prevent collapse and offer more protection. But people are seldom calm; they do not stand in doorways. They run into the streets and behave hysterically. In short, they panic, and casualties are far larger than they need to be. While casualties are high in Iran and Turkey, the highest casualty rate per seismic event may be in Chile. This statistic may correlate with a higher seismic risk in Chile, but general communications and specific warning systems are poor in both areas. The casualty rate in Chile is lowest at noon, rises in the afternoon and evening, reaches a peak just before midnight, and declines into the daylight hours.

The extent of property damage brought about by earthquakes is incalculable, and we shall make no attempt at overall estimates. Discussion of particular earthquakes later in this chapter, will, however, give some figures for particular cases. Suffice it to say that no phenomenon has brought about such abrupt changes in the surface features of the earth, and no natural disaster can approach the havoc earthquakes wreak on human habitations and other man-made structures.

Major Seismic Belts

Earthquakes are often associated with volcanic areas. There are three major volcanic

Figure 9-4

Earthquake and volcanic belts of the world.

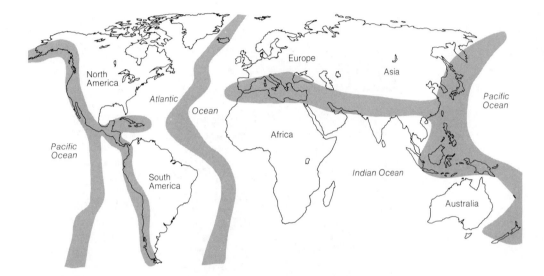

Figure 9-5

Seismic risk map for the conterminous United States. (Source: Uniform Building Code, International Conference of Building Officials)

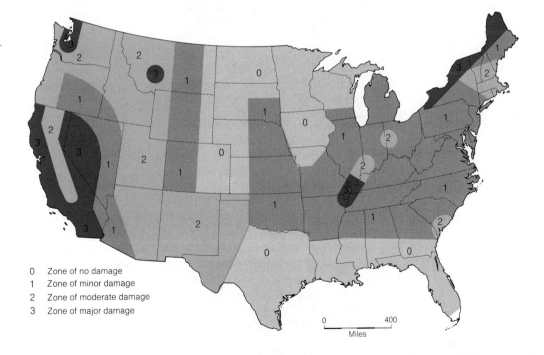

0   Zone of no damage
1   Zone of minor damage
2   Zone of moderate damage
3   Zone of major damage

0   400
Miles

Figure 9-6

Detailed seismic risk map for part of the San Francisco area.

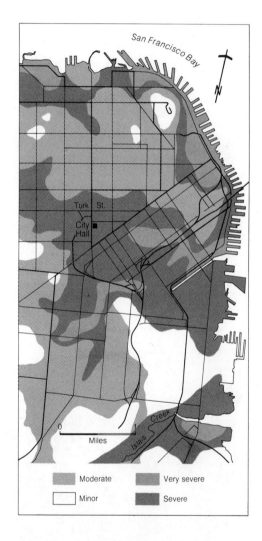

adjacent areas of Asia Minor, the west coast of the Americas as far north as California, and much of the arc from Alaska southwest through Japan and the East Indies.

If one studies these seismic belts in detail, it is possible to draw seismic frequency maps and seismic risk maps. These two types of maps are usually identical, but are given different names depending on the emphasis of the author. Seismic risk maps of the United States (Figure 9-5) show reasonably stable areas along the Gulf of Mexico and in the north-central part of the continent. Along the Gulf a thick wedge of unconsolidated sediment behaves plastically, and although there is active faulting, there are few if any seismic events. Central Texas and the northern midcontinent are associated with shield areas, which are essentially stable (seismically inactive) throughout the world. The active areas around Helena, Montana; Puget Sound, Washington; Nevada and California; and the St. Lawrence River Valley, show up very clearly in these figures. A seismic risk map of California would be on a larger scale and show more detail. Severe seismic risk areas are in the southern part of the Great Valley, around Los Angeles, and along the northern coast. The active areas around and south of San Francisco Bay (Figure 9-6) appear less severe than those just mentioned.

Proper planning against earthquakes requires even more detailed maps. Figure 9-6 illustrates in detail the seismic risk along the southern and western shores of San Francisco Bay. The most sensitive areas are on unconsolidated sediments. By comparing geologic and seismic maps, it is possible to extrapolate and predict the seismic risk in other areas on the basis of

areas and seismic belts: (1) the Tethyan or Mediterranean, (2) the circum-Pacific, and (3) that of the midoceanic ridges (Figure 9-4). Certain areas within these belts have long been noted for severe earthquakes, among them the eastern Mediterranean and

Figure 9-7

Axis of maximum seismic risk for the eastern Mediterranean Sea and the Iranian Plateau.

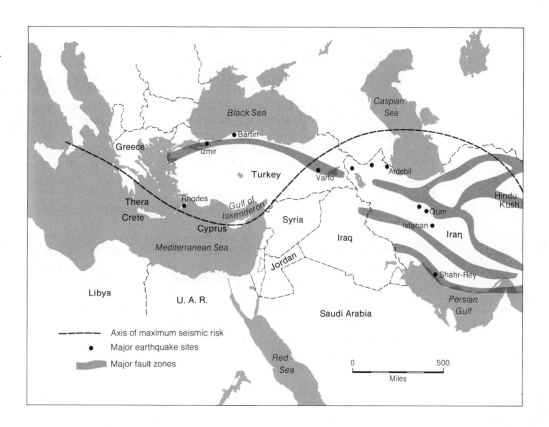

the lithology. A map of San Francisco Bay, showing filled areas and those that can be filled for development, may be read as a gross seismic risk map, since areas that rest on bayfill will be most severely affected by earthquakes (see Figure 4-8).

Disaster in the Middle East

Since prehistoric times, the Middle East has been an area of disastrous earthquakes and volcanoes. The most active seismic area (Figure 9-7) follows the marginal zone of the Alpine-Himalayan geosyncline, across the Peloponnesian Peninsula, through the island of Thera, almost through Cyprus, northeast through the Gulf of Iskenderun, and east toward the southern margin of the Caspian Sea, until it joins the trend of the Hindu Kush. The folded mountains, the disrupted bottom of the Mediterranean Sea, and the great fracture zones running across the Iranian Plateau all testify to the long history of faulting and earthquakes. But volcanic events can be dated more easily than earthquakes, and the earliest records from this area concern volcanoes. The volcano Santorin (from which the island of Thera remains) had the greatest eruptions.

The earliest known eruption of Santorin is recorded by a great ash fall 23,000 years B.C., while Neolithic man was living in slab stone houses and caves along the Mediterranean. Earthquakes, as well as volcanic activity, must have been almost continual.

Almost every ancient city in the area has a long history of earthquakes. The island and city of Rhodes, powerful in ancient history, suffered a severe earthquake in 227 B.C. The Colossus of Rhodes, the giant statue designed and erected by Chares of Lindus and considered one of the Seven Wonders of the ancient world, was destroyed by an earthquake in 224 B.C. A series of earthquakes in 155 A.D. and the following years completely destroyed the city.

Records of deaths over the last 1500

years are more complete in Iran and Turkey than in most areas. Although earthquakes may have been more severe in the eastern Mediterranean, the records there are less accurate. There is some evidence of cycles of more and less severe activity, but this may only appear to be so because records are more complete for some periods than others.

Some hazardous events are closely spaced over wide areas, though it is hard to prove from ancient records that any single event was geographically wide-spread. For example, in 856 A.D., approximately 45,000 people were killed by earthquakes in each of the widely scattered cities of Qum and Khorāsān (Iran) and Tunis (Tunisia). These are among the earliest records that list numbers of deaths. During a single year, from March 893 to March 894, it is estimated that there were 100,000 deaths from earthquakes in the vicinity of Ardabil, Iran. Records show that

Table 9–2

Disastrous earthquakes in Turkey, Iran, and Tunisia in the last 1500 years. Some of the localities are given in Figure 9-7.

Earthquakes

Year (A.D.)	Deaths	Location	Year (A.D.)	Deaths	Location
668	20,000	Izmer region, Turkey	1758	several thousand	Tunis, Tunisia
856	45,000	Qum, Iran	1853	10,000	Isfahan, Iran
	45,000	Khorāsān, Iran	1855	1,600	Bursa, Turkey
	45,000	Tunis, Tunisia	1872	1,800	Antakya, Turkey
872	20,000	Dar-I-Shar, Iran	1903	1,700	Malazgirt, Turkey
893	100,000	Ardabil, Iran, and vicinity	1939	40,000	Erzincan Basin, Turkey
1268	15,000	Erzincan and Erzurum, Turkey	1943	4,000	Lâdik, Turkey
1458	30,000	Erzincan and Erzurum, Turkey	1944	4,000	Marmara and Aegean region, Turkey
1509	13,000	Istanbul, Turkey	1957	1,130	Hamadan, Pakistan
1549	3,000	Qayen, Iran	1966	2,529	Varto, Turkey
1668	2,000	Izmir, Turkey	1968	7,000–12,000	Dasht-E-Bayaz, Iran
1755	40,000	Kashan, Iran	1970	1,086	Gediz, Turkey

nearly 500,000 people were killed by earthquakes in the Near East from 668 A.D. to 1970 (Table 9-2).

For many earthquakes the number of destroyed houses was recorded. Thus it is said that 150 villages were destroyed at Taligh, Iran, in 1957; 18,000 buildings in the Varto area of East Turkey in 1966; 40,000 buildings in the Ladik area in 1943, and 50,000 in the Boly sector (both in Turkey) in 1944. Records of seismic events on the island of Cyprus are concerned mainly with the destruction of temples and churches over the last 2500 years.

Between 1909 and 1967 there were 30 earthquakes of magnitude 6 or greater along the Anatolian fault zone of northern Turkey. Severe earthquakes also occurred in this area between the ninth and the thirteenth centuries. One large city, Shah-Rey, near Tehran, was destroyed five times. During the decade ending June 30, 1967, over 1400 earthquakes of magnitude 4 or greater occurred in the Iranian Plateau and the Hindu Kush.

Western North America

Two violent earthquakes that occurred on the West Coast in the last few years—at Anchorage, Alaska, in 1964 and San Fernando, California, in 1971—are extensively documented. As yet, the Alaska earthquake has been described in print more completely than any other, but the San Fernando earthquake is the most fully documented of all time, and in a few years all the documentation will be published.

The Alaska earthquake of 1964

Most events related to the Alaska earthquake are known. For instance, subsidence or emergence is well documented; there occurred from 2 to 6 feet of subsidence at Homer, about 5.3 feet at Whittier, and varied amounts at Anchorage and other localities. Movements ranged from a maximum uplift of 10.5 feet at Port Chalmers and Sawmill Bay to a maximum subsidence of 6.6 feet at Chance Cove.

There were 13 casualties at Whittier,

Land slippage occurred in Anchorage as a result of the Alaska earthquake, 1964

but casualties throughout the earthquake area were light because (1) the populace exhibited pioneer independence and the ability to survive, and (2) the entire country was lightly populated and some of the areas most severely struck were uninhabited. Damage was severe—$5 million in property damage was reported in Whittier and $22 million in Seward.

Landslides occurred everywhere, and probably caused the greatest damage to cities and towns. The earthquake reactivated geologically old slides, perhaps produced by one or more earlier earthquakes. A study of geologically old slides would yield information about areas in the Alaska earthquake belt where men should not settle or build. Severe submarine landslides, resulting in part from quick clay—a mixture of clay, silt, and water that liquifies when shocked and then usually flows as a viscous liquid—destroyed dock and port facilities at Homer, Whittier, and Anchorage. At Seward alone, 86 houses were destroyed and 269 houses were badly damaged by landslides and by ground fracturing. Resulting fires ignited oil in storage tanks. Slide-generated ocean waves washed onto the shore and damaged structures. As in most earthquakes, buildings on less consolidated rocks were most severely damaged. For example, a 14-story building of reinforced concrete on sand and gravel was moderately damaged, but a 6-story building of reinforced concrete on bedrock was undamaged.

Hydrologic effects accompanied the Anchorage earthquake. Horizontal movement offset and dammed streams, whereas vertical movement changed the elevation and gradient of the water table and the gradient of streams. Any movement caused snow avalanching and increased melting or freezing, depending on the temperature. Among the Anchorage earthquake's immediate but temporary effects were increases in stream discharge because of melting snow and, in areas where the land rose, greater drainage of groundwater. Wave action on lakes and fluctuation in groundwater levels occurred. Water tables were generally lowered because of increases in elevation and water table gradients. Water supplies to towns and cities were temporarily disrupted by snow slides into the streams. There may have been sanding or turbidity in water wells, and salt water encroachment into the fresh water systems will probably increase in areas where the land subsided.

Highway damage was extensive, largely where the ground fractured and destroyed bridges and asphalt. Total damage to highways probably amounted to $46 million.

The Anchorage earthquake had a Richter magnitude of 8.5, which is equivalent to exploding 31,550,000 tons of TNT (see Table 9-1).

San Fernando, a study in luck

Early on the morning of February 9, 1971, at one minute after six, most of the Los Angeles area was still asleep. Suddenly everything in the San Fernando area began to shake and roll. A few buildings and many freeway interchanges collapsed. Had the

A hospital damaged by the
San Fernando earthquake,
1971

intense ground motion lasted ten seconds
instead of seven, damage would have been
multiplied many times. The San Fernando
earthquake had a magnitude of only 6.6,
but the population explosion in the Los
Angeles area in recent decades has made
earthquakes much more serious than before.
With a death toll of 65, the San Fernando
earthquake caused the third most serious
loss of life resulting from earthquake in the
conterminous United States in the last
200 years, following San Francisco's 1906
earthquake and one in Long Beach in 1933.

The San Fernando earthquake was the
second worst in the United States with regard to property damage ($5 billion), after
the San Francisco disaster.

Within an hour of the earthquake, ten
geologists from the California Division of
Mines and Geology were in San Fernando
to appraise damage and search for further
hazards. They discovered a broad east-
west trending zone from 20 to 50 yards
wide and a mile long, in which water and
gas mains were ruptured, curbs disrupted,
and pavements damaged. Some east-west

Figure 9-8

Index map of the geology and localities of the San Fernando area, California.

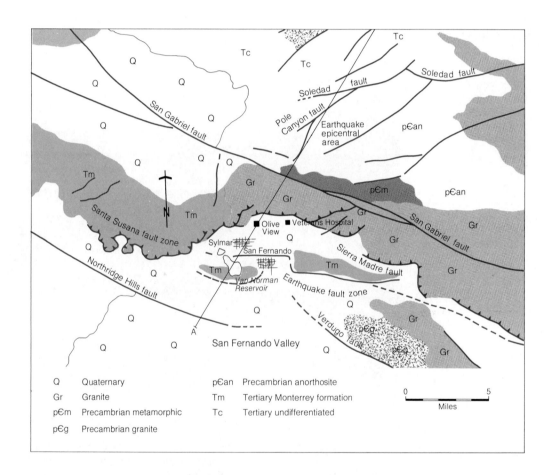

streets were offset as much as five feet. In a direction that was more north-south, other streets were shortened by several feet.

The most severely damaged buildings were two hospitals. The main buildings of the Olive View Hospital, completed the previous year at a cost of $25 million,

were damaged beyond repair. The greatest loss of life occurred at the Veterans Administration Hospital, where vertical acceleration lifted the hospital with the groundswell. The lower floor was not strong enough to support the upper floors on the rebound and collapsed, killing 48 patients.

Some schools were also badly damaged,

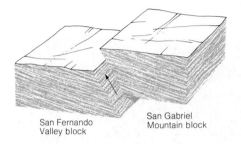

Figure 9–9

Relationship of the San Gabriel Mountain block to the San Fernando Valley block.

San Fernando Valley block

San Gabriel Mountain block

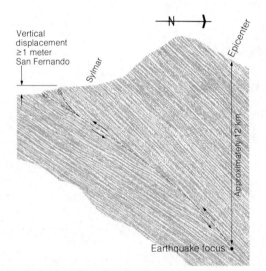

N

Vertical displacement ≥1 meter San Fernando

Sylmar

Epicenter

Approximately 12 km

Earthquake focus

due less to location than to construction. The Field Act, passed by the state legislature after the Long Beach earthquake in 1933, had established standards for school construction to prevent earthquake damage. All of the eleven schools in the San Fernando area that had to be demolished after the 1971 earthquake were built before 1933. One school building erected after the Field Act was severely damaged. The lesson of the San Fernando earthquake is that hospi-

tals, at least, should be constructed in accordance with the same standards as schools.

Transformers and electrical switching gear at the Sylmar converter station were almost destroyed. The Van Norman Reservoir was badly damaged but, fortunately, the water level was 15 feet below maximum. Most of the water was withdrawn from the reservoir and a flood was avoided, but not until after 80,000 people had been evacuated from a 36-square-mile area below the dam.

The greatest economic damage was to freeways. Open, above-ground spans of most freeway interchanges were of pre-stressed concrete. There were about 70 such structures within a 10-mile radius of the rupture zone. Of these, 40 suffered considerable damage and many completely collapsed. Because the quake occurred at six A.M., the freeways were virtually empty, and in this respect only could San Fernando be called lucky. One can only guess at the casualty list had the shock occurred two hours later during rush hour.

As the continent of North America overrides the Pacific Basin in its westward drift, parts of the Pacific Ocean floor are descending and sliding (subducting) under its western border. At the same time the continent is slowly moving northward relative to the Pacific Basin. The San Andreas fault zone in one of the fracture zones near the boundary between the northward-moving continent and the subducting and southward-moving Pacific Ocean Basin.

San Fernando is a highly faulted area where a twisting of fault trends accompanies the general subduction pattern (Figure

Effects of the San Fernando
earthquake on the Los
Angeles Freeway

9-8). The city of San Fernando is on the valley side of, but near, the San Gabriel Mountains, a range of granitic rocks just north of San Fernando. The San Gabriel fault, a major fault, is about five miles north of the city, and the Verdugo and North Ridge Hills faults are two to three miles south of it. The San Fernando rupture at ground surface did not occur at the old faults, but produced its own new fault. The north side of the fault moved west relative to the south side, offsetting streets; the north-south streets were shortened by movement which caused the San Gabriel

Mountains to override the north side of the San Fernando Valley (Figure 9-9).

Future earthquakes can be expected in the area, but nobody knows where or when.

Tsunami

Tsunami (singular and plural), or seismic sea waves, are ocean waves generated by earthquakes, (Figure 9-10). They are sometimes caused by sudden displacement of the sea floor, and sometimes by submarine slides produced by earthquakes. The giant tsunami associated with the eruption of Krakatoa in 1883 is said to have been vol-

canic, but a violent earthquake certainly accompanied the volcanic explosion. Any earth event that results in a sudden displacement of large volumes of ocean water will produce a tsunami.

In the open ocean tsunami range from 5 or 6 centimeters to one meter in height, and are not discernible to the casual observer. When such waves reach shore, however, the energy dispersed through 9000 or more feet of ocean depth concentrates in only a few feet of water in the nearshore area. As a result, great waves override the shore. This phenomenon, called runup, may be only a few feet or hundreds of feet high. The Krakatoa tsunami of 1883 raced across the Sunda Strait to drown the coastlines of Java and Sumatra. Great tsunami were generated by the Lisbon earthquake, and by Aleutian earthquakes, Chilean earthquakes, and major earthquakes all over the world.

Tsunami can be predicted, and as soon as an earthquake is located, a warning can be broadcast. A time plot can be made to predict the time at which the wave will reach any particular locality. Figure 9-11 is a time plot for Hilo, Hawaii, from various parts of the Pacific Ocean. For example, a tsunami produced by an earthquake at the southern end of Chile would require 16 hours to reach Hilo. This is the reason the tsunami from the Chilean earthquake on the afternoon of May 22, 1960 reached Hawaii on May 23. A tsunami originating at

Figure 9–10

Origin of a tsunami.

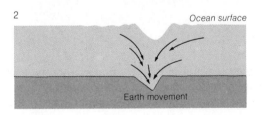

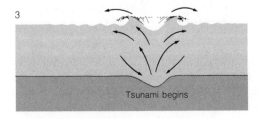

Figure 9-11

Elapsed time plot at Hilo, Hawaii for tsunami originating in the Pacific Ocean Basin. (From "Tsunamis" by Joseph Bernstein. Copyright © 1954 by Scientific American, Inc. All rights reserved.)

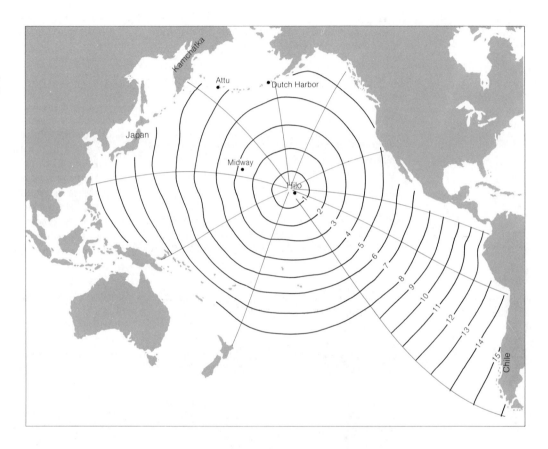

Dutch Harbor, Alaska, would require only five hours to reach Hawaii; one from the east coast of Kamchatka, six and one-half hours; and one from the east coast of Japan, eight hours. Such plots can be constructed for any locality.

The story of a tsunami

A great earthquake centered near the coast of central Chile with the epicenter near Concepcion, occurred at 6:04 A.M. on May 21, 1960, and generated seismic waves 20 to 30 centimeters high. The central headquarters for the Pacific tidal wave alarm system, in Honolulu, Hawaii, was alerted. The seismographs at Honolulu recorded the earthquake, and the warning system began operating at 6:45 A.M., but no tsunami developed.

At approximately 3:10 P.M. the following day, a second great earthquake occurred in Chile, with the epicenter south of Concepcion between there and Chiloe. This earthquake devastated the Chilean coast. Honolulu received word of it within 30 minutes, and alerted the Pacific alarm system in two messages, the first preliminary:

> This bulletin is a tidal wave alert. A violent earthquake occurred in Chile, the third in that area in the last 36 hours. It is possible that it has generated a large tsunami. Although we have as yet no data, we are awaiting information from Valparaíso and from Balboa. If a tidal wave has been originated it should arrive about midnight Hawaiian time today at the Island of Hawaii and 30 minutes later at the Island of Oahu. New information will be given as soon as more data is available.

The second bulletin read:

> This is a tidal wave alert bulletin. A violent earthquake in Chile has caused a tidal wave that is radiating in all directions over the Pacific Ocean. It is estimated that the first wave will reach the Island of Hawaii at midnight Hawaiian time and 30 minutes later at the Island of Oahu. Its destructive effects will last several hours. The intensity of the wave cannot be predicted. The southern part of Hawaii will be the first to be affected and will be the first indication of the damage that might be produced in other parts and other islands of the Hawaiian group. The times have been calculated for arrival at other Pacific islands based on the best data obtainable; they are not very exact: Tahiti, 0230 hours; Christmas Island, 0400 hours; Samoa, 0500 hours; Fiji and Canton, 0600 hours; Johnston, 0700 hours, and Midway, 0830 hours.

Actual arrival times appeared to be 30 to 45 minutes later than predicted. Because of local configurations of the sea floor, arrival times may vary up to two hours from estimates.

About ten minutes after the earthquake, observers at the lighthouse on the island of Guafo off the coast of Chile saw the sea receding. When the sea returned, the water reached a height of 10 meters (about 33 feet) above sea level on the cliffs below the lighthouse. The island was lifted three or four meters, displacing a tremendous volume of water. So much water was forced out by the rising island and the adjacent sea floor that the momentum of the returning water produced the tsunami.

About two hours after the earthquake, the tsunami began climbing the coast of Chile, first on the south-central coast near the epicenter of the earthquake, and then to the north and south, moving away from the epicenter. There were three successive waves, the second one hour and 20 minutes after the first, and the third about two hours after the second. The number of deaths in Chile caused by these waves may never be known. Destruction was most severe up and down the coast from the epicenter. Ships were destroyed, port facilities demolished, and thousands of buildings flooded, damaged, or destroyed.

Meanwhile, the tsunami was heading across the Pacific. It entered French Polynesia just before dark. Although the greatest runup there was 3.4 meters (13.2 feet) above low water, these islands are fairly well protected by barrier reefs that break up waves at the reef front and dissipate their energy before they arrive at the main island. Tahiti, with barriers only partly developed on the south side, and the Marquesas, with few barrier reefs, were more vulnerable. The greatest runup on Tahiti was 3.4 meters, and the average was 1.7 meters. Because the distance from Chile to Tahiti is so great, and because much of its energy was expended crossing the ocean, the wave did little damage and caused no loss of life. Two houses and a bar were destroyed on Tahiti. The bartender received radio warning only 10 minutes before landfall, but safely evacuated his patrons.

The tsunami reached Hilo, Hawaii, at 7 A.M. In Hawaii the runup ranged from 2 to 17 feet, except at Hilo, where a bore was formed that reached a height of 35 feet. Sixty-one people were killed and 282 injured in Hawaii; the greatest damage was in the Hilo area, where about 600 acres were flooded, frame buildings destroyed, and dozens of automobiles wrecked. Rocks weighing up to 22 tons were plucked from a seawall and moved as much as 600 feet inland. Destruction elsewhere in Hawaii was light.

Meanwhile, the great wave continued across the Pacific and spent itself against the shores of New Zealand, Australia, Japan, Kamchatka, and other bordering lands, where it was so far from its source that it did little harm. The area where it did greatest damage was closest to its point of origin—the coast of Chile.

Man-made Earthquakes

Petroleum geologists have known for many years that pumping liquids under pressure into reservoir rocks with water flooding programs, or expanding rocks so that more oil could be obtained (hydrofracking) would produce microearthquakes. This phenomenon had been experienced for a number of years at the Rangely Oil Field near the Colorado-Utah border. These geologists, then, were probably not surprised at the production of earthquakes, some of them not minor, by nuclear explosions in the vicinity of Benham, Nevada. These explosions were accompanied not only by earthquakes, but also by movement on old faults and the development of new faults.

A proposal to create a harbor by nuclear blasting at Kerauden, in northwestern Australia, was cancelled because of the fear of producing a major earthquake. But fear did not stop the U.S. Atomic Energy Commission from detonating a nuclear blast on Amchitka, in the Aleutian Island chain, a very active segment of the circum-Pacific volcanic and earthquake belt. When the nuclear event was carried out without an accompanying major earthquake, the Commission suggested that there had really been no danger. But basing such a conclusion on a single sample is not statistically valid.

The greatest attention has been paid to manmade earthquakes in the Denver area. In the past decade the South Platte River Valley has received a greater variety of pollutants than most river valleys in the United States, if not the world. It has suffered all the normal types of pollution, as

Figure 9–12

Correlation between frequency of earthquakes and injection of waste into the subsurface through the Rocky Mountain Arsenal well near Denver, Colorado.

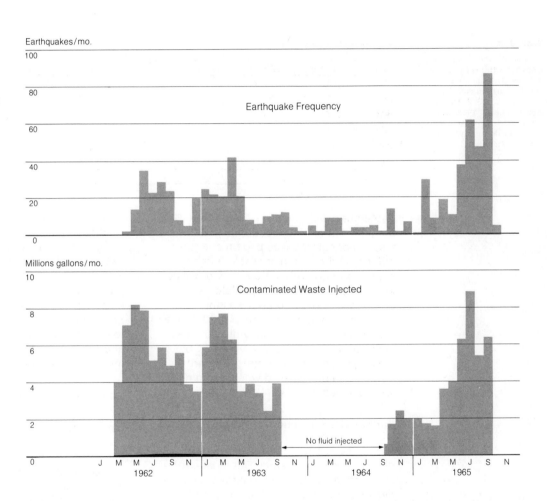

well as radiation pollution from the Rocky Flat Thorium Concentrating Plant and chemical warfare pollution from the Rocky Mountain Arsenal. For a number of years the Arsenal had been burying waste CBW products in the sanddunes along the southeast side of the South Platte River downstream from Denver. It was feared that these products—if they had not already— would eventually leak into the underground water system of the South Platte River Valley. Consequently, the Arsenal drilled a deep well and pumped its chemical warfare wastes into basement rocks at a depth of over 12,000 feet. The pumping of liquid wastes began early in 1962 and continued until September 1963. Fluid was injected in September 1964 and continued to

Table 9-3

Magnitude and frequency of Denver earthquakes. From J. H. Healy, *et al.*, "The Denver earthquakes." *Science* 161:1302, Table 1, September 27, 1968. Copyright 1968 by the American Association for the Advancement of Science.

Magnitude (Richter scale)

Year	1.5-1.9	2.0-2.4	2.5-2.9	3.0-3.4	3.5-3.9	4.0-4.4	4.5-4.9	5.0-5.4
1962	72	29	4	2	1	1		
1963	89	34	9	3	1	1		
1964	26	8	6					
1965	168	64	25	6	4			
1966	61	18	3	2	1			
1967	62	29	15	4	4	2		3
Total	478	182	62	17	11	4		3

September 1965 (Figure 9-12). Within a month of the first injection in 1962, a series of earthquakes began, and their number and magnitude (Table 9-3) correlated directly with the quantity of waste pumped underground. From late 1963 to late 1964, when no fluid was injected, the number and magnitude of earthquakes was minimal, although they did not cease because pressure had been built up by previous pumping. Magnitude of earthquakes again increased when pumping was resumed in 1964.

Earthquake Prediction

In the past, correct prediction of earthquakes was not possible. But most earth scientists have considered earthquake prediction a legitimate field of research since about 1964, and a few estimate that some predictive success will be possible by 1985. Richter has said that predicting an earthquake may be compared to "the situation of a man who is bending a board across his knee and attempts to determine in advance just where and when the cracks will appear."

Since there are thousands of earthquakes every year, and many large ones, almost anyone can predict an earthquake: if you predict an earthquake in March, the odds are that you will be right. Predicting the exact magnitude, time, and place, however, is beyond the capability of even the best professional seismologist at the present time. But some precursor phenomena have been noted to have a consistent relationship to earthquakes, and careful observation and recording of these will unquestionably improve the ability to predict.

Shortening or extension

One of these precursors is the shortening or extension of distances between points on the earth's surface that have been accurately measured. At Matsushiro, Japan, it has been noticed that the rate of such movement increases before the occurrence of an earthquake. Close and accurate geodetic surveying and continual monitoring for changes in the rate of land movement may someday make it possible to predict quakes.

Tilting

At Matsushiro it was observed that although there was longterm, very slow tilting, sharp changes occurred just before a moderate earthquake. Continuous tilt was recorded before and during the Danville, California earthquake swarm in 1970, and sharp changes in tilt preceded major events. A constant change in tilt and a one-day accelerated change preceded earthquakes. Russian seismologists are also conducting tilt studies and are optimistic about the future use of tiltmeters to predict more accurately the time of an earthquake.

Microseism swarms

It was also noticed at Matsushiro that short swarms of microseisms occurred in the vicinities of epicenters of larger shocks that followed several months later. This phenomenon might make it possible to predict earthquakes more than a few days or hours in advance. On the other hand, microseism swarms alone do not indicate the probability of major earthquakes. The intensity of microseisms varies greatly, and there are "hot spots" of such activity at different sites along the San Andreas fault zone. Although the frequency of microseisms for a 29-year period correlates with the frequency of larger earthquakes, microseism activity can also be stable for many decades. Some seismologists believe that foreshocks (seisms of lesser intensity preceding earthquakes) may have different characteristics than normal continuing microseisms, and thus be identifiable. But some earthquakes are not preceded by foreshocks.

Anomalous magnetic fluctuations

At Matsushiro it was further observed that anomalous magnetic fluctuations occurred both immediately preceding earthquakes, and in the areas of much later earthquakes. These fluctuations are not yet understood, but may be associated with shifts of magma bodies at depth.

Quiescence

In 1857 there was a severe earthquake in California between Cholame and Valyermo—one of the three largest earthquakes in California in the last 200 years. At present there is very low microseismic intensity along the 300-kilometer segment of the San Andreas fault involved in the 1857 earthquake. This may mean that movement along this segment of the San Andreas fault is not to be expected in the near future.

Density changes in rock

When rocks are stressed, their density changes slightly. As a result, the speed of transmitted pressure waves is increased. Seismologists are now trying to perfect devices to measure and record these changes in speed and, therefore, in stress at depths in the earth. If such buildups in stress can be determined, they may aid in prediction.

Increase in tectonic stress

As a corollary of density changes, tectonic stress builds up in rocks where earthquakes are common; most of the phenomena discussed above are associated with changes in stress. The various underground nuclear explosions in Nevada released stress. Accompanying the Benham ex-

Figure 9-13

Damaging earthquakes in the San Francisco area from 1800 to 1966.

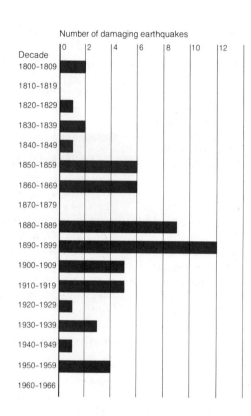

Number of damaging earthquakes

developing. Withdrawing water from the rocks at these depths could release the stress. But this suggestion presupposes a permeability that usually does not exist in stressed, fine-grained rocks at such depths. The well density necessary to release stress for sufficient distances would make them very costly.

Cyclicity of earthquake phenomena

There may be cycles in the severity of earthquakes. Some research has suggested that earthquakes are more severe during the peak of solar activity in the 11-year solar cycle, and less severe when such activity is low. Figure 9-13 shows the number of damaging earthquakes in the San Francisco Bay area per decade for the last 160 years. Even if there is an 11-year cycle, there also seem to be periods of high and low activity on a much longer time scale. Table 9-4 shows the number of earthquakes of different magnitudes from 1904 to 1946. With different rates of stress accumulation in the rocks, cycles of magnitude are expected to be different in different areas, and prediction according to cycles will be extremely difficult.

Predictions: a summary

The correlation of data on shortening and extension, tilting, microseism swarms, and anomalous magnetic fluctuations was used to predict earthquakes at Matsushiro, Japan. Warnings at Matsushiro have been considered a scientific success, although motel operators and others dependent on

plosion, not only were there increases in microseismicity, but there was active movement along old faults, and new faults were opened. Some think earthquakes of large magnitude might be prevented by periodically setting off small atomic explosions in active fault zones, thereby triggering small amounts of movement before great stress developed. It has also been suggested that wells be drilled into areas where stress is

Magnitude	Number
7.7–8.6	2
7.0–7.7	12
6.0–7.0	108
5.0–6.0	800
4.0–5.0	6,200
3.0–4.0	49,000
2.5–3.0	100,000

the tourist trade were not particularly pleased to have earthquakes predicted. Furthermore, if one is to predict earthquakes in any seismically sensitive area, the population must be taught how to behave when an earthquake is expected, in order to prevent hysteria and frantic activity. Our present knowledge of human behavior does not suggest a very high probability of success for such a program.

There is considerable reason for hope in learning to predict earthquakes. Since earthquakes can be produced by injecting fluids, adding weight or load to an area by impounding water, or setting off explosions, it may be possible to control earthquakes in some areas by releasing energy at a steady rate through some of these methods. It must be emphasized, as a result of the Danville, California, microseism swarms, that areas without known faults cannot be eliminated as seismic risk areas. Danville is east of San Francisco and had not previously been known to be active. A federally funded ten-year research program devoted to learning to predict earthquakes is now in progress. In addition, a cooperative U.S.-Japanese study is in process, and

the Russians are also heavily involved in such investigations.

Construction in Earthquake Areas

The ideal way to prevent earthquake damage is to prevent earthquakes. Since this is impossible, the next best alternative is to avoid living in sensitive areas. Yet, the Oakland, California, freeway follows or overlies the Hayward fault for about a mile, and the Oakland emergency warning station was once located on the same fault.

Stronger zoning ordinances and the application of present knowledge to urban development could greatly reduce seismic risk. The juggling and displacement of loose grains and other particles in unconsolidated sediments create greater seismic risk to buildings than does solid rock, and have been said to magnify effects of subsurface movement much as the motion of gelatin in a saucer is a magnification of the saucer's movement. The large-scale seismic risk map of the San Francisco Bay area (see Figure 9-6) reflects the degree of rock consolidation.

Similarly, houses and villages on soft ground in Turkey suffer the greatest damage. Although large buildings in developed countries can be designed and constructed to be resistant or tolerant to earthquakes, this does not mean that they are "earthquake-proof." The developer or building contractor can also construct residences that are relatively tolerant of ground motion. Whether in California or in Turkey, wooden buildings, being more resilient, invariably withstand ground motion better than brick, tile, rock, or stone buildings.

Designing buildings to tolerate horizontal

Example of poor landuse planning: aerial view of housing developments built along the San Andreas fault. The vertical white line indicates the fault line.

acceleration is a different problem from designing them to withstand vertical acceleration. In the San Fernando earthquake of 1971, vertical acceleration was much greater than had been experienced in previous California earthquakes, and much of the damage was therefore unanticipated by engineers. In many buildings the lower floors were not strong enough to support the increased load exerted by the upper floors during acceleration. The effect of such acceleration is much the same as that of jumping on a can that supports one's normal standing weight. Engineers can design to protect against vertical acceleration, and after the earthquake they asked why they had not been told that vertical acceleration could be so great.

It is seldom the engineers who are to blame for design failures. Instead, it is the individuals, societies, and governments that choose to meet only the minimum legal standards in order to minimize the cost of construction. Where zoning laws are weak— and they usually are—buildings are not well-constructed.

Earthquake insurance

Earthquake insurance has been a prime problem even in affluent California, and in nonaffluent areas such insurance does not exist. In 1926, an earthquake department was established by the Board of Fire Under- writers of the Pacific. This department found zoning laws and regulations to be so in- adequate that insurance was out of the question, because there was little or no uniformity or supervision of site location, construction materials, or building design. With the help of many organizations, zoning codes were studied and proposals for the reduction of seismic damage developed. But premium rates are generally still too high, especially for small businesses and home- owners who can afford earthquake in- surance only as part of an insurance pack- age. Small businesses' and homeowners' lack of insurance and planning results in massive bills for disaster relief being paid from tax monies. After the Long Beach earthquake of 1933, though, the state of California did pass a set of rigid require- ments for school construction. As a result, schools built since 1933 have suffered less damage than buildings that are not re-

quired to meet these specifications. Com- pared to New Zealand's all-inclusive earth- quake insurance, the accomplishment of a more socialist government, earthquake in- surance in California is meager. Yet in most of the world such insurance does not even exist.

Summary

Most earthquakes occur in the Tethyan and circum-Pacific volcanic and earthquake belts, and along the midoceanic ridges. The western United States, including the southwestern coast of Alaska, is in the circum-Pacific belt. Although the United States has been rather mildly affected by earthquakes, damage and death have been not inconsiderable. Including tsunami originating outside the United States, 37 major earthquakes from 1865 to 1966 have caused 1506 deaths and over $1.3 billion in damage (Table 9-5).

Buildings can be made earthquake-resist- ant, if not earthquake-proof. The prediction of earthquakes is steadily improving. In- creased financial support for prediction studies is desirable, but will not be com- pletely justifiable without better public education in behavior during earthquakes. Such a program could also include educa- tion in behavior during such other natural hazards as hurricanes, floods, and volcanic explosions. The program should probably be introduced in earth science classes in elementary schools. Earthquakes cannot be prevented, although their effect may be mitigated by proper building and street design and by better prediction.

Earthquakes

Table 9-5

Estimates of death and damage from 37 earthquakes that occurred during the century from 1865 to 1966, including tsunami originating outside the United States. From Lloyd S. Cluff and Bruce A. Bolt, "Risk from Earthquakes in the Modern Urban Environment, with Special Emphasis on the San Francisco Bay Area." In E.A. Daheny, ed., *Urban Environmental Geology in the San Francisco Bay Region*. Special Publication, San Francisco Bay Section, Association of Engineering Geologists, 1969, Table IV-1.

Year	Location	Property damage	Persons killed
1865	San Francisco, Cal.	$ 500,000	
1868	Hayward and San Francisco, Cal.	350,000	30
1872	Owens Valley, Cal.	250,000	27
1886	Charleston, S.C.	23,000,000	60
1892	Vacaville, Cal.	225,000	
1898	Marc Island, Cal.	1,400,000	
1899	San Jacinto, Cal.		6
1906	San Francisco, Cal.	524,000,000	700
1915	Imperial Valley, Cal.	900,000	6
1918	Puerto Rico	4,000,000	116
1925	San Jacinto and Hemet, Cal.	200,000	13
1926	Santa Barbara, Cal.	8,000,000	1
1933	Long Beach, Cal.	40,000,000	115
1934	Kosmo, Utah		2
1935	Helena, Montana	4,000,000	4
1940	Imperial Valley, Cal.	6,000,000	9
1941	Santa Barbara, Cal.	100,000	
1941	Torrance-Gardena, Cal.	1,000,000	
1944	Cornwall, Canada-Massena, N.Y.	2,000,000	
1946*	Hawaii	25,000,000	173
1949	Puget Sound, Wash.	25,000,000	8
1949	Terminal Island, Cal.	9,000,000	
1951	Terminal Island, Cal.	3,000,000	
1952	Kern County, Cal.	60,000,000	14
1954	Eureka-Arcata, Cal.	2,100,000	1
1954	Wilkes-Barre, Pa.	1,000,000	
1955	Terminal Island, Cal.	3,000,000	
1955	Oakland-Walnut Creek, Cal.	1,000,000	1
1957	Hawaii	3,000,000	
1957	San Francisco, Cal.	1,000,000	
1958	Lituya Bay, Alaska		5
1959	Hebgen Lake, Montana	11,000,000	28
1960*	Hawaii and west coast of U.S.	25,000,000	61
1961	Terminal Island, Cal.	4,500,000	
1964	Alaska and west coast of U.S.	500,000,000	131
1965	Puget Sound, Wash.	12,500,000	7
1966	Dulce, N.M.	200,000	

*Damage resulted from tsunamis generated by earthquakes occurring outside the United States.

Selected Readings

Algermission, S. T. 1969. Seismic risk studies in the United States. Paper presented at Fourth World Conference on Earthquake Engineering at Santiago, Chile. (preprint).

Ambrasyes, N. N. 1968. Early earthquakes in north-central Iran. *Seismological Society of America Bulletin* 58:485–496.

Anderson, Don L. 1971. The San Andreas Fault. *Scientific American* 225(5):53–64.

Barazangi, Muawia, and Dorman, James. 1969. World seismicity maps compiled from ESSA, Coast and Geodetic Survey, epicenter data, 1961–1967. *Seismological Society of America, Bulletin* 59:369–380.

Bernstein, Joseph. 1954. Tsunamis. *Scientific American* 191(2):60–64.

Creighton, W. H. 1972. A centennial . . . the great Owens Valley earthquake of 1872. *California Geology* 25(3):51–54.

Davies, David. 1973. Monitoring underground explosions. *Nature* 241:19–24.

Evans, David M. 1966. Man-made earthquakes in Denver. *Geotimes* May–June:11–18.

Greensfelder, Roger. 1971. Seismologic and crustal movement investigations of the San Fernando earthquake. *California Geology* 24(4–5):62–68.

Hamilton, R. M.; McKeown, F. A.; and Healy, J. H. 1969. Seismic activity and faulting associated with a large underground nuclear explosion. *Science* 116:601–604.

Ilhan, Emin. 1971. Earthquakes in Turkey. In *Geology and history of Turkey*, ed. Angus S. Campbell, pp. 431–442. Tripoli: Petroleum Exploration Society of Libya.

Pakiser, L. C., et al. 1969. Earthquake prediction and control. *Science* 166:1467–1474.

Pecora, William T., chairman. 1968. Proposal for a ten-year national earthquake hazards program. Washington: Ad Hoc Interagency Work Group for Earthquake Research of the Federal Council for Science and Technology.

Rikitake, T. 1972. Problems of predicting earthquakes. *Nature*. 240:202–204.

Sievers, C.; Hellmuth, A.; Villegas, Guillermo C.; and Barros, Guillermo. 1963. The seismic sea wave of 22 May 1960 along the Chilean coast. *Seismological Society of America Bulletin* 53:1125–1190.

Steinbrugge, Karl V. 1968. *Earthquake hazard in the San Francisco Bay area: a continuing problem in public policy.* Berkeley: University of California Press.

Wallace, R. E. 1970. Earthquake recurrence intervals on the San Andreas Fault. *Geological Society of America Bulletin* 81:2875–2890.

Wood, J. H., and Jennings, P. C. 1971. Damage to freeway structures in the San Fernando earthquake. *New Zealand Society for Earthquake Engineering Bulletin* 4 (3):347–375.

Chapter 10
Beware the Volcano

If, as Plato records, the morphology of Athens was so drastically changed by rainfall, it would seem that the central basin of Crete must have been changed to, at least, a similar degree.
Galanopoulos and Bacon, *Atlantis: The Truth Behind the Legend,* 1969

The best introduction to the hazards of volcanism is to review the effects of both recent and older volcanic activity. Volcanoes have been a threat to the peoples who live near them since the dawn of history (Table 10-1). Although deaths from volcanic events have not been as numerous as those from earthquakes, casualties have sometimes been extensive. Thus 92,000 people were killed during the eruption at Tambora, Indonesia, in 1815, and 30,000 by that of Mont Pelée on the island of Martinique in the West Indies in 1902. It is not necessary to discuss all the volcanoes listed in Table 10-1; Krakatoa is a fairly representative example and is worth a brief account.

Krakatoa was once a very large volcanic mountain, fragments of which remain in the form of three small islands. It is located in the middle of the Sunda Strait between Java and Sumatra (Figure 10-1). The earliest historical eruptions at Krakatoa were flows of magma that produced a rock called andesite in 1680. But an older, extremely large volcano is thought to have existed on the site at one time, and to have collapsed or blown out to form a large central basin, called a caldera, in which vents later developed (Figure 10-2). In 1877 earthquakes began to occur rather frequently in the area, culminating in a violent explosion and eruption in 1883. Approximately 12 cubic miles of earth and volcanic ash were blown into the air, and fell over a huge triangular area about 3300 miles long in one direction and about 2000 miles long in each of the other two—an area of approximately 5.4 million square miles. The explosion was heard 3000 miles away in west-central Australia, in Ceylon, and in the Philippines. Although the casualties caused by the

Figure 10-1

The geography of Krakatoa and the history behind the 1883 eruption.

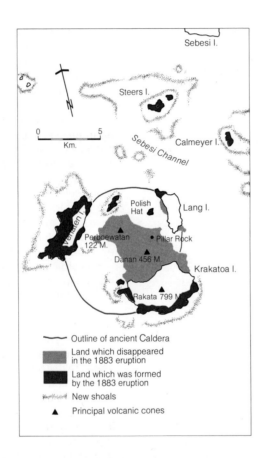

maintain, and still contributes to, the atmosphere and the hydrosphere (pages 53–54) and to the various cycles of the ecosystem, such as the hydrologic, nitrogen, carbon, and sulfur cycles (pages 16–18). Without volcanism this earth would be a very different and much less habitable place.

Volcanism is an outpouring of molten rock, heated rock debris, gases, and water vapor; most volcanism occurs at the boundaries of those plates in the earth's crust believed to be associated with drifting continents (see Chapter 1 and Appendix B). Along the midoceanic ridges, where ocean basins widen, new rock is formed from magma. Faults, fractures, and vents at volcanic sites allow the escape of gases, liquids, magma, and rock debris. Such areas ring the Pacific Ocean and extend through the Caribbean Sea, the Mediterranean Sea, and along the islands of the East Indies.

Types of volcanism

Volcanism takes different forms. Sometimes magma flows from long cracks known as fissures, opening to the surface of the earth. Examples of fissure flows are the lava fields of Iceland; the Columbia River basalts of Idaho, Oregon, and Washington; and the Deccan Plateau basalts of India. Some volcanoes are essentially sheets of lava thicker at the ends nearest to a single vent source. These are low, gently sloping mountains called shield volcanoes.

Other volcanoes are composed of rock fragments (ejecta) or semimolten to molten blobs (tephra) that pile up around a vent in the form of a cone. These volcanoes are called tephra cones, or cinder cones, and

eruption of Krakatoa were estimated at 36,000, most of the deaths resulted from the tsunami that flooded the north and northwest coasts of Java and the south coast of Sumatra.

Why Volcanism?

Volcanic activity is a natural part of the ecosphere and of ecosystems. By outgassing volcanism has established and helped to

Table 10-1

Dangerous volcanoes in man's history.

Volcano	Location	Known start of activity	Rate of activity	Known casualties
Stromboli	Sicily	Pre-600 B.C.	More or less continuous	?
Vesuvius	Italy, Bay of Naples	Quiet for centuries before 63 A.D.	More or less continuous since 63 A.D.	Thousands
Santorin	Between Greece and Turkey	Pre-2500 B.C.	Violent off and on since 2500 B.C.	Thousands, especially from 2500 to 1200 B.C.
Hekla	Iceland	18 eruptions since 1100 A.D.	Regular	?
Pelée	Martinique, Lesser Antilles	?	Irregularly violent	30,000 in 1902
Etna	Sicily	Pre-Homer, pre-900 B.C.	More or less continuous	Thousands, 20,000 in 1669
Krakatoa	Indonesia, Sunda Strait	1680 and earlier	Irregularly violent	Thousands, 36,000 in 1883

Figure 10-2

Cross section of Krakatoa. (a) Outline of ancient caldera and pre-caldera cone; (b) prior to the 1883 eruption; and (c) following the 1883 eruption.

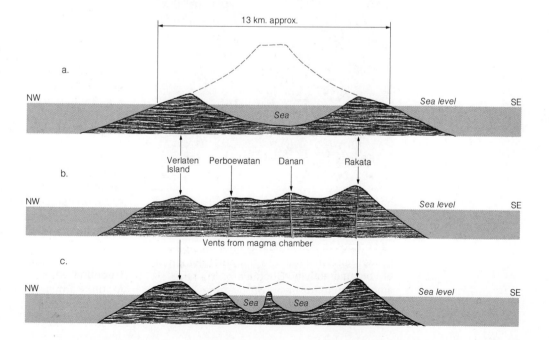

Table 10–2

Greatest volcanic eruptions since 1815 in terms of volume of rock ejected. From *Anatomy of the Earth* by André Cailleux. Translated by J. Moody Stuart. © André Cailleux, 1968. Translation © George Weidenfeld & Nicolson Limited, 1968. Used with permission of McGraw-Hill Book Company. (New York: McGraw-Hill, 1968), table 9, p. 47.

Date	Volcano	Location	Volume (cubic kilometers)	
			Lava	Ash
before 870	Veidivatna	Iceland	43	
930	Eldgja	Iceland	9	
1783	Laki	Iceland	12	
1815	Tambora	Indonesia		150
1883	Krakatoa	Indonesia		18
1888	Bandaisan	Japan		1
1912	Katmai	Alaska		30
1956	Bezymianny	Kamchatka	3	

Geyser emitting steam, Yellowstone National Park, Wyoming

have steeper sides than do shield volcanoes. Volcanoes composed of layers of both lava and tephra, or cinders and ash, are known as composite volcanoes. Sometimes gases escaping from vents are the chief evidence of volcanic activity. Because of the large amount of sulfur given off by some of these vents, they are called solfataras, after an area in Italy. Similar vents, which primarily emit steam, are called fumaroles; fumaroles frequently appear before eruptions.

Geyser fields are usually interpreted as resulting from the heating or superheating of groundwater that has penetrated to the heated rocks at depth. Most geyser fields, such as those at Yellowstone National Park, are thought to represent the end stages of volcanic activity. But some geyser fields, such as those in Iceland, are in volcanically active areas. Geyser fields are sources of geothermal energy for the future, and several are already partially developed.

Magnitude of volcanic events

Despite its ejection of 12 cubic miles of material, Krakatoa is rated only fourth in volume of ejected material among volcanic eruptions since 800 A.D. First in volume is Tambora, estimated to have ejected 150 cubic kilometers (100 cubic miles) of ash in 1815. Second is a volcano in Iceland called Veidivatna, which before 870 A.D. extruded about 43 cubic kilometers of lava. Mount Katmai, in Alaska, ejected about 30 cubic kilometers of ash in 1912. Eruptions of lesser magnitude are listed in Table 10-2.

Table 10-3

Relative energy released by five major volcanic eruptions and by hydrogen and atom bombs.* From *Anatomy of the Earth* by André Cailleux. Translated by J. Moody Stuart. © André Cailleux, 1968. Translation © George Weidenfeld & Nicolson Limited, 1968. Used with permission of McGraw-Hill Book Company. (New York: McGraw-Hill, 1968), table 8, p. 46.

Event	Year	Relative release of energy
Tambora, Indonesia	1815	4200.0
Sakurazime, Japan	1914	23.0
Bezymianny, Kamchatka	1956	11.0
Krakatoa, Indonesia	1883	5.0
Asama, Japan	1783	4.4
hydrogen bomb		1.0
atom bomb		0.0007

*This comparison refers not to the energy of the explosion, but to the total energy liberated during the eruption. Tambora released 4200 times as much energy as a hydrogen bomb and six million times as much as an atom bomb.

Kilauea Iki erupting (1959) in the Hawaii Volcanoes National Park, Hawaii

The energies released by several volcanoes have exceeded those of either the atom bomb or the hydrogen bomb (Table 10-3).

Kinds of Volcanic Activity

Although the intensity of volcanic activity has been classified in several ways, the system most widely accepted, with some modification, is that proposed by A. Lacroix, a French volcanologist, in 1908. His classification contains four categories named after individual volcanoes, ranging from less to more explosive: Hawaiian, Strombolian, Vulcanian, and Peléan. That there are valid differences, and yet overlaps, among the types, will be apparent.

Hawaiian

The Hawaiian type of volcano, named for the island, is the least violent. The magma is mafic—that is, the lava and ejecta contain more iron- and magnesium-rich minerals and less silica than the Peléan type. Magmas of this composition cool less rapidly and allow gases to escape more easily. Because escaping gases do not exert great pressures in a crater, chamber, or vent, Hawaiian volcanoes are quieter and less explosive than other types.

The great fissure flows, sometimes called Icelandic volcanism, are related to the Hawaiian type of activity. Hawaiian-type activity produces fissure flows, shield volcanoes, and composite cones. Fissure flows also occur in areas where the ocean floor is spreading, or in areas of subduction, where one plate glides beneath another. Hawaiian-type volcanoes are characterized by greater lava flows, less pumice or ash and ejecta,

Small crater cones within
the gigantic crater
Haleakala, Maui

Figure 10-3

The Aeolian Islands, including Vulcan and Stromboli.

and in comparison to other types, only
mild explosions.

Strombolian

As mixtures richer in silica and lower in
iron and magnesium appear in the lava,
gases escape from the magma less easily,
because it's more viscous and congeals
more quickly. Volcanoes that emit magmas
richer in silica tend to be more explosive
than those of the Hawaiian or Icelandic
types. Strombolian volcanoes usually form
composite cones.

Stromboli, one of the Aeolian Islands just north of the eastern end of Sicily (Figure 10-3), lends its name to this type of volcanism. Strombolian volcanoes are constantly active; Stromboli itself has been active since ancient times, and has more or less regular explosions of moderate intensity. These explosions are produced by gas in the crater where the magma barely crusts over, and only slightly impedes its escape. Recent studies indicate that the tidal cycle disrupts this delicate balance. Some scoria, a frothy rock resulting from the entrapment of small bubbles of gas in the surface of the cooling and solidifying magma, is emitted.

Vulcanian

Vulcanian volcanic activity is named for the volcano Vulcan, also in the Aeolian Islands. Vesuvius, which has been thoroughly studied, is the model of Vulcanian activity. A volcano of the Vulcanian type has more viscous magma than one of the Stromboli type, and so is more explosive. Strombolian eruptions are not strong enough to break the cone, whereas Vulcanian explosions sometimes do. They also eject scoria and pumice, a very finely-frothed rock more silicic than scoria. Ash, gases, and other ejecta form great clouds that disperse and drop dust and ejecta from the sky. The most violent explosions of Vulcanian volcanoes have been termed Plinian, after Pliny the Elder, the Roman natural historian who was killed at Stabiae in 79 A.D. during an eruption of Vesuvius. Vulcanian volcanoes may be composite, but most are tephra cones.

Peléan

Peléan volcanoes are the most violent. (Mont Pelée, on the island of Martinique, is discussed on pages 257–260.) There is very little magma flow, and nearly all of the very viscous lava is blown out as ash and other kinds of small ejecta. The normal volcanic neck or vent may become so solidly blocked that the side of the mountain is blown out as great rolling clouds of hot gases and incandescent ejecta and tephra. These horizontal clouds are called nuées ardentes and the rocks they produce are called ignimbrites or welded tuffs, so named because the heated and glowing ejecta and tephra weld together when they finally come to rest. Peléan volcanoes are almost always tephra cones.

Volcanic classification and volcanic risk

Lacroix's system for classifying volcanoes is, of course, imperfect. The more violent explosions of a Vulcanian volcano may be stronger than the less violent explosions of a Peléan one. The most violent eruptions occur at volcanoes extruding rock of greater silica content, whereas the less violent extrude products lower in silica and higher in iron and magnesium. This does not mean that Hawaiian or Icelandic events cannot be dangerous. The 1783 Laki fissure eruption in Iceland killed thousands of people and animals, some directly and others by starvation as a result of crop destruction, probably by sulfur dioxide. (Sulfur dioxide pollution was discussed in Chapter 3.)

Historically, the violent eruptions of more silicic volcanoes have been the greatest killers. The eruption of Mont Pelée caused 30,000 deaths directly and several tens by mudflow. The explosion of Krakatoa was responsible for 36,000 deaths, most of which were indirectly caused by tsunami. At Pompeii, a relatively violent eruption of a Vulcanian volcano killed tens of thousands, many directly, although the deaths at Herculaneum were by mudflow. The violent explosions at the island of Thera in the eastern Mediterranean in ancient times must have resulted in much greater kills, directly by tsunami, and indirectly by famine (since all the crops on Crete were probably destroyed more than once, and crops may have been destroyed as far away as northern Egypt).

In summary, Hawaiian volcanism is less dangerous than Peléan because it is much easier for people to escape from a lava flow moving at one to eight miles per hour than from a nuée ardente moving at 100 miles per hour. It is also possible to divert small lava flows, but not nuées ardentes.

Recent Volcanism

Many volcanoes pose an unending threat due to continual, frequent or intermittent, activity. Examples are Mount Etna in Sicily and Mount Vesuvius in Italy. Some volcanoes in Central America are continually dangerous. Mount Irazu, near San Jose, Costa Rica, erupted in March 1963. By 1964, 15,000 people had been evacuated from a 25,000-acre dead zone around the volcano; 2000 cattle had been killed; and 100,000 acres of farmland, mostly planted with coffee, had been blighted by the ashfall. The economic loss was estimated at $160,000,000, about two-thirds of the Costa Rican gross national product. Mount Arenal, also in Costa Rica, erupted in 1964 for the first time in over 450 years and killed 78 persons.

On February 19, 1964, Mount Agung in Bali, Indonesia, erupted for the first time in over 100 years. Further eruptions occurred from March 17 to March 21 and on May 16. Total deaths approximated 1610, but 78,000 people were left homeless.

Much volcanism in the United States is so recent that it is mentioned in the legends of several Indian tribes. The western United States had a large number of volcanic eruptions in the late Tertiary and the Quaternary (Figure 10-4). These occurred in Trans Pecos Texas; several areas in New Mexico, Arizona, Colorado, Nevada, and Utah; Yellowstone Park; the Snake River area of Idaho and Oregon; and some of the mountains of Oregon, Washington, and northern California.

There are about 460 active volcanoes in the world, of which 38 are in the United States. But only five are in the contiguous United States—the other 33 are in Hawaii, the Aleutian Islands, and part of the mainland of Alaska. Of those in the contiguous United States, only one has been active since European colonization. But several others, though sometimes called dormant, may awaken and become dangerous. Lava flows west of Albuquerque, New Mexico, are said to be between 600 and 700 years old, and the ever-dangerous Mount Vesuvius has been known to be quiescent for a much longer period than that. Sunset Crater, near Flagstaff, Arizona, erupted about 1067 A.D., and volcanoes in northeastern New Mexico were active only 10,000 years ago. Malpais (fields of rough, blocky lava) north of Alamagordo, New

Figure 10-4

Tertiary and quaternary volcanic areas in the western United States.

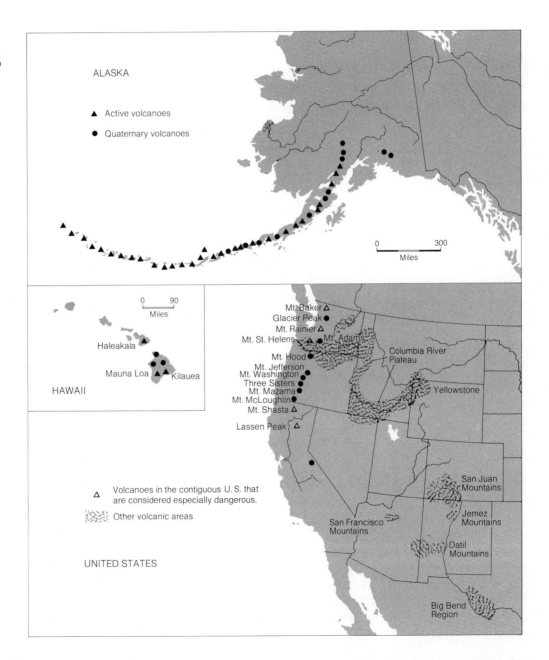

ALASKA

▲ Active volcanoes

● Quaternary volcanoes

0   300
Miles

0   90
Miles

Haleakala

Mauna Loa ▲ △ Kilauea

HAWAII

Mt. Baker △
Glacier Peak ●
Mt. Rainier △
Mt. St. Helens △ ● Mt. Adams
Mt. Hood ●
Mt. Jefferson ●
Mt. Washington ●
Three Sisters ●
Mt. Mazama ●
Mt. McLoughlin ●
Mt. Shasta △
Lassen Peak △

Columbia River Plateau

Yellowstone

San Juan Mountains

Jemez Mountains

San Francisco Mountains

Datil Mountains

△ Volcanoes in the contiguous U. S. that are considered especially dangerous.

Other volcanic areas

Big Bend Region

UNITED STATES

Figure 10-5

Postglacial mudflows on Mount Rainier, Washington.

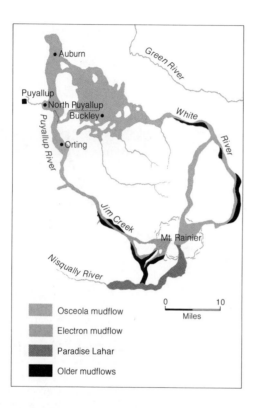

Osceola mudflow

Electron mudflow

Paradise Lahar

Older mudflows

duction is still an active process along the western margin of the United States, these volcanoes—which some geologists believe to be related to subduction—should be considered active.

Potential Volcanism

Like seismic activity, volcanoes are distributed along the Tethyan belt, the circum-Pacific belt, and the oceanic ridges (see Figure 9-4). The volcanic part of the Tethyan belt includes Indonesia and the Mediterranean; the latter has been called the Mediterranean belt. The activity of volcanoes, along these belts in particular, is a constant reminder that they are a hazard and an environmental problem. But human behavior is such that much of the damage and loss of life cannot be prevented. After every catastrophe people begin rebuilding their villages and tilling their farms on the sides of volcanoes that will wipe them out once again.

Any of the volcanoes of California, Oregon, and Washington (see Figure 10-4) could erupt at any time. Some appear more active than others; at present the U.S. Geological Survey is most concerned with Mount Rainier. This concern centers on three phenomena: (1) there has been a two- to sixfold increase in the frequency of microearthquakes at Mount Rainier since 1968; (2) aerial infrared images taken in 1969 reveal that several new hot spots on the mountain have developed since 1966, which probably means that the temperature of the volcano's interior is rising; (3) although glaciers normally change very little from year to year, in the last five or six years the glaciers of Mount Rainier have

Mexico, lack soil and vegetation, and the similarity of their appearance suggests that they cannot be much older than those near Albuquerque. Mount Shasta, Mount Rainier, Mount Mazama, and other volcanoes in Oregon and Washington were active in the last ice age, if not more recently. Mount Rainier has erupted repeatedly in the last 10,000 years, most recently between 2000 and 2500 years ago. Fifty-five mudflows have originated at Mount Rainier in the last 10,000 years, and some of these reached the lowlands near Puget Sound. If sub-

been cracking, slipping, sliding, and perhaps even melting, producing greater quantities of water than normal, probably as a result of the increasing temperature of the underlying rock.

New villages are being developed around the foot of Mount Rainier, and geologists are less worried about a volcanic eruption than about the villages being inundated with water and mud as the warming mountain allows the newly saturated and thawed ashbeds to move down its slopes in great mudflows (lahars). That such flows have occurred before is attested to by the many fossil mudflows around the base of the mountain (Figure 10-5). About 30,000 people live in five towns in the Rainier lowlands, all within reach of the mudflows of the past. If a mudflow were to discharge into one of the reservoirs in the valleys above these villages, 40,000 people would be endangered by a disaster similar to, but much greater than, the Vaiont disaster (see Figures 11-7 and 11-8).

Volcanic Prediction and Engineering

Prediction

Because of potential danger in the western United States, the U.S. Geological Survey has been monitoring some "dormant" volcanoes, particularly Mount Rainier and Mount St. Elias. If their changing features can be understood, prediction may someday be possible. The present status of prediction rests primarily on studies of Hawaiian volcanoes, particularly Kilauea. Methods of predicting volcanic events are similar to those for predicting seismic events, and include studies of volcanic cycles, monitoring of seismic events, magnetic studies, and dilatation studies.

Volcanic Cycles. Volcanic cycles, like earthquake cycles, are irregular and do not provide consistent data for predicting a particular event, even though it is possible to chart the frequency of events of different strengths and magnitudes.

Seismicity. Swarms of minor seismic events, or a series of earthquakes, often precede volcanic eruptions. Earthquakes began to increase in number and magnitude at least six years before the eruption of Krakatoa in 1883. Earthquakes predated the eruption of Vesuvius in 79 A.D. by at least sixteen years, and swarms of minor seisms have preceded eruption at Kilauea. Such phenomena may suggest impending volcanism without indicating when volcanism may occur.

Magnetic Studies. As with earthquakes, magnetic changes appear to precede volcanic eruption. But no correlation has been made between the magnetic change and the length of time preceding eruption.

Dilatation. Perhaps more important than the studies mentioned above are dilatation studies. Dilatation is swelling or expansion. Figure 10-6 illustrates the changes in elevation of the top of Kilauea, Hawaii, from January to July 1966. Increases in the altitude of the summit in conjunction with lesser increases down the slopes indicate movement within the crater. Figure 10-7 illustrates the increase in slope of the flank

Figure 10–6

Changes in elevation of Kilauea before an eruption (figures are in millimeters).

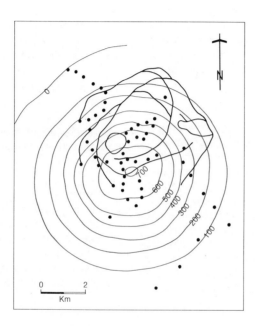

of Kilauea before each eruption and the collapse during eruption. It may eventually be possible to say that eruption is imminent when the slope reaches a certain angle.

Problems of prediction

All these potential methods of prediction need much more research, and the budget for research into volcanic hazards appears to be insufficient, not only in the United States but throughout the world. Furthermore, the data on Kilauea, for example, although potentially sufficient for prediction there, may not pertain to another volcano. Each may need to be documented separately.

Thus the only way to protect oneself from the dangers of volcanic eruption at the present time is to be somewhere else. Yet

Figure 10–7

Tiltmeter data for Kilauea, Hawaiian Islands from 1964 to 1966.

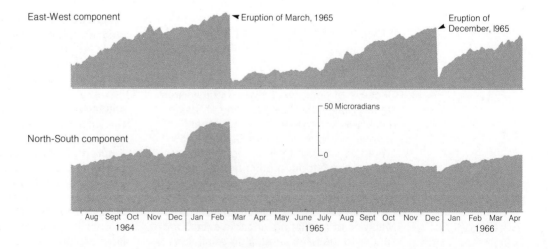

people ignore this recommendation, now as always. The complete disregard of history is one of the remarkable aspects of human behavior manifested at all active volcanoes. Some towns at Mount Vesuvius have been destroyed five or six times, but people continue to return. Volcanoes seem to be addictive.

Engineering

Nuées ardentes and black clouds are clearly beyond the control of man. Nor can massive lava and mudflows be controlled by engineering methods. Smaller lavaflows have often been diverted by walls constructed for other purposes. During the 1669 eruption of Mount Etna, when 20,000 people were killed by an earthquake, a lavaflow that had destroyed 14 towns was directed toward Catania, by the sea. To protect their homes, the people of Catania raised the city's wall, which stopped the lava until it had piled up to a depth of 60 feet, at which point it flowed over the wall and entered the city. This was the first attempt to divert a lavaflow.

Lava builds its own walls by congealing along the edges of the flow. Such a wall is a low transmitter of heat, and protects the internal lava from cooling rapidly, so that it flows more freely. Thus the walls that congeal along the sides of the flow contain the lava in the direction of flow. At one point these same Catanians covered themselves with the hides of freshly butchered cattle for protection against the heat, and with iron bars and other tools dug a tunnel through the side wall of a flow and diverted at least a part of it from their town. But their success was not celebrated, for citizens of Paterno, the town toward which the flow was redirected, drove the Catanians away, and the unattended tunnel soon closed.

In December 1935, when a lavaflow was moving toward Hilo, Hawaii, Dr. T. A. Jaggar, the outstanding American student of Hawaiian volcanoes, convinced the United States Army Air Corps to bomb the flow. The flow stopped, but some volcanologists have argued that it did so coincidentally, of natural causes.

Engineering methods are not likely to prevent or cure many eruptions. The primary hope of people living in volcanic risk areas lies in excellent warning systems, dependable transportation out of the area, and maybe future prediction.

Examples of Major Volcanic Eruptions

The nuée ardente—Mont Pelée

Mont Pelée is on the French island of Martinique (Figure 10-8), which is 30 miles long and a dozen or so miles wide. Pelée means bald, and the volcano at various times has had a new crown of unvegetated rock, giving it the look of a bald mountain. It is roughly circular and has a single peak with a crater and lake.

Ruins of Mont Pelée on
Martinique, West Indies

Before 1902 there were two known eruptions of Mont Pelée, one in 1792 and another
in 1851. In 1851 ash fell on St. Pierre, some
six miles from the summit. In 1902 the first
activity was noticed on April 2, when steam-
ing fumaroles appeared. This activity con-

Figure 10-8

Mont Pelée, at the northern end of Martinique.

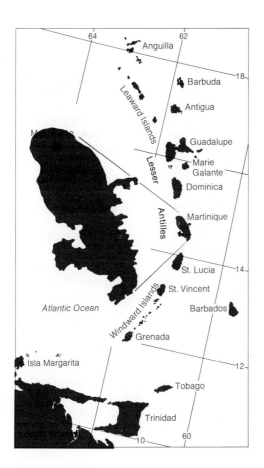

larger, and the constant roar of the volcano became deafening. Birds and small animals were killed by the ash.

By May 3 ash had been falling continually on St. Pierre for almost a week. Many businesses were closed and large houses were shut. But a commission appointed by the governor reported no immediate danger. The governor had scheduled an important election for May 10, and he did not want alarmed voters to evacuate his largest city. To impress the voters that danger was not imminent, he and his wife went to St. Pierre to stay until after the election.

By May 5 ash had been falling for almost two weeks and had accumulated in considerable thickness on the flanks of the mountain. Hot moisture-laden air was rising high enough to condense and create torrential rainfalls. Water quickly saturated the porous, newly fallen ash and created mudflows that raged down the valleys, started by earthquakes and heated to the boiling point by escaping steam. The first casualties, between 30 and 150 persons, occurred on May 5, when a boiling mudflow rushed down the valley of the Riviére Blanche and inundated a sugar mill.

People did not understand the volcano's behavior; they were expecting a violent earthquake, or a tsunami. On May 6 the governor ordered the militia to prevent citizens from leaving the city.

Eruptions on May 7, and the never-ending rains, continued to produce great floods of muddy ash-laden water. On the night of May 7 the main vent of the volcano was clogged and permanently blocked. May 8 dawned bright and sunny, with only a column of vapor rising above the summit. But suddenly, violently, at 7:50 A.M., the

tinued until April 23, when there was a slight fall of ash and a strong odor of sulfur in St. Pierre. On April 25 explosions sent up clouds of ash. On April 27 a small cinder cone formed in the center of the lake. A large column of steam arose, probably from groundwater coming in contact with heated rocks, and there were boiling noises as ashfalls continued. Explosions became

volcano blew up. Four deafening, cannon-like explosions produced two great black clouds, one directed upward and the other to one side. The lateral cloud was blown out of the side of the mountain directly at St. Pierre, and within two minutes the mass of hot gases and incandescent ash and rock fragments had completely engulfed the city of 30,000 people. This cloud of super-heated steam and dust was blown out so forcefully by expanding gases that it traveled at about 100 miles per hour. All the houses in the city were unroofed by its force. Trees were stripped of branches and bark to the bare trunk. The entire city burst into flames. Three and one-half hours later the heat was so intense that a ship could not approach the shore. Of the many ships in the harbor only two escaped, and most members of their crews were dead when they reached another port to the south.

The destruction of the city was complete; no building remained intact. The ash was not very deep, drifting to only three or four feet, but death had been instantaneous. According to most reports all but two of St. Pierre's inhabitants were killed, including the governor and his wife. One of the two survivors was said to be a prisoner in solitary confinement in a room so poorly ventilated that the air did not circulate enough for the hot gases to suffocate him. There were five more nuées ardentes, in June, July, August, and December of 1902, but there was nothing left to destroy after May 7. Mild activity continued at Mont Pelée until 1904.

Violence and ash—but little damage

The volcano Coseguina is on the cape of the same name that juts into the Pacific Ocean from the northern shore of Nicaragua. In 1835 Coseguina produced the most violent eruption known in the western hemisphere—in other words, more violent than Mont Pelée.

The crater of Coseguina is 1500 to 2000 feet deep and one to one and one-half miles across. Around the mountain lies an old crater ring, indicating a tremendous explosion in the geologic past, and suggesting the ever-present danger of volcanic explosion. In the 1835 eruption of Coseguina the most violent explosion occurred during the night of January 22. The explosion was heard at Bogotá, Colombia, and in Jamaica, over 800 miles away, where ash also fell. Because no cities or villages were located near the volcano, casualties were minimal. Without inhabitants there is no danger and no drama.

Violence and ash—and death

The shores of the Bay of Naples in southern Italy have been plagued by volcanism for thousands of years. About 800 B.C. the Greeks first settled the bay and the island of Ischia at its western margin. The solfataras and volcanic cones of the Phlegraean Fields, west of Naples, have been active most of the time since then (Figure 10-9). The old volcanic craters underlying Naples itself testify to the active volcanism of the area from times remote.

Figure 10-9

Mount Vesuvius and some important localities around the Bay of Naples.

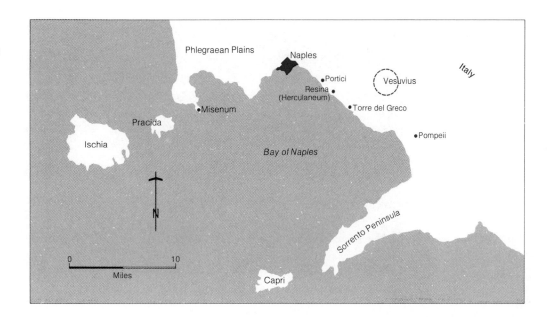

Vesuvius lies across the bay from Naples, about six miles away, and is responsible for a number of firsts in volcanology. The Vesuvius Volcanic Observatory, established in 1845, was the first of its kind. Professor L. Palmieri announced in 1872 that the patterns of Mount Vesuvius' eruptions were sufficiently cyclic to allow for rough prediction.

The volcano of Ischia had erupted early in the period of Greek settlements and forced the inhabitants of the island to move to the mainland. Ischia last erupted in 1398. Solfatara, northwest of Naples, last erupted in 1198, but since that time has been continuously emitting gases. Monte Nuova, also northwest of Naples, is a volcanic feature that appeared in 1598.

Vesuvius itself probably began as a submarine volcano shortly after the last ice age. It is located near the western end of the Mediterranean (or Tethyan) volcanic belt, and represents to Italy the same problem that Santorin (of which Thera is a remnant) did to the civilizations of the eastern Mediterranean, particularly in the Bronze Age. There are no records of eruptions at Vesuvius before 79 A.D., and it had apparently then been dormant for many centuries— since well before the Greek settlements of 800 B.C.

On February 5, 63 A.D., severe earthquakes did great damage to Pompeii and Herculaneum. Earthquakes continued intermittently for 16 years. On August 24, 79 A.D., tremendous earthquakes rocked the entire

Human bodies found in
Pompeii, petrified by lavaflow from Vesuvius

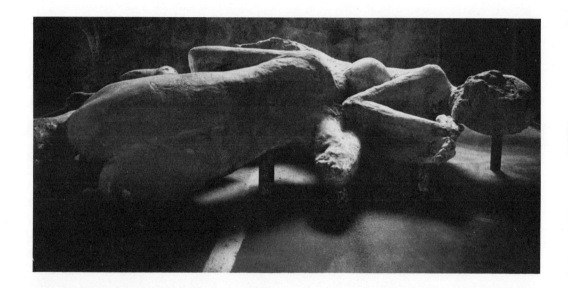

Ruins of ancient Pompeii,
Italy

area, and most of the cone known as Monte Somma (the precursor of Vesuvius) was blown away. Pliny the Younger and his mother lived with his uncle, Pliny the Elder, at Misenum, west of Naples on the northwest side of the bay. The elder Pliny, a natural historian, lost his life in an attempt to rescue friends from the villa of Stabiae, north of Pompeii. Much of the reconstruction of events at Vesuvius is due to excellent archeological work, but Pliny the Younger also left highly descriptive, if unscientific, letters.

At about 1 P.M. on August 24, a cloud of unusual size, in the shape of a pine tree, rose above Monte Somma. The earth had trembled for several days, and at one time the sea withdrew, although there was no subsequent tsunami.

On August 25 ashes fell on Misenum, 20 miles across the bay from Monte Somma, and it became dark early in the day. Before evening, the glare from the volcano made the day brighter, even at a distance of 20 miles. Earthquakes continued. Pompeii, an old city even in 79 A.D., was covered with ash and other ejecta to a depth of 15 to 25 feet. People were asphyxiated, crushed by falling buildings, and trapped in their homes. The manner of death can be interpreted from the locations and positions of casts of bodies recovered from the ash. Many people died holding their hands or cloths over their noses and mouths. Many escaped to the seaside, only to die later when there were no boats to evacuate them.

Herculaneum, another city on the west flank of Monte Somma, suffered a different fate. With the tremendous amounts of ash and pumice and the torrential rains that always accompany the black clouds at this stage of volcanic activity, mudflows were frequent. Herculaneum was engulfed by a mudflow, at some places to a depth of 65 feet. Most of the inhabitants, however, had evacuated the city prior to the flow. By 1100 A.D. a new town, Resina, arose on the site of Herculaneum, and its citizens did not know that they lived on top of a buried city.

Although Vesuvius wrought its greatest destruction in the explosion of 79 A.D., it has been active ever since. There were eruptions in 203, and a great eruption in 472 spread volcanic ash all over Europe and aroused great fear even at Constantinople, 700 miles away. Eruptions followed in 512, 685, 993, and 1036. Although there had been lavaflows in prehistoric days, the first in historic times occurred in 1036. These were massive flows down the flanks of the volcano and into the Bay of Naples. Others followed in the eruptions of 1049, 1138, and 1139.

By 1631 Vesuvius had been quiescent almost 500 years, and a people occupied the sides of the mountain who had not heard of its previous eruptive history. On December 16 of that year there were violent explosions, and another pine-shaped cloud shot into the air, accompanied by cinders and ash. Thousands fled the villages into the city of Naples. Again there were lavaflows, and fissures opened on the flanks of the cone. Lava flowed over the villages of Pugliano, La Scala, Torre del Greco, San Georgio, and Portici, reaching the sea, but most of the inhabitants escaped to Naples. On the evening of December 17 it rained mud in Naples, and mudflows enveloped villages on the northern and western slopes of Vesuvius, including the village of Resina on the exact site of Her-

Figure 10–10

The area influenced by outfalls from eruptions of the volcano Santorin (Thera).

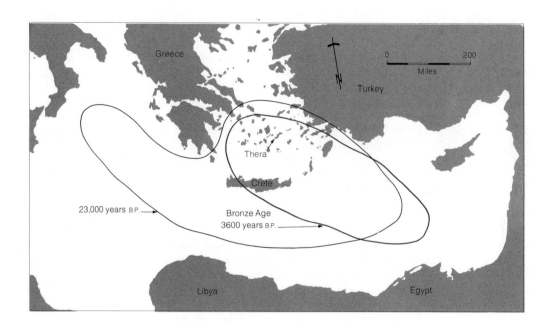

culaneum, which had been engulfed in a mudflow 1552 years earlier. The explosions declined in January 1632, but only after 4000 deaths and the destruction of six towns. Vesuvius erupted many times in the last half of the eighteenth century, and also in 1822, 1838, 1850, 1872, 1893, 1906, and 1941.

The Lost Atlantis

Thera is an island in the eastern Mediterranean (Figure 10-10), about 80 miles north of Crete and 120 miles east of the southern end of the Peloponnesus, most noted as the site of the ancient volcano Santorin. In fact, Thera and Therasia are parts of the rim of a collapsed volcanic crater, about seven miles north-to-south by four miles east-to-west. Small volcanoes of more recent origin form islands within the drowned, collapsed crater of Santorin.

Although there are older rocks on a small island not far from Thera, the first volcanic activity at this site dates from the late Pliocene. The island began as a submarine extrusion, and gradually grew. It is known that Santorin was violent in the Bronze Age and again in 365 A.D., and that in 726 A.D. pumice from the volcanoes of the island reached the shore of Asia Minor, some 200 miles away. In 1650 there was a severe earthquake on Thera, and a destructive tsunami struck the island of Ios, 10 to 12 miles north, with a runup of over 50 feet. Another severe earthquake occurred in 1672.

Cycene, a town built on Thera in 651 B.C., has since been destroyed and excavated. There were also eruptions at Santorin in 196 B.C., 1570, 1707, and 1806. It is impossible to estimate the loss in human life and property caused by this volcano, since the greatest destructions took place in pre-Hellenic times and left no dependable record.

Not until submarine cores were taken in the eastern Mediterranean in the 1950s and 1960s did the extent of the volcanism of ancient Santorin begin to be realized. Some volcanic ashes can be dated radiometrically, and an ash that extends over a large part of the eastern Mediterranean is dated at 23,000 B.C. This ashfall, which covered 200,000 square miles (see Figure 10-10), occurred in the time of Neolithic man, so the only record is geologic. Following this great explosion there was apparently a long quiescence—or, if there were eruptions, no trace of them has been discovered.

Again, in the fifteenth and sixteenth centuries B.C., during part of the Bronze Age, Santorin deluged the eastern Mediterranean with so much ejecta that archeologists and oceanographers can once more identify it. That there were earthquakes in the sixteenth century B.C. is attested to by stone buildings from that period with wooden pegs pinning adjacent stones together; these pegs provided structural resiliency, but held the buildings together and prevented too much shifting of stone on stone. Similar construction is still used with some success in parts of Greece today.

Beneath the volcanic layers deposited in the sixteenth century B.C. are remains of the inhabitants of Thera, who were fishermen, farmers, and growers of olives, wheat, and other cereals. They wove cloth and fishing nets. Their buildings were of stone, and they carved stone for ornament. Interiors were stuccoed and painted or frescoed with colorful murals depicting the life of the time. The inhabitants of Thera were pottery makers, and traded with neighbors for gold, silver, and copper, though most of their tools were still made of stone. Altogether an advanced culture had developed.

Then at some time in the sixteenth century B.C. a storm of pumice, ash, lapilli, bombs, tephra, and hot volcanic gases enveloped the island. The thickness of the deposits testify to many eruptions. Much as at Pompeii, the people had no time to leave their island or even their homes. The pale colors of the sixteenth-century ejecta contrast strongly with the darker underlying tephra of the older eruptions on Thera. These later deposits range from 100 to 150 feet in depth, and bury the entire world of these people.

If ejected deposits had not destroyed the culture of the island, accompanying events would surely have done so. It is thought that the series of geologic events began with a huge submarine volcanic explosion and the emission of tremendous amounts of gases, ejecta, and tephra, far exceeding anything that has occurred in more recent times. It was ash from these emissions that covered the city and the area circled in Figure 10-10. The explosion also left a vast empty chamber under the sea, some seven miles across at its greatest dimension. This chamber then collapsed, drowning much of the city and island of Thera, and producing a gigantic tsunami. Evidence of this seismic wave has been obtained near Tel Aviv, Israel, 500 miles distant, where there was over 20 feet of runup. On islands 15 miles from Thera, runup was 150 feet, and some estimates are of four times that, as tsunami bores ran far up the narrow ravines on these volcanic islands. The deposits that provide evidence of these bores still lie at the heads of the ravines, some 600 to 650 feet above sea level.

The old crater, or caldera, is now a bay of 32 square miles, and its waters are deep enough to indicate that many cubic miles of ejecta were spread over the eastern Mediterranean. Crete is only 80 miles to the south, and at the time of the explosion its High Minoan culture was at its peak. Fine pottery and exquisite sculptures, splendid buildings adorned with rich frescoes, a sophisticated economy, and "world" trade characterized this culture. It appears to have been destroyed by the violence of Santorin. It is estimated that the first tsunami destroyed all nearshore structures. Not less than four-inch, and probably much thicker, deposits of ash covered eastern and central Crete, if not all of the island. It has been determined in Iceland that four inches of ash are enough to make a land barren, and to remain so until the ash is (1) eroded away, (2) leached, (3) weathered to a good soil, or (4) some combination of these processes. While Crete was being buffeted by tsunami and covered with ash, severe earthquakes were toppling its buildings and crushing its inhabitants.

In two of his dialogues, the *Timaeus* and the *Critias*, Plato speaks of Solon's visit to Egypt, where he was told of the great island empire referred to as "Atlantis," which after violent earthquakes and floods "in a single day and night disappeared beneath the sea."

Although speculation has placed the lost "Atlantis" at many different localities around the world, including the center of the Atlantic Ocean, Greek archeologists, headed by Spyridon Marinatos, now believe that they have discovered the ancient land and are salvaging its relics from beneath the sea adjacent to the island of Thera.

Plato also mentions that the entire Peloponnesus was once good agricultural and forest land and had a thick mantle of soil. Before discussing the destruction of Atlantis, he alludes to the long period of great rains that washed the good soil of the Peloponnesus into the oceans and left only the thin soil and rocky surface that it still has today.

Any volcanic eruption, long or short, and especially a black cloud eruption, is accompanied by torrential rains. At Parícutin, Mexico, a relatively small volcanic phenomenon caused torrential rains every afternoon. Torrential rains also occurred at Vesuvius, Coseguina, Mont Pelée, and other ash-emitting volcanoes. Plato's writings suggest that the prelude to Santorin's final eruption was a long period, perhaps decades, when skies were darkened by volcanic ash, and torrential rains eroded and destroyed thousands of square miles of good agricultural land.

Summary

Volcanoes have existed since the cooling crust of the once-molten earth began to solidify, and are thus a link with our most ancient past. Volcanoes are also a part of the ecosystem; after the loss of the earth's original atmosphere, it was through outgassing of water vapor and other substances that much of our present hydrosphere and atmosphere were produced. This process is continuing.

The three major belts of volcanic activity coincide with those of earthquake activity. The four types of volcanoes differ largely in the chemical composition of the lava, and more violent volcanic activity is associated with more silicic magmas. Prediction of volcanic explosions is seldom possible, and engineering methods of minimizing damage are usually ineffective.

Volcanoes have been a hazard to life and property throughout human history, and the losses they have caused are incalculable. Increasing population, migration into volcanic areas in greater numbers, and the growing uncertainty about which volcanoes are inactive and which are only sleeping suggest the importance of intensified research into prediction.

Selected Readings

Bullard, Fred M. 1962. *Volcanoes in History, in Theory, in Eruption.* Austin: University of Texas Press.

Chesterton, Charles W. 1971. Volcanism in California. *California Geology* 24(8):139–147.

Crandell, Dwight R. 1971. Postglacial lahars from Mount Rainier volcano, Washington. U.S. Geological Survey Professional Paper no. 677.

Crandell, Dwight R., and Mullineaux, Donald R. 1967. Volcanic hazards at Mount Rainier, Washington. U.S. Geological Survey Bull. no. 1238.

Fiske, Richard S., and Kinoshita, Willie T. 1969. Inflation of Kilauea Volcano prior to its 1967–1968 eruption. *Science* 165:341–349.

Galanopoulos, A. G., and Bacon, Edward. 1969. *Atlantis: the Truth Behind the Legend.* Indianapolis: Bobbs-Merrill.

Luce, John Victor. 1968. *Lost Atlantis: New Light on an Old Legend.* New York: McGraw-Hill.

MacDonald, Gordon A., 1972. *Volcanoes.* Englewood Cliffs, N.J.: Prentice-Hall.

Thorarinsson, Sigurdur. 1958. The Oraefajökull eruption of 1362. *Acta Naturalia Islandica* 2(2):1–95.

———. 1970. *Hekla, a Notorious Volcano.* Reykjavik: Almenna Bokafleagid.

Chapter II

Unstable Lands

Two dry seasons in the White Mountains, in New Hampshire, were followed by heavy rains on the 28th August, 1826, when from the steep and lofty acclivities which rise abruptly on both sides of the river Saco, innumerable rocks and stones, many of sufficient size to fill a common apartment, were detached, and in their descent swept down before them, in one promiscuous and frightful ruin, forests, shrubs, and the earth which sustained them.
Charles Lyell, *Principles of Geology I*

In the course of populating and developing the earth, man continually encounters less favorable areas of the ecosphere. Some are undesirable because of unstable soil, rock, or crust. Most people who hear the word "unstable" immediately think of landslides or slumps. But instability also occurs where and when permafrost melts, the earth rises or sinks due to internal forces, or subsidence occurs because of resource depletion. Some of these geologic hazards are described in this chapter.

Surficial Earth Movements

Landslides and other earth movements destroy dwellings and public buildings, disrupt transportation and public utilities, and wreck or bury natural resources. Some slides move a tremendous volume of earth (Table 11-1). The Saint Albans slide in Canada moved 25 million cubic yards. The 1909 Gros Ventre slide in Wyoming, just south of Yellowstone Park in Jackson Hole, moved about 50 million cubic yards.

Earth movements kill many people and leave many more homeless. In Gunong Kloet, Indonesia, in 1926, a lahar or mud-flow (page 272) traveled 38.25 kilometers (about 25 miles) and killed over 5000 people. Landslides started by an earthquake in Peru in 1970 caused 20,000 deaths. Slides in the province of Quebec, Canada, periodically result in a number of deaths.

Landslide Terminology

Slide surfaces and rocks are commonly described in terms of their morphologic features (Figure 11-1). The crown of a slide is that part of the slope from which the slide

Table 11–1

Age and volume of some large landslides. Data from Douglas M. Morton and Robert Streitz, "Landslides, Part I." *California Division of Mines and Geology Mineral Information Service* 20(10):127, 1967.

Landslide	Date	Volume (millions of cubic yards)
Riviere Blanche, Quebec, Canada	1898	3.5
Elm, Switzerland	1881	10–11
St. Albans, Quebec, Canada	1894	25
Frank, Alberta, Canada	1903	35–40
Gros Ventre, Wyoming	1909	50
Blackhawk, California	Prehistoric	370
Vaiont, Italy	1963	390
Tin Mountain, California	Prehistoric	2,350
Flims, Switzerland	Prehistoric	15,000
Soid Marreh, Iran	Prehistoric	21–27,000

breaks away, whereas the head is the upper part of the slide. The toe is the distal, or lower, end of the slide. Above the head of a slide there may be cracks in the ground, called lunar cracks, which usually form early and help in the prediction of impending movement. Enechelon cracks extend out and downslope from the lateral margins of a slide, and radial or longitudinal cracks develop at high angles to the margin or toe. Sometimes ridges and cracks cross the slide from one side to the other; these are called transverse ridges and transverse cracks. If there is a scarp at the head of the slide, it is called the main scarp.

Some land movements do not move on a distinct slide surface, the surface of rupture (Figure 11-1), but those that are called landslides usually do. In some slides the upper part of the surface of rupture, where it breaks away from the slope, is concave. Downslope, this surface is usually the land surface, although some downslope soil may be incorporated into the slide.

Figure 11–1

Terms used in describing landslides.

Figure 11-2

Creep.

Figure 11-3

Hummocky topography on the Frontier Formation, Bighorn Mountains, Montana.

Mass Movement

Movements of earth and soil range from a very slow creep to the almost instantaneous airborne avalanche. The types blend into each other, and of the following types some are slow, some rapid, and some characterize both movements. It is not clear where one type leaves off and another begins. Creep (Figure 11-2) is the gradual movement of surficial material under gravitational forces acting on the surficial zone, including parts of the soil and the C horizon. Creep may also occur in unstable rocks underlying weathered bedrock. It may be accelerated by the loosening of material by the hooves of animals; particles kicked out horizontally fall back vertically to the surface and gradually move downhill. Earth material is also moved downslope by freezing and thawing. Freezing creates expansion at right angles to the slope, and thawing allows debris to drop vertically. Expansion and contraction, caused by temperature changes, produce creep in the same manner as freezing and thawing, as do the expansion and contraction of expandable clays due to wetting and drying. If the rate of creep is slightly increased, slow earthflows may develop, and many undulating or hummocky landforms have been created not by rapid slides but by slow earthflows moving at different rates (Figure 11-3). Other types of slow earthflow occur on steep slopes, where a competent bed resting on an incompetent bed is gradually pulled down the slope by gravity (Figure 11-4). Slumps may be slow or fast, and a particular type of slump with backward rotation is known as a toreva block (Figure 11-5). Other slumps are earthflows that move downslope in a matter of minutes or, at the most, hours. They occur along arcuate (concave upward) slip planes. Large earthflows may travel hundreds or even thousands of feet.

Solifluction is a mass movement phenomenon characteristic of cold regions. Most movement occurs during the spring thaw, when saturated soil and debris move over

Figure 11-4

Crumbling of the Madison Formation, Bighorn Mountains, Wyoming, is the result of slow sliding on the underlying Amsden Formation.

Figure 11-5

Toreva blocks. (From "Quick Clay" by Paul F. Kerr. Copyright © 1963 by Scientific American, Inc. All rights reserved.)

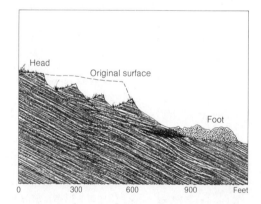

Figure 11-6

"Fallen City," a massive rock slide near Dayton, Wyoming.

still frozen ground. The depth of the movement depends on the depth of thawing, and therefore increases as spring advances.

Rock slides

Rock slides are landslides in which large blocks or areas of rock move down steep slopes for great distances, causing extreme disruption of the land and soil surface. West of Dayton, Wyoming, in the Bighorn Mountains on Highway 16, are the remnants of such a rock slide (Figure 11-6), called "Fallen City," that occurred many decades ago.

A rock slide in Italy in 1963 resulted in the Vaiont Reservoir disaster, in which 390 million cubic yards of rock suddenly broke loose from the wall of a valley containing the Vaiont Reservoir and slid into the water, forcing it out over the lips of the dam (Figures 11-7 and 11-8). The Vaiont Valley was flooded and 2600 lives were lost.

It would take about 100 million trucks of 2 1/2-ton capacity to remove that much material, provided they were overloaded; or one-half billion half-ton pickup trucks. If all the motor vehicles in the United States were half-ton pickups, it would require more of these than have been built since the automobile was invented.

Debris slides are rapid movements of rock, soil, and whatever other material is present, that cover distances from a few hundred feet to several miles (Figure 11-9).

Mudflows (lahars) are another type of mass movement; they may be quite rapid if the mud becomes so saturated with water that it liquefies and moves rapidly downslope, crossing highways or drowning villages. Two small villages were covered and 3500 lives lost by just such a mudslide in Peru in 1962.

Figure 11-7

The Vaiont Reservoir and slide area, Italy.

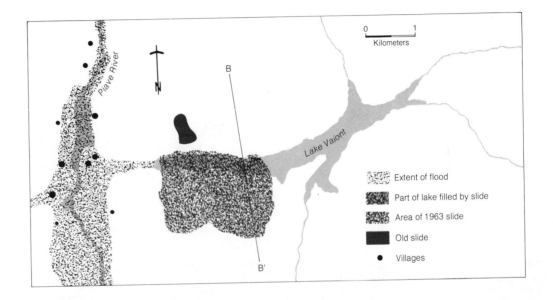

Glaciated mountain valleys

Figure 11–8

Cross section of the Vaiont Reservoir and slide.

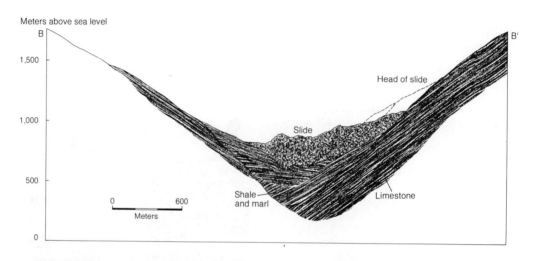

Figure 11–9

Debris (mostly rock, soil, and trees) from slide above Yankee Jim Canyon, Yellowstone River, Montana.

Avalanches

The most rapid slides—avalanches—move so fast that pockets of air become trapped beneath the debris, which then rides on the air with the full force of gravity and very little friction. They are borne up by the air much like the new refrigerators that can be moved by connecting to them the air outlet of a vacuum cleaner. The base of the refrigerator is pumped full of air, so that much of its weight is borne by the air instead of the floor. Only a few such airborne slides have been described. The Sherman slide (Figure 11-10), initiated by the Alaskan earthquake of 1964, moved so rapidly down the mountain south of Sherman Glacier that it was carried over two miles across the nearly flat surface of the glacier and up the north wall of the glacial valley for 75 feet.

Conditions That Promote Slides

Slides are promoted by the presence of unstable rock or soil, overloading, saturation, collapse, structural weakness, and structural slopes.

Unstable rock or soil

Any montmorillonitic clay that contains a little water will be unstable enough to produce downward movement, even on a

Avalanche in Squaw Valley,
California

Figure 11-10

The Sherman slide, Alaska.

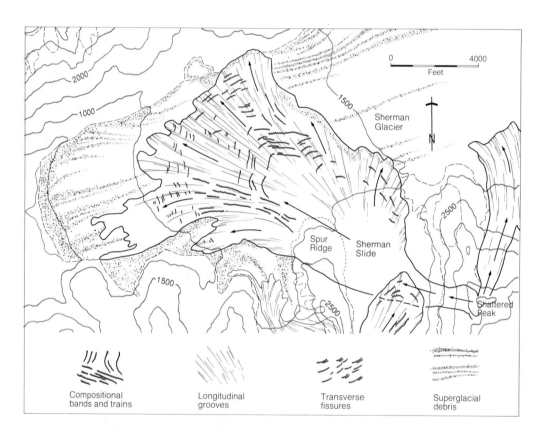

| Compositional bands and trains | Longitudinal grooves | Transverse fissures | Superglacial debris |

very shallow slope. Other types of clay are also unstable, but to a lesser degree. Quick clays are those that can liquefy when jarred or shocked. In the Alaskan earthquake of 1964, much of the damage was caused by landslides on quick clay. Weak seams or beds of unstable clay or soil on slopes may produce sliding.

Overloading

A slope is said to be overloaded when the weight of its load exceeds the strength of the rock forming the slope. The same effect can be produced by cutting the toe out from under a hill of rock of minimal strength. More often a house or other structure is built near the edge of a hill composed of rock with little strength; over the years the formation fails and the building moves down the slope or, during a particularly wet season, the formation may yield so that the structure slides more rapidly.

Saturation with water

Decreasing the strength of rock by saturating it with water also results in overloading.

Foundation of house damaged by landslide and mudflow, Bel-Air, California

tonite is a weathered volcanic product composed largely of montmorillonitic clay, which takes on water, expands, and becomes even more unstable.

Saturation lessens the strength of most clay-bearing rocks and soils and increases the probability of sliding. Water also has a flotation potential. Rocks normally held in place by friction are partly buoyed when saturated with water, and thus move much more easily. Many slumps attributed to other causes do not occur until the ground is saturated. In recent years the most dramatic examples have occurred in such parts of Los Angeles as the Hollywood Hills and along the coastal highway. Although always subject to sliding, as is indicated by many ancient landslides, and underlain by layers and formations of montmorillonitic clay, these areas are most hazardous during and immediately following the rainy season. Thus, a combination of slide-producing factors increases the hazard of an area.

Structural weakness

Where there are caves or unconsolidated rocks, collapse into the caves or subsidence into the unconsolidated rocks may occur. But structural weakness usually involves joints or faults near slopes, where the formation is already broken. When the fault or joint is overloaded and becomes filled with water, collapse and sliding are facilitated.

Figure 11-11

Weak layer of rock in a slope.

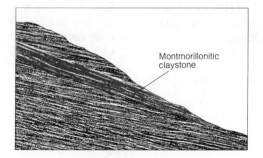

Montmorillonitic claystone

Structural slopes

Structural slopes (Figure 11-11) are structurally weak when dip parallels slope and provides a glide plane down which a mass

The Frontier Formation illustrated in Figure 11-3 develops a stable slope, but as meteoric water seeps into and saturates a bentonite bed, the slope above it becomes overloaded and movement occurs. Ben-

of earth or rock can move. Unstable strata are much less hazardous when they dip into the slope.

Mass Movement: Some Examples

A highway or street with an improper base usually has a bump in the roadbed where a stable formation meets an unstable clay formation. East of Rapid City, South Dakota, long stretches of Interstate 90 are rough and undulating, because much of the highway is constructed on and of the Pierre Formation, a claystone containing considerable shrink-swell clay. Such undulations, which suggest the instability of the formation, can be prevented by using great quantities of gravel or by stabilizing the clay. But in this area hauling gravel great distances, or stabilizing clay, is more costly than building new highways.

Another way to estimate the stability of a slope on unstable formations is to see whether the telephone poles or trees lean. Likewise, a fence will have small sags or curves where the downslope movement has varied within the formation.

More rapid slides—and greater damage—occur in Los Angeles along the Hollywood Hills, the Pomona Freeway, and the coastal highway. In this area unstable formations allow tremendous movement, especially during or shortly after rainy seasons, when those areas underlain by layers of montmorillonitic clay take on the greatest amounts of water. In such areas the value of homes and transportation facilities destroyed by landslides has been estimated at millions of dollars. The courts have ruled that, because Los Angeles has zoning ordinances and presumes to control development, the city is liable for damages. Over the last 20 years the city has passed stricter and stricter zoning ordinances to protect itself from damage suits. These ordinances require engineering reports, soil reports, and geological reports before construction may begin. The design engineer, the soils engineer, and the engineering geologist then become liable if their reports are inaccurate. Or if their reports were adverse before construction began the builder becomes responsible. The overall effect of such stringency has been beneficial. James E. Slosson, a California geologist, has shown that the continually stricter zoning codes have resulted in continually reduced damage. This is certainly a high recommendation for strong and effective zoning ordinances (Table 11-2).

Table 11-2

Decrease in destruction to building sites from landslides due to successively stricter building codes in Los Angeles. Modified from James E. Slosson, "The role of engineering geology in urban planning." In *Colorado Governor's Conference on Environmental Geology*, ed. John B. Ivey *et al*. Colorado Geological Survey Spec. Publ. No. 1, pp. 8–15, 1970.

Building sites	Before 1952	1952 to 1962	1963 to 1970
Approximate number of sites constructed	10,000	27,000	11,000
Approximate damage	$3,300,000	$2,767,000	$182,400*
Approximate number of sites damaged	1040	350	17
Average amount of damage per site constructed	$330	$100	$7.00
Predictable failure percentage	10.4%	1.3%	0.15%

*Over $100,000 of the $182,400 was incurred on projects where grading was in operation and no residences were involved.

A case study

In central Texas several years ago, part of a dam being built to contain the water supply of the city of Waco slid into the Bosque River. The heart of the problem was an outcrop, about 1000 feet along the dam, of the most unstable plastic clay formation in Texas, the Pepper Shale. During the construction of the dam, but—fortunately—before there was any water in the reservoir, a section of the dam slid out on the Pepper Shale. The outcrop was associated with an old fault, but it was the instability of the shale, not the fault, that was responsible for the slide.

This happened even though all the necessary information was available. The fault was known. The formations, including the unstable claystone, had been mapped and cored. Samples were available. Yet all this information was ignored by most of the personnel of the agency involved. Even after this fiasco the dam was completed, and ten years later the Waco dam is one of the most heavily monitored sites in the United States, as engineers continue to study the reaction of these clays to water, load, and use.

Prevention and Cure of Slides

Engineers can prevent and cure shallow slides, but with deep slides the load exceeds the strength of the remedies. Shallow slides can sometimes be prevented by bolting potentially dangerous flat slabs of rock to solid rock that will not move. Rock bolting is also used to prevent the scaling of roofs and walls in mines.

If sliding occurs in thin beds resting on stable beds (Figure 11-12), a strong abutment set solidly into the underlying stable formation may hold the potential slide area, provided the beds are not too thick. In some areas retaining walls are used to hold back unstable clays and soils if the slope is low and the unstable area is not too deep (Figure 11-13). This does not work if the retaining wall is deadmanned (anchored) to the moving slope (Figure 11-14). One sometimes sees expensive retaining walls of reinforced concrete built to hold back limestone formations that would never have moved in the first place.

In areas where surface water flows into the soils in the vadose zone, seeps along the slope, and saturates the earth, the water

Figure 11-12

An abutment bedded in solid rock to prevent sliding on an overlying weak stratum.

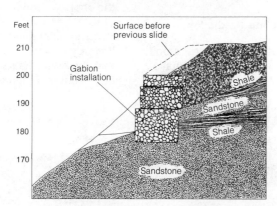

may be the primary cause of slides. Such sliding may be prevented by draining the slope. So far attempts to prevent sliding by the use of spaced piling have been unsuccessful. It bears repeating that shallow slides may be cured or prevented, but that for deep and massive slides there is little hope of cure or prevention.

Slope Designs and Slide Prediction

Engineers usually describe empirical slope designs in ratios of vertical to horizontal distance. A one-on-one slope is a 45° angle, with one unit of horizontal distance for every one of vertical distance. A one-on-seven slope has seven feet horizontal for each foot vertical. A seven-on-one slope, on the contrary, is very steep, with one foot horizontal for every seven feet vertical. The slopes in Figure 11-15 are too steep for construction to be recommended by geologists, except in extremely stable rock. In some areas houses on shrink-swell clay have moved though the slope was less than one-on-twelve.

Little slide prediction has been attempted. In areas where frost and ice split, pry, and loosen hard rock, prediction of rock slides may be feasible. The Norwegian Geotechnical Institute, for example, has stationed sensing extensometers along the edges of the fjords. These instruments detect the very slight shifts in rock that precede rock slides or rock falls. In the spring, as ice slides or melts, water circulates down through the crevices and joints. Slides of rock masses into a fjord often produce huge waves that damage villages along the shore. Some movement and adjustment of rock frequently precedes such a slide. Sensing extensometers detect this preliminary movement, record it, and warn that the entire nearshore area of the fjord has become dangerous.

Buda
Limestone

Del Rio

Claystone

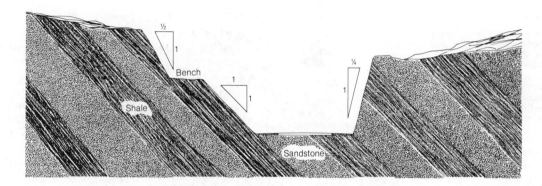

Through geologic studies, including analyses of formations for instability, geological engineers can predict areas of potential sliding. Russian geologists have for years mapped ancient slides, and used them to predict areas of potential sliding. The increasing temperature of so-called dormant volcanoes creates a danger of saturation of unstable volcanic clays with water from melted ice, resulting in mud-flows (pages 254-255).

Changes in Elevation

Areas such as Long Beach, and Tulare and Kern Counties, California, and Houston and Beaumont, Texas, have in recent years experienced local sinking of the land, in some places to elevations below sea level. This phenomenon is particularly hazardous to cities near the coastline, where after flooding or major storms it becomes necessary to pump floodwaters back into the oceans. Some coastal inundation results not from land sinkage but from the rise in sea level that accompanies the melting of the Greenland and Antarctic ice caps. Other areas are sinking as the result of isostatic adjustments that so often accompany rebounds of nearby areas once covered by major icesheets. Still other areas are sinking or rising as the result of tectonic activity, and many are sinking as a result of the withdrawal of fluid from the subsurface. Because sinking or rebound is the result of isostatic adjustment, it is considered a variety of tectonic subsidence; thus two types of subsidence are recognized: tectonic, and that which results from withdrawal of fluids, either water or hydrocarbons.

Tectonic subsidence and emergence

Eustasy is either worldwide constancy or the lack of constancy of sea level. This is in contrast to the changes in sea level that are not worldwide. Eustatic change can be caused by removal of water to form icecaps, which lowers the sea level, or by the melting of icecaps, which raises the sea level. Sea level can also be lowered (negative eustatic change) by a lowering of the ocean floor or by the depression of a large geosyncline. Positive eustatic change, or a general rise in sea level, can also be

Figure 11-16

Uplift of the Scandinavian Shield as a result of glacial rebound. (© Andre Cailleux, 1968. Translation © George Weidenfeld & Nicolson Limited, 1968. Used with permission of McGraw-Hill Book Company.)

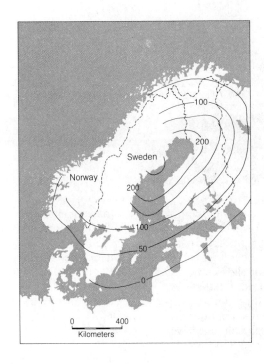

lowering of the ocean floor, also produce eustatic, or worldwide, changes, eustatic changes are included with tectonic subsidence, even though melting of ice and deposition of sediments may produce eustatic changes that are not tectonic. Since the cause of eustatic change cannot always be ascertained, all eustatic changes are discussed under tectonic subsidence or emergence.

As a result of isostatic adjustment, Iceland, England, Scotland, Northern Ireland, and the Scandinavian shield (Figure 11-16) have been rising since the last glaciation, and the North Sea basin, the Rhône basin, and the head of the Adriatic Sea have been sinking. Sometimes this combination of processes produces peculiar results. From Figure 11-17 it can be seen that the hinge line, the line of zero movement, cuts across Denmark. It has been shown that northern Denmark is rising slightly and that southern Denmark, particularly the southwest, is sinking. Part of this sinkage is tectonic, and part is only apparent, actually resulting from a rise in sea level caused by melting ice over the past several centuries. It is estimated that about two-thirds of southern Denmark's sinkage is tectonic, and that the other third is the result of eustatic change. Since the North Sea off Denmark receives no sediments from great rivers to build out the shoreline, the total effect is that the southern part of the western shore of Denmark is being eroded away. This is a serious problem for a country with so little land area.

North America is adjusting isostatically with the rebound of the Canadian shield after the melting of glacial ice, just as the Scandinavian shield is rebounding in Europe. The portion of North America last

caused by gradual erosion of land to base level, filling parts of the ocean basins and displacing water.

Isostasy is the state of balance between masses of the earth's crust. The surfaces of those parts of the earth's crust underlain by rocks of greater mass, such as the oceans, lie at lower elevations than the surfaces of those parts underlain by rocks of lesser mass. As the mass underlying an area changes by deposition or erosion, there is isostatic adjustment due to the slow transfer of rock deep in the earth's crust; this process balances the masses of the areas altered by erosion or deposition. Since some isostatic adjustments, such as the

Figure 11–17

Eustatic and tectonic subsidence in the vicinity of Denmark.

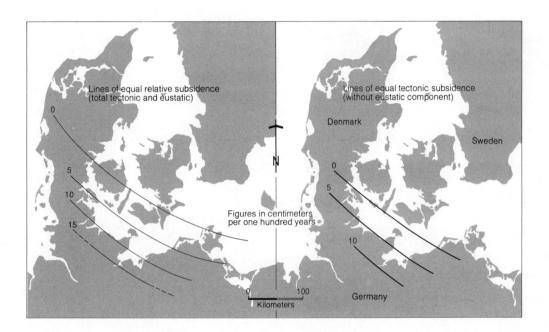

glaciated is rising as a result of the disappearance of the load of ice in the past 20,000 years (Figure 11-18). And there is evidence that the eastern and central part of the United States is tilting to the south: (1) late Pleistocene beaches are well above sea level along the shore of Labrador; (2) postglacial beaches that once represented the shore of Lake Michigan are tilting southward; and (3) the flow of the Illinois River has gradually reversed, since it flowed into Lake Michigan in the 1830s when Chicago was settled, and now flows southwest toward the Mississippi River. Figure 11-18 also shows the expected position of the shoreline along the Gulf and Atlantic Coasts of the United States if the remainder of the ice on Greenland and Antarctica were to melt. The Atlantic and Gulf Coasts are

undergoing a complex of (1) tilting as a result of isostatic adjustment accompanying rebound, (2) eustatic rise in sea level from melting ice, (3) isostatic adjustment with depositional loading (see pages 155–177), and (4) the fact that rate of deposition at some sites (e.g., the Mississippi delta) exceeds rate of sinkage. Thus some combinations of forces produce land whereas others destroy it. It is apparent that rebound accompanying isostatic adjustment prevents that part of the Atlantic shore nearest the Canadian shield from being drowned by eustatic rise in sea level. In glacial periods, because great quantities of seawater were taken up as ice, the Atlantic coastal plain extended much farther to sea than it does now.

Another example of uplift occurred in

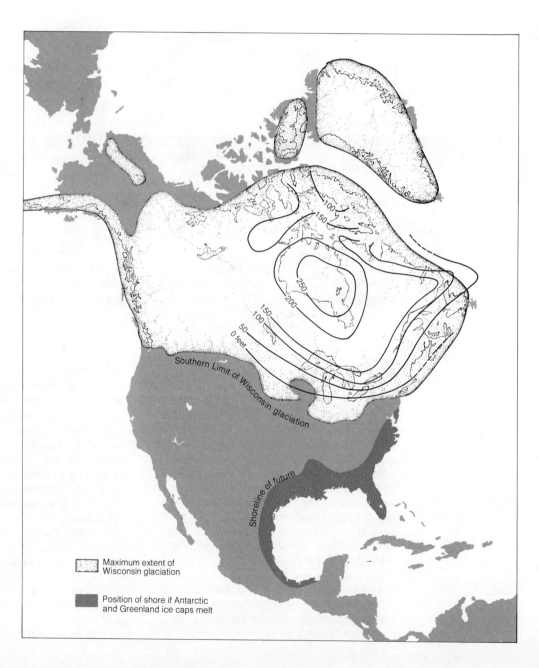

Figure 11-18

Maximum extent of Wisconsin glacial ice in North America, and glacial rebound since its retreat.

Southern Limit of Wisconsin glaciation

Shoreline of future

Maximum extent of Wisconsin glaciation

Position of shore if Antarctic and Greenland ice caps melt

the spring of 1970, when parts of the main street of the small city of Pozzuoli, on the north side of the Bay of Naples, rose as much as three meters within a month. Because the city is practically in the shadow of Vesuvius (see pages 260–264), it was expected that volcanic activity was about to occur. But it did not. The area of the Bay of Naples has experienced many relative changes of shore and sea level within historic times, and it is assumed that the tectonism that raised the streets of Pozzuoli is a phenomenon associated in some way with this volcanic province.

Tectonic changes in elevation similar to that at Pozzuoli have occurred in the earthquake zones of California. The California Department of Water Resources has developed one of the most intensive seismic monitoring systems in the world to determine as rapidly as possible the effects of tectonism on the California aqueduct. It has already been found that some areas are sinking and others are rising (see page 118). Some areas are sinking because of withdrawal of groundwater; others are sinking or rising because of tectonism along California fault systems.

Subsidence from withdrawal of liquids

Subsidence through withdrawal of liquids from the subsurface is quite common. Some areas are sinking because groundwater is withdrawn for irrigation, industrial uses, and municipal consumption; others because of the withdrawal of liquid hydrocarbons. Typical areas subsiding because groundwater is used for irrigation are Tulare and Kern Counties in California. In such areas there is a correlation between rate of subsidence and rate of withdrawal of liquids (Figure 11-19).

The Wilmington Field near Long Beach, California, is a good example of subsidence caused by removal of hydrocarbons. Subsidence in the port areas has been over 20 feet, and has been abetted by a series of faults along which movement allows the formations to adjust to the compaction of underlying sediments being deliquified. Subsidence can be prevented or lessened by pumping water into the oil horizon so that total liquid content remains constant.

In the greater Houston area, a complex pattern of subsidence has occurred because of withdrawal of groundwater for industrial and municipal uses, and withdrawal of hydrocarbons from oil fields. Certain parts of this area, such as the Port Arthur region, are subsiding because of the withdrawal of groundwater, whereas other parts, such as Spindletop and Goose Creek, increase this subsidence through hydrocarbon withdrawals. Closer to Houston there is more extreme subsidence along the Houston Ship Channel, a highly industrialized area between Houston and Galveston Bay that withdraws large amounts of groundwater.

Causes of Subsidence from Withdrawal of Liquids. Subsidence caused by withdrawal of liquids results from compaction of sediments; compaction, in turn, results from some combination of (1) reorientation of mineral or sediment grains, (2) deliquification, and (3) faulting. Extant faulting partly determines the sites of subsidence.

Underground liquids are usually confined by overlying and underlying impermeable layers. The liquid itself may hold these layers apart, so that the grains in the intervening aquifers or producing horizons

Figure 11-19

Subsidence in the Delano Area (Kern and Tulare Counties), California from 1902 to 1940.

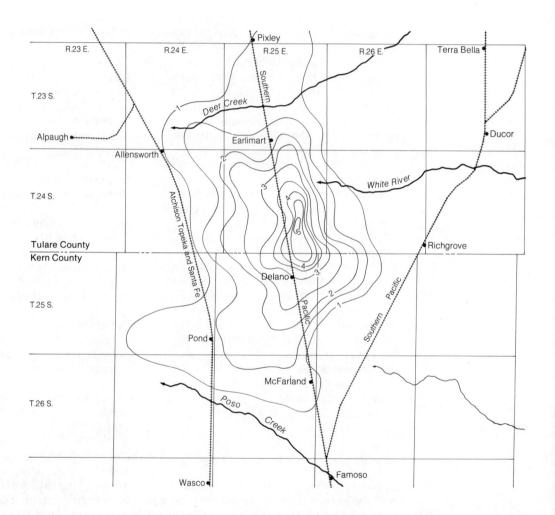

are not as closely compacted as they would otherwise be (Figure 11-20). With the removal of the liquid, the sand grains rotate into an arrangement that occupies less space, and the containing formation is not as thick as it was.

Clay particles in a formation, on the other hand, are usually compacted more by de-

liquification than by rearrangement of the particles (Figure 11-21). Although the particles can be rearranged, they seldom take up less space as a result. When it is deposited, clay usually contains water between its layers amounting to from 40 to 90 percent of its total bulk. Impermeable clay usually retains much of this water,

Figure 11-20

Compaction of poorly consolidated rocks after the removal of liquids. (From "Geological Subsidence" by Sullivan S. Marsden, Jr. and Stanley N. Davis. Copyright © 1967 by Scientific American, Inc. All rights reserved.)

Figure 11-21

Compaction of clay-rich beds after the removal of liquids. (From "Geological Subsidence" by Sullivan S. Marsden, Jr. and Stanley N. Davis. Copyright © 1967 by Scientific American, Inc. All rights reserved.)

and shrink-swell clays even expand to contain more water between their layers. If a nearby aquifer is deliquified by withdrawal, water in adjoining clay may be subjected to enough pressure to force it out of the clay and into the aquifer in which pressure has been reduced. As this water moves through the clay and eventually into the aquifer, the layers of the clay are pressed closer together by the load of overlying strata, and the formation becomes less thick (Figure 11-21).

Other kinds of subsidence

In Mines. In many mining areas from which large amounts of coal, salt, or ore have been removed, tremendous cavities have been gouged in the earth. Pillars are ordinarily left at regular intervals to support the roofs of such cavities. When this is not done, or when the pillars are too far apart, the ground surface may gradually subside into the cavern. This sinking of the earth is called "subsidence" if the areas are large, and "collapse" if they are small. Recent antisubsidence laws in Pennsylvania have forced coalmining companies to leave 50 percent of the finest coal in North America, the Pittsburgh Coal, in the ground as pillars. It would have been preferable to zone the surface above mined areas for recreation, agriculture, or forestry, so that damage from subsidence would have been minimal and the coal could have been used.

Limestone Sinks. A similar type of collapse or subsidence occurs in limestone terrain, where the roofs of natural caverns fall in and leave sinkholes. In Florida houses have collapsed into ancient caves, and in Alabama parts of an interstate highway built over a karst area (a limestone terrain with extensive cave development) have subsided. In some limestone areas a high water table strengthens the rocks by flotation. But when the water table is lowered the limestone is weakened and collapse is initiated. Light geophysical equipment is now being used in some areas to locate caverns, so that they can be avoided in construction.

Overloading. Other areas have subsided, though slightly, as the result of overloading and isostatic adjustment similar to that which accompanies the accumulation of ice on an area of continental glaciation. This type of overloading is frequently accompanied by swarms of microseisms.

The leveling data collected in studies of Lake Mead on the lower Colorado River show that it has subsided a few millimeters per year since the filling of the lake. This phenomenon has been attributed to subsidence due to overloading by water; from the standpoint of lowering the land, the amount of subsidence is too small to be serious. Lakes behind the Kariba and Aswan dams in Africa have produced similar swarms of microseisms.

As populations increase, urban areas create heavier loads, and more hydrocarbons and water are withdrawn from underground, subsidence will increase as an environmental problem.

Permafrost

Permafrost is usually defined as permanently frozen ground; the term is sometimes applied to all ground permanently below zero degrees Centigrade (that is, below freezing) regardless of moisture content. Some Americans first became acquainted with permafrost in World War Two, when metal linked runways were laid on the permafrost to create landing fields in the Arctic. In summer the metal transmitted enough heat from the sun to the permafrost to melt it and produce cavities. The linked metal bases of the runways sagged into these holes, and planes could not land safely. Since that time, as the Arctic areas have been developed, permafrost has become more of a problem and its study more important.

Distribution of permafrost

Permafrost is distributed across the northern part of North America and Eurasia (Figure 11-22). It is extensive in the deglaciated areas of Antarctica, and occurs in patches on Tierra del Fuego and the higher parts of the southern Andes. But most problems with permafrost occur in the northern hemisphere, where these areas of Alaska, Canada, and the U.S.S.R. are being developed. Russian specialists have developed the greatest number of permafrost areas in the Siberian Arctic. Permafrost areas are divided into two regions, a northern region in which the permafrost is continuous and a southern region in which it is discontinuous.

Engineering in permafrost areas

Special precautions must be taken in construction on permafrost. When it melts, permafrost usually becomes a soupy mud with no strength. A structure built on it gradually sinks into the mud and becomes useless. Thus structures must be designed to prevent the transfer of solar or other heat to the permafrost. When drilling oil wells in permafrost, a part of this problem is avoided by building the structure and drilling the hole in winter when there is little danger of melting. Highways must have a gravel base thick enough so that the permafrost will rise into the base and hold it solid (Figure 11-23). The road base should not be so thin that heat can be transmitted through the surface and base to melt the subbase and allow the highway to collapse. The same concept can be applied to the building of other structures. That is, there must be a firm base thick enough that the

Figure 11–22

Distribution of permanent
(dark) and discontinuous
(very dark) permafrost areas
in the northern hemisphere.

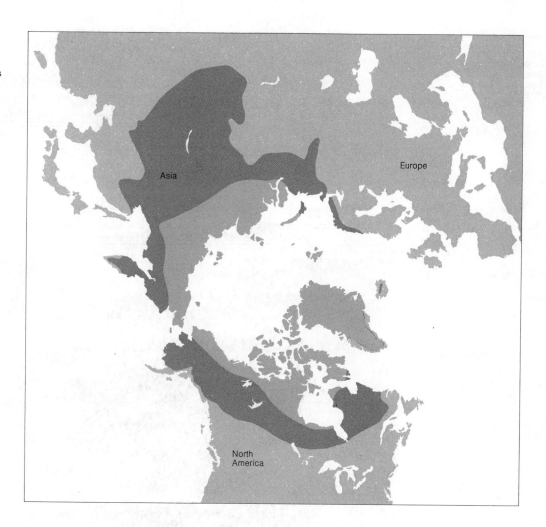

permafrost will rise into it and provide a
firm foundation.

To bury a heated pipeline in permafrost
would require a blanket beneath the pipeline much too thick to be economically
feasible. Otherwise the pipeline would melt
the permafrost and sink out of reach. For
this reason pipelines in areas of permafrost will have to be suspended in air.

It appears possible to transport hydrocarbons across the permafrost without
serious destruction of the environment if

Figure 11–23

(a) Properly constructed highway with thick base on permafrost, and (b) improperly constructed highway with thin base on permafrost.

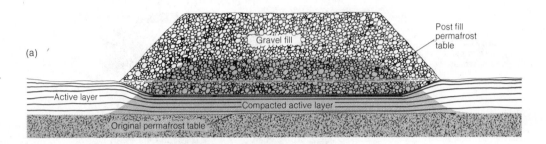

(a)

Gravel fill

Post fill permafrost table

Active layer

Compacted active layer

Original permafrost table

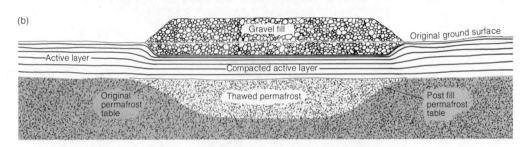

(b)

Gravel fill

Original ground surface

Active layer

Compacted active layer

Original permafrost table

Thawed permafrost

Post fill permafrost table

Landslide caused by permafrost meltout, Alaska

Water pipeline north of
Fairbanks, Alaska

blankets of sufficient depth are used under supporting highways, and if the pipelines are suspended. Little disruption of reindeer migration patterns was caused by similar development in northern Scandinavia; there should be no more disruption of caribou in the North American Arctic. The real danger to ecosystems in these areas will be caused not by pipelines but by the highways and power facilities—not to mention the people—that accompany the production of natural resources.

Summary: What Can Be Done?

Wherever landslides and subsidence occur, there are deaths, injuries, and economic loss. To counter both successfully, we need more geological research into the properties of rocks in different areas, and more geological mapping on a scale useful in planning. Further, the public needs to be educated to recognize unstable sites when development is being planned. It is important also that consumers and the environment be protected. Experience in Los Angeles has demonstrated that rigorous zoning laws can be designed to protect the consumer, the municipality, and the environment from the risk of landslides by

requiring engineering and geological reports on all development sites. More rigorous land-use planning ordinances can provide further protection.

We do not yet have satisfactory legislation for the control of subsidence and potential subsidence. Since we now have enough knowledge to help solve the problems of subsiding lands, the main bottleneck at present is legislative. Legislation should include proper zoning and protection to govern both the development of subsiding areas and the removal of liquids from the underground.

One question that needs to be answered is whether the cost of subsidence to major nearshore municipalities makes the expense of desalination plants preferable to continued subsidence due to the withdrawal of underground water.

Achieving a right and just balance between the need for resources and degradation of the ecosystem is difficult and delicate. As yet neither the environmentalists nor the resource developers have approached these problems in a thoroughly rational and dispassionate way. Until they do, legislation to achieve the most workable compromise is impossible—and so is the prospect of improvement.

Selected Readings

Cleaves, Arthur B. 1961. Landslide investigations, a field handbook for use in highway location and design. Washington: U.S. Department of Commerce Bureau of Public Roads.

Crandell, Dwight R. 1971. Postglacial lahars from Mount Rainier Volcano, Washington. U.S. Geological Survey Prof. Paper 677.

Ferrians, Oscar J., Jr.; Kachadoorian, Reuben; and Greene, Gordon W. 1969. Permafrost and related engineering problems in Alaska. U.S. Geological Survey Prof. Paper 678.

Kerr, Paul F. 1963. Quick clay. Scientific American 209 (5):132–143.

Kiersch, George A. 1965. Vaiont Reservoir disaster. Geotimes May-June 1965:9–12.

Köster, R. 1968. Postglacial sea-level changes in the

western Baltic region in relation to worldwide eustatic movements. In *Means of Correlation of Quaternary Successions,* Proceedings of the 7th Congress of the International Association for Quaternary Research, Roger G. Morrison and Herbert E. Wright, Jr., eds., pp. 407–419. Salt Lake City: University of Utah Press.

Lofgren, B. E., and *Klausing, R. L.* 1969. Land subsidence due to groundwater withdrawal, Tulare-Wasco area, California. U.S. Geological Survey Professional Paper 437-B.

Marsden, Sullivan S., Jr., and *Davis, Stanley N.* 1967. Geological subsidence. *Scientific American* 216(6): 93–100.

Morton, Douglas M., and *Streitz, Robert.* 1967a. Landslides, parts I and II. *California Division of Mines and Geology Mineral Information Service* 20(10):123–129; 20(11):135–140.

Shreve, Ronald L. 1966. Sherman landslide, Alaska. *Science* 154:1639–1643.

———. 1968. The Blackhawk landslide. Geological Society of America Special Paper 108.

Slosson, James E. 1970. The role of engineering geology in urban planning. *Colorado Governor's Conference on Environmental Geology,* ed. John B. Ivey, *et al.,* pp. 8–15. Colorado Geological Survey spec. publ. no. 1.

Zolotarev, G. S. 1961. Engineering-geological study of shore slopes and the significance of the history of their development in evaluating stability. In *The Stability of Slopes,* vol. 35, L. V. Popov and F. V. Kotlov, eds., pp. 9–29. Trans. Consultants Bureau, New York.

And here we find that the Flood that destroyed this land of the gods was the Flood of Deucalion, and the Flood of Deucalion was the Flood of the Bible, and this, as we have shown was the last great Deluge of all, according to the Egyptians, which destroyed Atlantis.
Donnelly, 1898

Floods have been the bane of mankind since man first settled and farmed near streams in order to irrigate his crops. Great floods are recorded throughout history. One such was Noah's flood described in the Old Testament. This flood, or another at about the same time and place, has been authenticated by archaeologists as having occurred in the valleys of the Tigris and Euphrates Rivers (Figure 12-1). It covered thousands of square miles, if not the "entire world." There was a tremendous flood in the Pleistocene in the Snake River plain of Idaho, and an ancient flood of a magnitude not since experienced in that area occurred in California. More recently, floods of great destructive violence occurred at Florence, Italy, in 1966 and at Genoa, Italy, in 1970. At Florence several hundred people and thousands of animals were killed, 1500 miles of streets and highways were submerged, and parts of the city were immersed in debris to above the height of an automobile. Invaluable artistic and historic treasures were destroyed. In Genoa, a flood almost as severe, but with little natural sediments, piled up automobiles in the streets until they produced an "automobile conglomerate."

Some Phenomena Related to Flooding

The frequency with which a river floods depends on the size of the floodplain, the hardness of the bedrock, the amount of rainfall, and the length of time between torrential rainfalls.

Type of climate

Some areas, such as Mexicaltitlan on the Rio San Pedro in Sinaloa, Mexico, are

Figure 12-1

Part of the Tigris and Euphrates River Valleys, showing the extent of a flood that may have been from the time of Noah.

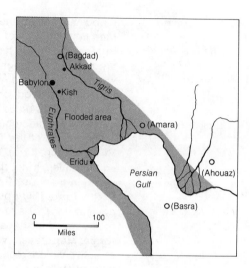

100,000 cubic feet per second (cfs) covers it completely. In contrast, the Pecos River below Pandale, Texas, has a fairly narrow channel cut in hard limestone, and its floodplain is completely covered with a flood of perhaps 25,000 cfs. Yet because of periodic high rainfall from hurricanes, the Pecos has floods of 100,000 cfs more often than does the South Platte.

Severity and distribution of rainfall

Areas frequently hit by hurricanes suffer disastrous floods more often than areas characterized by less violent storms. Some areas are flooded annually in the spring when the accumulated snows of winter melt. Most areas of high rainfall are flooded more often than areas of low rainfall. For example, the tradewind areas are flooded every year in the monsoon season. There are a few regions in which warm, moist air rises and is cooled suddenly; exceptionally large amounts of rain result.

Streamflow

A stream fills only about 10 percent of its channel about one-half of the time (Figure 12-2a). At average flow somewhere between 10 and 50 percent of the channel is filled, depending on the climate. Streams overflow their banks when precipitation is higher than normal. This may occur once every two years, twice a year, or several times a year, also depending on climate. In most streams moderate flooding occurs about every ten years (Figure 12-2d). A peak flood is any flood that covers the entire floodplain, and such extensive flooding occurs on the average about every 50 years

flooded part of every year. In this small village the streets are used as canals and transportation is by boat in the months of flooding, whereas in the dry months the streets are open to bicycles and automobiles. In an equitable oceanic climate, such as that of the Meuse River in Belgium, rainfall is consistent, gradual, and not in great amounts. Because there is little flooding, there are many intact old houses almost at the water's edge.

Hardness of bedrock

Rivers cut in hard rock usually have narrower channels than rivers flowing over soft rock. Thus a river in an area of hard rock has a narrower floodplain than one of same flood magnitude in a soft rock area. The floodplain of the South Platte River at Denver is fairly wide, and a flood of about

Figure 12-2

Classification of flow in streams: (a) low flow, (b) average flow, (c) bank full, (d) moderate flooding, and (e) peak flooding.

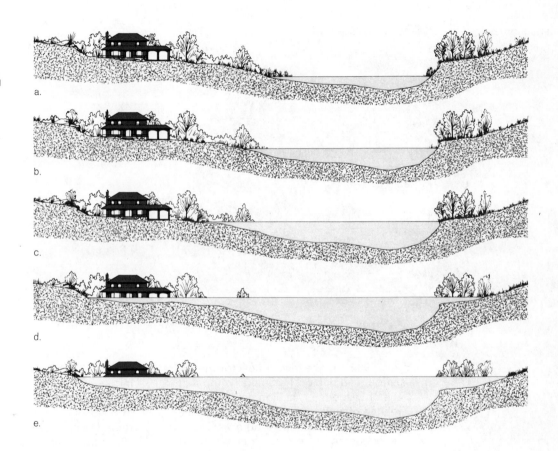

a.

b.

c.

d.

e.

(Figure 12-2e). Thus floods are normal events in the lives of rivers.

Planning for Water Storage

Since the average flow of a river can be calculated over a certain length of time, planning for the storage of floodwater to be used in times of drought is often based on average flow. Yet a hydrograph of the

Colorado River of the West, at Lee's Ferry, Arizona (Figure 12-3), demonstrates that variations are so extreme that there is no such thing as a dependable average flow, at least in areas of continental climate. Over the 70 years from 1895 to 1965 shown in this record, there were only nine in which the flow approached the average.

Planning should therefore be based on maximum and minimum average flows. The

Figure 12–3

Hydrograph of the Colorado River at Lee's Ferry, Arizona.

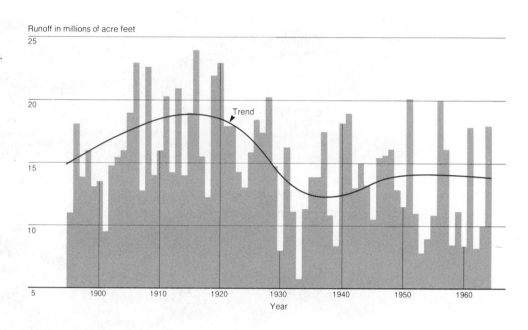

Runoff in millions of acre feet

Year

year 1933 had the lowest flow for the entire period, and demonstrates that without storage only about 5½ million acre feet (the volume of water that will cover 5½ million acres to a depth of one foot) a year can be depended on; in years of greater flow, such as the wet period from 1910 to 1927, water not immediately used was simply allowed to run off. By 1960 there was less waste of water in the Colorado River Basin, because large reservoirs had been constructed in the watershed to store excess water.

Flood Recurrence

Engineers use the term recurrence interval to describe the average length of time between floods of a particular magnitude.

Figure 12-4 is a flood frequency curve showing the recurrence interval for floods of given magnitudes on the Navasota River at Navosota, Texas. The greatest flood recorded on the Navasota River occurred in 1899, but there have been many small floods. A 20,000 cfs flood occurs about every three years, a 40,000 cfs flood about every ten years, a 60,000 cfs flood about every 20 to 40 years, and an 80,000 cfs flood about every 100 years. Figures above the 40,000 cfs flood level are not very accurate, because there are too few large floods to be significant. On this particular curve, the 1899 flood would normally be called the 100-year flood, although if accurately plotted such floods come closer to having a frequency of 125 years. Thus, using curves

Erosion and deposition
from flooding have left
their mark on this now dry
river bed

to illustrate the recurrence of flooding,
engineers can designate the 100-year flood,
the 50-year flood, the 20-year flood, and
so forth, depending on how often a flood
of a given magnitude occurs on a particular
river. It should be emphasized that floods
are labelled on the basis of average interval
between occurrences. For example, 20-year
floods can occur in each of three successive years, and not again for 40 years, and
still be called 20-year floods, for if averaged
out over 40 years, a flood of that magnitude
occurs every 20 years.

Exceptions to the pattern

But flooding on some rivers does not corre-
spond to a curve on which magnitude can be
correlated with frequency. For example, the
Pecos River in Texas (Figure 12-5) has a
very eccentric flood curve. In the 50 years
from 1905 to 1955, there were about 10

Figure 12-4

Flood recurrence curve for
the Navasota River. The
dots represent floods of
known magnitude.

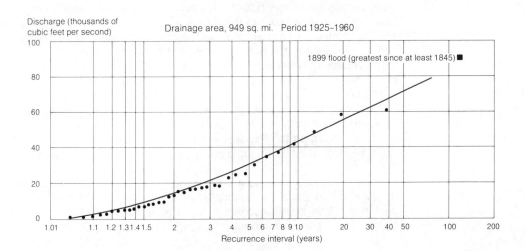

Figure 12-5

Yearly peak floods on the Pecos River, Texas.

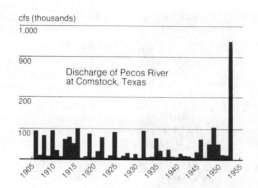

floods that ranged between 80,000 and 120,000 cubic feet per second. Such floods appear to occur about once every five years; by the usual scheme each would be called a 5-year flood. But it is impossible to arrange these events in typical frequency curves because there are almost as many "5-year" floods as "2-year" floods. Indeed, there are more 100,000 cfs floods than 80,000 cfs floods, and twice as many 100,000 cfs floods as 50,000 cfs floods. This condition exists because most of the floods on the Pecos result from hurricane outfall. Indeed, in 1953 the Pecos flowed about 960,000 cfs, a volume of water equivalent to mid-flood stage on the Mississippi River at Baton Rouge, Louisiana.

But if one defines a peak flood as any flood that completely fills the floodplain, the 5-year flood on the Pecos River is a peak flood because the floodplain is so narrow. By contrast, the floodplain of the South Platte River at Denver is so wide that only the 100-year flood can be described as a peak flood. For these reasons much of the terminology used in discussing flooding and floodplains is confusing to the average citizen.

Flood recurrence has been studied primarily in areas where the floodplain is covered about once each century. In planning greenbelts for streams that do not behave according to this pattern, the floodplain has sometimes been defined as the 100-year floodplain, even though it is covered, for example, on an average of every 20 years. Thus, quite unintentionally, the magnitude and frequency of floods have been misrepresented, and planners, citizens, and public officials in charge of plans and expenditures for flood control projects have been confused. In planning for floods each municipality should calculate flood frequency curves for its own streams, because generalized flood frequencies cannot be safely applied to specific rivers.

Floodplains and Natural Levees

Many large streams build their own natural levees, which tend to confine the stream to the channel during normal flow. Natural levees are formed as part of the overbank deposits, which include levees and floodplain sediments. When, during flooding, water flows out of the channel and onto the bank, it suddenly loses velocity, and just as suddenly loses much of its ability to carry sediment. Most of the sediment, including all of the coarse material, is dropped along the edges of the channel. Only the very fine sediment, which can be carried by water of lower velocity, is spread across the floodplain. For this reason, the areas along the banks of the channel receive more sediment and grow more rapidly, producing the *nat-*

Figure 12-6

Floodplain, natural levees, and channel of a stream.

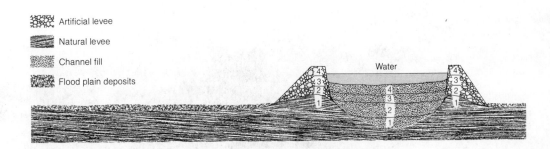

    Artificial levee

    Natural levee

    Channel fill

    Flood plain deposits

ural levees (stage 1 of Figure 12-6). The stream also deposits bed load in its channel, causing the bottom of the channel to slowly rise in elevation. When the stream bed approaches the elevation of the floodplain, floodwater may break through a natural levee; if the flood is sustained long enough, the stream may start a new channel on the floodplain.

People do not welcome streams that flow across floodplains and then their farms, houses, roads, and towns; so they construct artificial levees (stages 2-4 in Figure 12-6) on top of the natural levees. As the stream's channel rises ever higher above the floodplain, artificial levees are built correspondingly higher. The total effect is to make even more hazardous and destructive that severe flood that inevitably occurs and cannot be contained. Neither natural nor artificial levees contained the disastrous flooding from hurricane Agnes in 1972, or from the Mississippi River in April and May 1973, when flooding was the most severe in over 200 years.

Some artificial levees have caused flooding to be more severe than under natural conditions, because water that had broken out of the channel was prevented from returning to it. Such malfunctions of both artificial and natural levees suggest that the only certain way to prevent damage by floods is to avoid building on floodplains.

Floods and Urbanization

In many areas numerous accidents occur in the first few minutes of rainfall, due to the lack of nonskid pavement, and inexperienced driving. Many people need to see an accident or two before they become cautious enough to drive safely in a rainstorm. Moreover, newly wet asphalt suddenly becomes slippery as grease, and flotables are mobilized by the water. If the rain continues long enough, the flotables and grease are washed into gutters and the pavement becomes less slippery again.

Effects of paving and channelizing

It takes very little rain to start water running on streets. In a developed area, streets, sidewalks, parking areas, and the roofs of

Flooding in Missouri

buildings provide large areas of impermeability, and there is little vegetation to impede the flow of water. These factors prevent the seepage of water into the underground and increase the runoff. Since there is little vegetation and little soil, water runs off rapidly. The total effect of the increase in impermeable surface area with urban development, then, is to increase the total runoff and shorten the time of runoff for a given volume of water—and in these ways to increase the danger of flooding. Furthermore,

in most development, stream channels are straightened, causing water to be carried downstream much faster. This phenomenon, called channelization, is usually accompanied by the installation of storm drains and sometimes of concrete stream bottoms and banks. If channels are narrowed, water flows faster because there is less of a bottom to create friction and slow the flow. Also, channelization and storm drain installation cause water to flow a shorter distance, because meanders are removed,

Figure 12–7

Flood hydrographs for Brays Bayou, Houston, Texas before and after development.

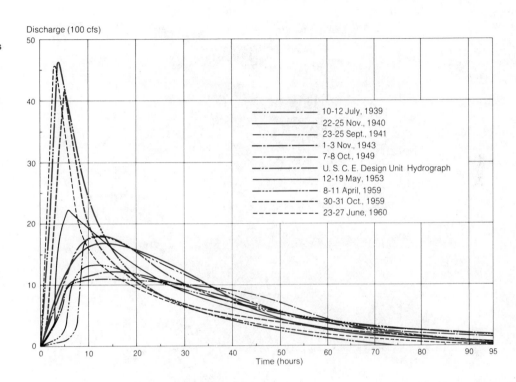

Discharge (100 cfs)

—·—·—·—·—·— 10-12 July, 1939
————————— 22-25 Nov., 1940
—·—·—·—·—·— 23-25 Sept., 1941
—··—··—··—·· 1-3 Nov., 1943
—·—·—·—·—·— 7-8 Oct., 1949
—··—··—··—·· U. S. C. E. Design Unit Hydrograph
————————— 12-19 May, 1953
—·—·—·—·—·— 8-11 April, 1959
— — — — — — 30-31 Oct., 1959
- - - - - - - 23-27 June, 1960

Time (hours)

and more rapidly, because there is less vegetation to hold back the flow. Artificial changes increase the danger of flooding; if there are constrictions in the channel, such as culverts or narrow bridges, or obstructions that almost dam the channel, local flooding is even more likely.

Storm hydrographs

Figure 12-7—hydrographs of Brays Bayou

in Houston, Texas, before, during, and after urban development—shows the change in flood runoff over about 20 years. For a given rate of rainfall, the peak discharge is over three times as great at the end of the period as at the beginning. From such studies predictions can be made for other communities with similar climates under specified conditions of development. Data from other areas indicate that storm runoff after development may increase as much as five times.

Figure 12-8

Flood map for Joes Creek and Bachman Branch, Dallas, Texas, 1962 and 1964.

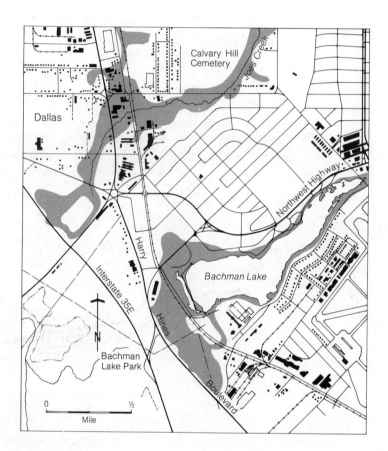

Flood maps

Flood maps now prepared by the Army Corps of Engineers and the U.S. Geological Survey Water Supply Branch shed further light on urban flooding. Maps like those for Joe's Creek and Bachman Branch in Dallas, Texas (Figure 12-8) are especially useful when a municipality is planning greenbelts, though future development must also be considered. As development becomes denser, and particularly as it moves upstream, the magnitude of the peak flood increases and more of the floodplain is covered by a given amount of precipitation than on natural terrains. Figure 12-9 is also a flood map.

Figure 12-9

Flood map for the 1935 flood on the Colorado River at Austin, Texas.

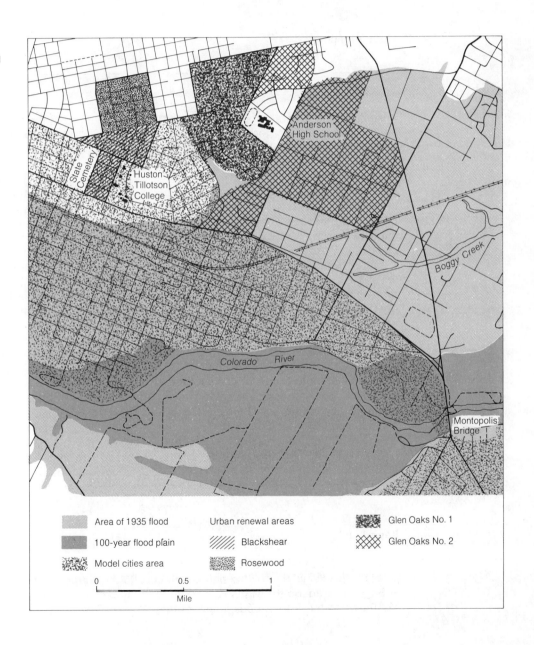

State Cemetery

Huston-Tillotson College

Anderson High School

Boggy Creek

Colorado River

Montopolis Bridge

	Area of 1935 flood		Urban renewal areas		Glen Oaks No. 1
	100-year flood plain		Blackshear		Glen Oaks No. 2
	Model cities area		Rosewood		

0   0.5   1

Mile

Flood Planning

Austin, Texas

Austin's urban renewal program is divided into five areas. Four of these—Blackshear, Brackenridge, Glen Oaks No. 1, and Glen Oaks No. 2—are near the Colorado River floodplain. Part of the Brackenridge tract near Waller Creek was underwater in the 1869, 1935, and 1937 floods. Blackshear and Glen Oaks No. 1 are not in the area covered by peak flood, but nearly all of Glen Oaks No. 2 was submerged by the three floods mentioned above. A director of the urban renewal program expected the engineers to remedy all the potential flooding problems before redevelopment actually began. The only structure now on the floodplain that was designed to withstand flooding is St. Joseph's Catholic Elementary School, which was built on pillars. Intended to protect it against the periodic floods of a nearby tributary, this planning may have proofed the building against most river floods as well.

Figure 12-9 also shows most of the area of the Austin Model Cities program, nearly all of which is within that area flooded three times in the first hundred years of record. This suggests that Glen Oaks No. 2 and much of the Model Cities area should either be left as greenbelt or undergo special flood planning if extensive flood damage is to be avoided. At present neither possibility is being considered. The danger of flooding in Austin will be better appreciated if it is pointed out that reservoirs above Austin on the Colorado River are either at constant level or are used for irrigation storage. There is little planned flood control.

This part of Texas is thought to be the area of most damaging rainfall in the country. It lies along the Balcones escarpment, a topographically elevated area that causes northward-flowing, moisture-laden, warm air from the Gulf of Mexico to rise and become suddenly cooled. The result is the most catastrophic precipitation events in the United States. The Balcones escarpment lies between the major cities and the dams that have been erected for a modest flood control program. Austin is one of the cities that lies within this area.

Baghdad, Iraq

Simple flood planning that alleviates major problems is exemplified by the plan in effect at Baghdad, Iraq, on the Tigris River (Figure 12-10). When the Tigris floods, much of the water can be diverted through canals to holding reservoirs or even into the Euphrates, thus diminishing flooding in the city. If the water is needed downstream, much of it can be diverted back to the Tigris through canals below the city. There are also diver-

Figure 12-10

Flood planning for the city of Baghdad, Iraq.

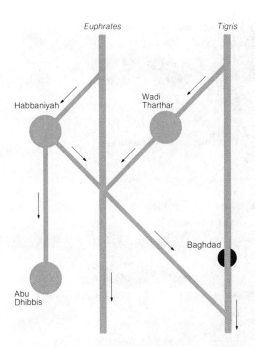

Euphrates

Tigris

Habbaniyah

Wadi Tharthar

Baghdad

Abu Dhibbis

artificial lake want a constant-level reservoir so that they will not have to worry about adjusting their boat docks to changing water levels. Cities downstream want a large storage pool for flood control, so that they will not need to undertake extensive flood planning or to pay out large sums for flood damage and relief. Farmers downstream prefer a large storage pool for irrigation, which produces high levels in lakes at the beginning of the irrigation season and low levels at the end, and leaves downstream cities without flood control during the early part of the growing season. Most reservoirs turn out to be single-, or at best dual-purpose, reservoirs.

On the upper Mississippi, particularly in Minnesota and Wisconsin, all reservoirs are at constant-level, primarily to maintain a water depth appropriate for barge traffic. Reservoirs above Prairie-du-Chien, Wisconsin, have had no effect on the flood problems of Prairie-du-Chien, LaCrosse, and other cities, because they are constant-level reservoirs. Despite all the reservoirs on the upper Mississippi, flood damage is more severe every year.

More flood-control reservoirs are needed in the United States, but it will be costly to build them on the upper Mississippi and other rivers where single-mission and constant-level reservoirs already exist. Under natural conditions deltas sink at approximately the same rate as sediment is added by rivers. Each dam on a stream prevents sediment from reaching a delta. When there are many dams on a stream, as on the Mississippi River, the delta receives so little sediment that it is gradually inundated by ocean. New reservoirs

sion canals and holding reservoirs on the Euphrates that can control some of the flooding downstream.

Flood control reservoirs

Multiple-use reservoir planning can provide for flood control. There are generally three approaches to multiple-use reservoirs (see Figure 6-16), since in addition to the sediment pool there are usually conservation, storage, and flood pools. Compromising among these uses is one of the major political problems of river authorities. People living on or above the easement of an

Mississippi River flood,
Illinois, 1965

should be designed to bypass silt or to be periodically cleaned (see Chapter 6).

The Tennessee Valley Authority

The Tennessee Valley Authority was established in the early 1930s as a government corporation to develop the Tennessee Valley and to operate certain government installations. The TVA is involved in power development, water resource development, fertilizer research and manufacture, and flood control. The TVA collects excess water in approximately 27 reservoirs during the rainy and melting seasons, shortly after the beginning of the year, and releases the excess in the dry parts of the year. The

Damage from flooding— a washed-out bridge

Tennessee Valley is not a hurricane area, but if a rare storm such as hurricane Agnes reaches the valley, flooding would not be completely controlled. At low reservoir level the TVA system has about 12 million acre feet of storage. A storm like Agnes would dump 26 million acre feet, and if the ground were already saturated the TVA's reservoir system could not prevent flooding; but the flooding would not be as serious as that on the Susquehanna River in June 1972 (pages 322–325). Furthermore, the TVA has no authority to zone the floodplains. Thus people settle on the lower part of the floodplain where flooding is caused by local rainfall before it reaches the storage system. But the TVA has saved Chattanooga several times from possible flooding due to abnormal rainfall and melting snow.

In retrospect, one probably cannot fault the TVA on flood control. It has provided enough flood control to earn its keep. More to be criticized is the official attitude, or the implicit faith of the public, that leads people to believe that protection is possible from all kinds and conditions of flooding.

Floodplain control

Planning for floods can go further than the construction of reservoirs or the maintenance of greenbelts. The safest of all flood plans is to zone a floodplain for farming or greenbelt only, ensuring that masses of people are not located there. If a floodplain and nearby areas are to be used by a municipality, it is possible, for example, to locate parking areas on the floodplain and adjacent buildings just off it (Figure 12-11). Another solution is to build artificial rises on the floodplain, construct buildings on

Figure 12–11

Varieties of flood planning:
(a) Development on natural topographic highs; (b) development on artificial topographic highs; and (c) buildings supported above flood level on the floodplain.

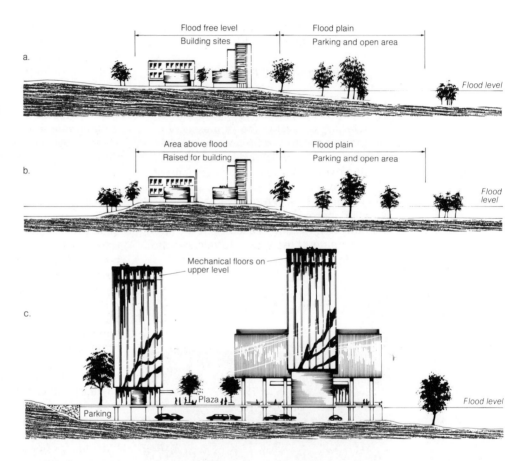

these, and use adjacent lower areas for parking or greenbelts. It is also possible to construct buildings on pillars, so that parking and open areas are at ground level but buildings occupied by people are above flood level.

It is also necessary to build pillars and bridges strong enough to withstand the pounding of flood-borne debris, such as branches, logs, boats, and even houses and house trailers. On the floodplain of the

South Platte River at Denver in 1965, there were a number of trailer parks. These were covered with water to such a depth that campers and house trailers floated downstream, lodged against bridges, and built dams that spread water even farther across the floodplain. If buildings on stilts or pillars had been located on the floodplain, such debris could have damaged the pillars and probably collapsed the buildings. Most planners have not considered the damage

Figure 12-12

Stream flow map for the United States, November 1971.

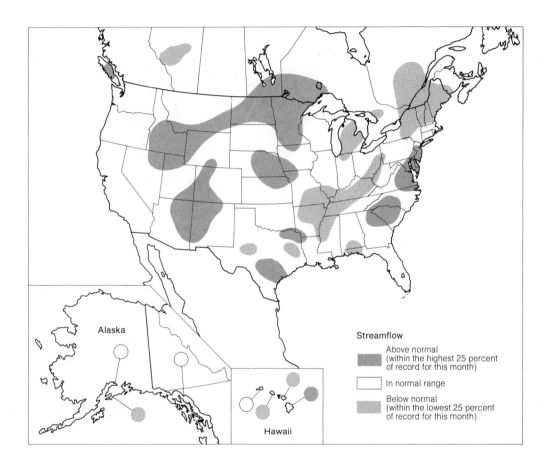

that can be done by debris that floats or saltates downstream and creates new hazards at flood time.

Flood Prediction

Once a month the Rivers Branch of the Canadian Geological Survey and the Hydrology Branch of the U.S. Geological Survey jointly publish streamflow maps (Figure 12-12). In winter they record snow accumu-

lation. Much flooding in the northern states results from rapid melting of snow, and a comparison of past streamflow maps and snowfall records can determine the contribution of snow to spring flooding. From such data flood prediction maps can be prepared (Figure 12-13) to warn citizens of potential flooding.

Prediction of flooding from tropical storms is not always accurate. In the summer of 1970, as a hurricane approached the

Figure 12-13

Flood prediction map for the United States, March 1969.

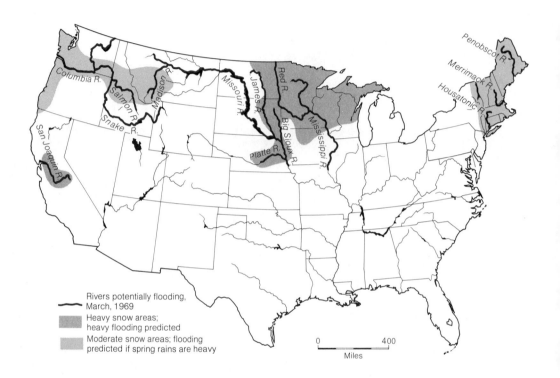

Rivers potentially flooding, March, 1969

Heavy snow areas; heavy flooding predicted

Moderate snow areas; flooding predicted if spring rains are heavy

0   400

Miles

Gulf Coast of Texas, it was predicted that there would be 10 or more inches of rain around San Antonio. San Antonio authorities, with an excess of untreated sewage, released a large dump of untreated municipal effluent into the San Antonio River, expecting it to be diluted by the flood. But the storm turned out to be a "dry" hurricane; it only sprinkled in San Antonio, and the river flowed raw undiluted sewage. The city was criticized by the Federal Water Quality Board and the Texas Water Quality Board, and felt the wrath of downstream neighbors. Prediction cannot always be accurate.

Flooding from hurricanes

Most of the floods along the Gulf of Mexico are caused by hurricanes, and extensive damage has given rise to attempts at control. Deaths from hurricanes (Figure 12-14) in the United States have steadily declined over the past few decades as the result of better communication and warning systems. But as urbanization has increased and more and more people have crowded the seashores, property damage has steadily increased during the same period. One proposal is to seed hurricanes with silver iodide or some other nucleating agent, to

Figure 12–14

Hurricane deaths and damage in the United States from 1915 to 1969.

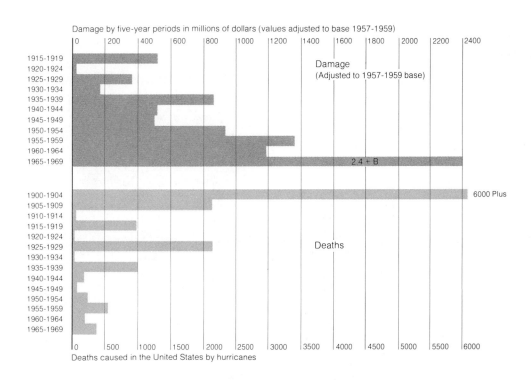

Damage by five-year periods in millions of dollars (values adjusted to base 1957-1959)

Damage
(Adjusted to 1957-1959 base)

2.4 + B

6000 Plus

Deaths

Deaths caused in the United States by hurricanes

decrease the energy of the wind and the amount of precipitation. Some attempts at seeding have been successful, but more research is necessary before seeding can be widely used.

To prevent hurricane damage the Army Corps of Engineers has proposed a great seawall along the Gulf Coast of Texas so that marine water cannot inundate the land. This would not prevent flooding from high rainfall on land, and it might even create bores of marine water in bays and estuaries when storm surges rush through restricted inlets (Figure 12-15). It's debatable if a seawall costs more than storm damage.

All areas should look cautiously at hurricane prevention plans before accepting them. Inland states along the Gulf of Mexico would receive an average of five inches less rainfall a year if hurricanes were prevented. Nearshore states would receive at least ten inches less. It is questionable whether Texas could economically withstand such a drop in annual precipitation, as the desert would move east of the Trinity River. An average of seven inches a year less surface water would enter the Edwards aquifer. This would be a very serious problem at San Antonio, where the limestone aquifer is replenished rapidly

Figure 12-15

Areas inundated by storm surge tides in Texas.

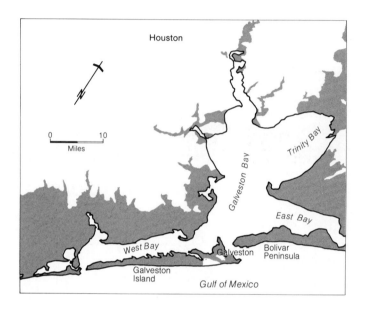

Hurricane Betsy, Florida, 1965

	Site	Dead and missing	Homeless	Damage (millions of dollars)	Damage per person
Table 12-1 Estimated flood damage, June 1972. Data from *The Floods of June '72.* New York: General Adjustment Bureau, Inc., 1972.	Northeast Appalachian states	132	thousands	1690	
	Rapid City, South Dakota	700†	more than 400	120‡	more than $2700
	Scottsdale, Arizona			more than 30	more than $2000
	New Braunfels, Texas	53	5000 evacuated	10	$ 500
	San Marcos, Texas*			2	$ 100

*San Marcos, Texas, had been flooded two years earlier. The area of the 1970 flood had never been rebuilt, and new greenbelt ordinances will prevent rebuilding. For this reason damage was minimal in the 1972 flood.
†Includes estimates of tourists and others on Rapid Creek who were not actually within the city limits.
‡Does not include outlying areas of damage.

after each drought, but might not be if there were less rainfall (see Chapter 5).

The gradual diversion of water from the Florida Everglades to urban areas along the east coast of Florida has caused the Everglades to become drier. Without the rain that comes with hurricanes the Everglades might have been destroyed before now. Tropical storms are a natural part of ecosystems, and the removal of other water for urban use has made them even more important to the Everglades' survival.

If hurricanes were successfully prevented, the economic loss in water might equal or even exceed the present cost in property damage. Yet if it were possible to abate the fury of the storms without losing the water they bring, hurricane prevention might be truly beneficial.

The Floods of 1972

The year 1972 was notable for disastrous floods in the United States. Buffalo Creek, West Virginia, was heavily flooded in February; New Braunfels, Texas, in May; Rapid City, South Dakota, early in June; and later in the same month eight eastern states from Florida to New York were ravaged by floods that accompanied the erratic and unusually destructive hurricane Agnes. Disastrous floods also occurred at Isleton, California; Scottsdale and Phoenix, Arizona; and San Marcos, Texas. Serious flooding continued into 1973, caused by rainfall and melting snow during the spring months. Figures indicating the extent and effect of some 1972 disasters are given in Table 12-1.

The 1972 floods fell into two categories. The first group, including those in Arizona and Texas and the floods associated with hurricane Agnes, were natural events. They could not have been predicted or prevented, and they were too great to be controlled. The only sure way to avoid the damage they caused would have been to stay off the floodplains—and this means floodplains of greater area than those of a 100-year flood. Some of the floods produced by hurricane Agnes were rated as being from 400-year to 800-year magnitude! The second group of floods in 1972 included natural events of only moderate magnitude whose effects were greatly augmented by the failure of manmade structures. At Isleton, California, a dike failed, and at Rapid City, South Dakota, and Buffalo Creek, West

Figure 12-16

Index map of Buffalo Creek Valley, Saunders, and Middle Fork, West Virginia.

Virginia, dams failed. We shall examine some of the floods in both categories.

Buffalo Creek, West Virginia

The Buffalo Creek Valley was opened to coal mining shortly before World War One, when the Chesapeake and Ohio Railroad constructed a spur line up the valley. Shortly after World War Two mines were opened along Middle Fork, a tributary to Buffalo Creek above Saunders, West Virginia (Figure 12-16).

The Buffalo Creek Valley is sinuous, curving around prominent spurs of hard bedrock. Its winding course impedes the flow of runoff from floods, because at each curve the valley narrows. The narrow floodplain, averaging only 400 feet in width, has steep valley walls, and is almost filled with farmhouses, villages, roads, and the railroad, all of which leave little space for floodwater. The soil is thin and mostly impermeable.

By February 26, 1972, rain had been falling for three days. The resulting 3.7 inches of rain had flooded streams at about the 10-year level. Then, shortly before 8 A.M. on February 26, the retaining dam above Saunders collapsed, and 132 million gallons of water were dumped into a tributary that drops 250 feet in about one-eighth of a mile. The town of Saunders was completely destroyed, and all or parts of 16 other mining villages, including Man at the junction of Buffalo Creek and the Guyandotte River 14 miles downstream, were also destroyed. One hundred eighteen people were killed, 500 houses were demolished, and damage exceeded $50 million.

One hundred thirty-two million gallons is not a great deal of water—less than 400 acre feet. The damage resulted less from quantity than from the sudden and complete collapse of the dam, which produced a 10- to 20-foot wall of water that rushed down the steep and narrow floodplain.

The history that precedes this flood does not excuse it but explains it. Mining practice for decades has involved the washing of newly mined coal with water from adjacent streams. The water was then returned

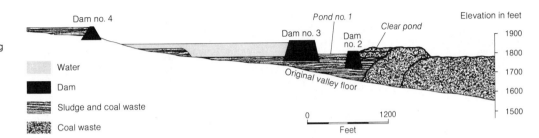

Figure 12-17

Cross section of dams, sludge deposits, and holding ponds along Middle Fork above Saunders, West Virginia.

Dam no. 4

Pond no. 1

Dam no. 3

Dam no. 2

Clear pond

Elevation in feet

1900

1800

1700

1600

1500

Original valley floor

0   1200

Feet

Water

Dam

Sludge and coal waste

Coal waste

to the streams. In the early 1950s an act of the West Virginia Legislature required coal companies to construct clearing ponds in which the undesirable coal dust and clay settled out as sludge before wash water was allowed to return to the streams. The clearing ponds were shallow and caused no problems. As new mines were opened along Middle Fork, new clearing ponds, using coal slag and waste for dams, were built on the sludge of older clearing ponds. After a disastrous mudslide from a large coal tailing pile (bone pile) in Wales in 1966, bone piles above Saunders were inspected by personnel of the U.S. Bureau of Mines. Their report stated that there was no danger of slides from the bone piles, but that the holding ponds were dangerous because the dams were constructed on unstable sludge.

After 1966, as mining increased above Saunders, the companies began to conserve water by retaining it behind dam Number 3 (Figure 12-17) to recycle it for washing. For the first time water accumulated to a depth of more than a few feet. Three days of rain had brought the water level to within one foot of overflowing by the night before the collapse of the dam at holding pond Number 3. There was no spillway. Only after the collapse of the dam was

it discovered that there was an overflow pipe, 24 inches in diameter, which was inoperative because it was filled with sediment.

The lesson learned—the hard way—from the Buffalo Creek flood is that dams for holding ponds should be correctly engineered and not constructed haphazardly; they should have firm foundations and proper spillways. The collapse of the dam on the tributary to Buffalo Creek turned a mild 10-year flood into a flood over 40 times that magnitude.

Water in the night—New Braunfels, Texas

Thursday, May 11, 1972, was a dark night in New Braunfels (Figure 12-18). The moon was on the other side of the earth. It began raining about 9:30 in the evening, but quietly, not with the gusty, boisterous violence that had been predicted. People were lulled to sleep by the quiet patter of raindrops, only to be awakened before midnight by water in their houses, seeping under doors, oozing around windows, or cascading to the floor in small waterfalls from the window air conditioners so widely used in the Southwest. Soon houses began to float downstream, some with their occupants still

Figure 12-18

Areas of flooding around New Braunfels, Texas, May 1972.

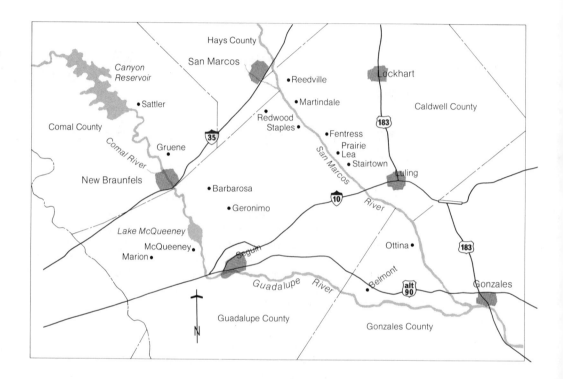

inside or on the roofs. Houses crashed into trees, into other houses, and into automobiles also being washed downstream.

About half an hour after it had begun to rain, the bottom had dropped out of the sky over Blueders Creek above the city, where ten inches of rain fell on an area of about twelve square miles. This amounts to 6400 acre feet, or a little over two billion gallons—enough water to supply a city of a million people for about two weeks. At the same time Comal Creek, which lies mostly within the city, got almost as much rain, and the Guadalupe River, into which both these streams converge at a point near the center of New Braunfels, was many feet above flood stage. At the Interstate Highway 35 bridge, at the southeast edge of New Braunfels, the Guadalupe had a 40-foot crest, almost 12 feet above the previous flood record of 1921.

Canyon Dam is on the Guadalupe River above New Braunfels, but most of the rain fell along the Balcones Escarpment between the dam and the city. There was no operative flood control. Water rushed through New Braunfels to do extensive damage to the towns of Seguin, McQueeney, and Gonzales, but these communities had at least three hours' warning.

Mud flats, Belgium

Figure 12–19

Preliminary flood map of New Braunfels, Texas.

At New Braunfels alone, a city of about 20,000, 53 people were dead or missing, and nearly 5000 were homeless (Figure 12-19). Some would return to damaged houses, but many returned to bare lots. Even brick and stone houses were swept away by the flood. A thousand telephones were out, and one-third of the city was without electricity. Incomplete assessment of damage was $10 million, or $500 for each

man, woman, and child in the city. Rescue work and relief were impeded because the Comal County Red Cross Disaster Center was under water.

Eighteen miles to the north, the city of San Marcos experienced less severe flooding. San Marcos had also experienced devastating floods in 1913, 1921, 1929, 1952, and 1970; of these the greatest was in 1921, though damage was less because the popu-

Flooding from the
Mississippi and Missouri
Rivers, Missouri, 1943

lation was smaller. The 1970 flood had stimulated the development of a flood control plan. On March 15, 1972, two months before the most recent flood, the city of San Marcos had voted a $500,000 bond issue to be used as seed money to obtain an additional $2 million from the federal government for a flood control system. One of the projects of this program is a massive park (greenbelt) along the San Marcos River, Texas.

The flood of May 11, 1972, at New Braunfels was the greatest flood of record. The Balcones Escarpment transects the western edge of New Braunfels, and because of the air patterns most of the rain fell between Canyon Dam and the city, but Canyon Dam farther up the Guadalupe River had given the citizens of New Braunfels a false sense of security.

Rapid City, South Dakota

Rapid City is a city of about 45,000 inhabitants. By late evening on 9 June 1972, there had been over ten inches of rain in 24 hours, and streams were expected to reach the 40-year flood stage. About midnight, Canyon Dam washed out on Rapid Creek, the principal stream flowing through the city. Shortly after midnight the mayor of Rapid City received a telephone call to the effect that the city had about 20 minutes to prepare for a disastrous wall of water moving down the creek. Although warnings were immediately broadcast over local radio stations, most residents were already asleep.

In Rapid City over 200 people died, and perhaps as many as 500 were missing. The number of missing persons is unknown because parks and greenbelts along rivers are inviting places for campers to stop overnight and few communities keep an accurate count of such visitors. On Rapid Creek, cars, tents, and trailers were swept away like kindling wood, in unknown numbers. Mud covered one-fifth of the city. Ten percent of the houses were destroyed, and 50 percent were damaged. Eight blocks of paving were ripped up. There was about $120 million in damage, about $2,700 per man, woman, and child, or about $13,500 for a family of five. The drinking water was polluted, power failed, and fires from broken gas mains and electric lines destroyed many buildings. Only a few telephone lines remained in operation, and these were jammed by incoming calls from people concerned over relatives and friends.

Hurricane Agnes

Hurricane Agnes was spawned in the northwest Caribbean in mid-June 1972. After touching eastern Yucatan and western Cuba, Agnes roared across the Gulf of Mexico on June 18, the Florida panhandle and Georgia on June 19, South and North Carolina on June 20, Virginia, Maryland and New Jersey on June 21, and Delaware, New York City, and Connecticut on June 22. A storm center and barometric low over the Great Lakes attracted Agnes back over southern New York and northern Pennsylvania on June 23 (Figure 12-20a). Rainfall was greater in New York and Pennsylvania than in other areas because in central New York the warm (hurricane) front encountered a cold (Canadian) front that produced even greater condensation and precipitation. All the major rivers of south-central New York, Pennsylvania, and northern West Virginia (Figure 12-20b) reached peak flood levels.

Cities along the Susquehanna River and its tributaries were hardest hit by flooding, all the way from Elmira, New York, through Wilkes Barre and Harrisburg, Pennsylvania, to Chesapeake Bay, where the tremendous influx of fresh water damaged brackish water fisheries, particularly clam and oyster beds. By the time hurricane Agnes reached New York and Pennsylvania, there had

Figure 12-20a

The path of hurricane Agnes from the Caribbean Sea to Pennsylvania, June 1972.

Figure 12-20b

Rivers in flood in the mid-Atlantic states as a result of hurricane Agnes.

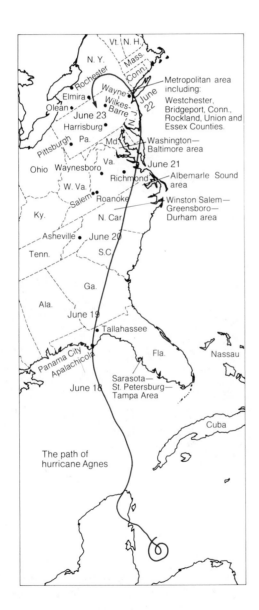

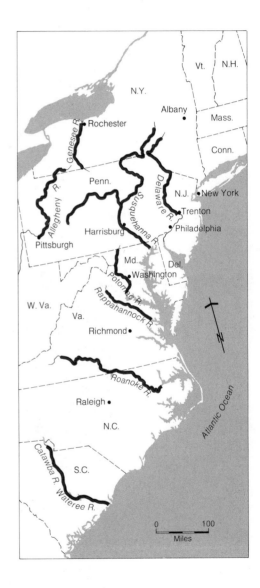

Hurricane damage

State	Deaths	Damage (millions of dollars)
New York	23	100
Pennsylvania	51	1250
Ohio	0	1
West Virginia	0	25
New Jersey	2	20
Maryland	21	55
Virginia	17	more than 210
Florida	16	25
Total	130	1,696

already been a week of rain. The soil was saturated and all rainfall ran off.

In the eight states hardest hit by hurricane Agnes, there were 130 deaths and damage in excess of $1,700 million (Table 12-2). Total damage from this hurricane was estimated at $2,200 million. By June 24 the hurricane had spent five days over Atlantic coastal states, and the President had declared Florida, Maryland, Pennsylvania, New York, and Virginia disaster areas eligible for federal funds. In Pennsylvania alone, 100,000 people were left homeless.

Hurricane Agnes taught us that some flooding cannot be controlled or prevented. Everything on floodplains in the areas flooded by Agnes was affected by water. Everything that could be damaged by water and mud suffered. Damage of such magnitude can only be avoided by zoning legislation that prohibits the construction on the floodplain of structures that can be damaged by water.

Flood Insurance

Title XIII of the Housing and Development Acts of 1968 and 1969, the National Flood Insurance Act of 1968 and its amendments provided for federal flood insurance at a cost of 40 cents for every $100 of insurance coverage. This means that a $20,000 policy costs just $80 a year; dollar for dollar, this is one-tenth the cost of automobile insurance. Title XIII provides insurance for floods from rivers; hurricane insurance is provided for by different legislation. As the June 1972 floods showed, most people apparently did not know that federal flood insurance was available. And many people, when they weigh a bill of $80 against the chance of a flood in that particular year, will take a chance and forego the insurance. In any given year the probability is in their favor; for a period of 20 years, if they live on the floodplain, the probability that each of 20 such decisions will be correct is not in their favor.

In the 15 states that suffered flood disaster in June 1972, there were only 92,500 active federal flood insurance policies. Nearly one-fourth of these were in Florida; Rapid City, South Dakota, was high on the list. There, of the total damage of $120 million, nearly $14 million, or about 12 percent, was covered by flood insurance.

Of the nearly $3 billion in flood damage in that disastrous month, only about $100 million, or about 3 percent, was covered by flood insurance. This means that the bulk of the $3 billion in disaster relief had to be paid for out of federal funds—that is, by taxpayers throughout the rest of the nation. This seems a high price to pay for the folly of those who build unsoundly on floodplains.

The lack of public participation in the federal flood insurance program has resulted in two new decisions. The first is the lowering of the cost of such insurance from 40 cents to 25 cents per $100 of policy. Second, federal flood insurance is now available only to those who live on floodplains that have been zoned. It is hoped thus to force more zoning, reduce the amount spent on developing floodplains, and, in turn, decrease the disaster relief bill paid for from tax monies. The zoning decision may of course backfire and result in the issuance of fewer policies. This may in turn increase the cost of the disaster relief, unless the zoning is very rigorous.

Summary

The year 1972 taught Americans a great deal about flooding. Geologists and hydrologists already knew the facts, but the public is not always convinced by argument alone; and, all too often, people only learn the hard way.

The 1972 disasters emphasized that, although we classify floods by recurrence intervals, the timing of floods cannot be predicted for "average" recurrence. The floods of 1972 also taught that flood control by dams on rivers has misled people into thinking they are protected when they are not, especially from heavy local rainfall. Flood protection systems are not designed to withstand more than the 40- to 80-year flood, depending on the design of the system. Floods like those brought on by hurricane Agnes, or those in Scottsdale, Arizona, and New Braunfels, Texas, discharge their wrath on everything on the floodplain. The two most concentrated floods of 1972 were at Rapid City, South Dakota, and Buffalo Creek, West Virginia. In each case, manmade dams washed out or collapsed to magnify many times what would have been only minor or moderate flooding. Obviously there should be legislation requiring proper engineering of all dams. Furthermore, a program of periodic monitoring of all dams should be initiated.

All these factors suggest that the best way to prevent death and devastation from flooding is to exercise much closer control of the building and development allowed on floodplains.

Selected Readings

Chemung Historical Journal. 1972. The flood of '72 in Elmira, Chemung County. *Chemung Historical Jr.* 18 (1):2141–2196.

Davies, William; Bailey, James F.; and *Kelly, Donovan B.* 1972. West Virginia's Buffalo Creek Flood. U.S. Geological Survey Circular 667.

Fisher, W. L., et al. 1972. Environmental geologic atlas of the Texas coastal zone—Galveston-Houston area. University of Texas Bureau of Economic Geology.

General Adjustment Bureau. 1972. The floods of June 1972, New York: General Adjustment Bureau.

Centry, R. Cecil. 1970. Hurricane Debbie modification experiments, August, 1969. *Science* 168:473–475.

Hayes, Miles O. 1967. Hurricanes as geological agents: case studies of hurricanes Carla, 1961, and Cindy, 1963. University of Texas Bureau of Economic Geology, Report of investigations 61.

Hoyt, W. G., and *Langbein, W. B.* 1955. *Floods.* Princeton: Princeton University Press.

Malde, Harold E. 1968. The catastrophic Late Pleistocene, Bonneville flood in the Snake River Plain, Idaho. U.S. Geological Survey Professional Paper 596.

Ruggles, F. H. 1966. Flood of October 8, 1962, on Bachman Branch and Joes Creek at Dallas, Texas. U.S. Geological Survey Hydrologic Investigations, Atlas HA-240.

Sheaffer, John R., et al. 1967. Introduction to flood proofing: An outline of principles and methods. University of Chicago Center for Urban Studies.

U.S. Department of Housing and Urban Development. 1971. National flood insurance act of 1968 as amended. Washington: the Department.

Vansickle, Donald. 1969. Experience with the evaluation of urban effects for drainage design. In Effects of watershed changes on streamflow, ed. Walter L. Moore and Carl W. Morgan, pp. 229–254. Austin: University of Texas Press.

Chapter 13
The Earth's Changing Climate

Therefore this one-time lake had still been a lake when Thaj was built. This was going to tie up with our speculations on the prehistoric climate of Arabia, for only a higher rainfall, or at least a higher water-table, could have held water in the lake bed.

Geoffrey Bibby, *Looking for Dilmun,* 1969

To early men droughts must have been little more than an inconvenience. Bands of men who were primarily carnivorous followed animals to new feeding grounds as the climate changed. Bands of foragers moved to those areas where the particular plants that were the main staple of their diet grew best. As climates changed, men moved from one place to another to stay within easy access of their food supply.

But when human beings began to live in communities and to depend on agricultural products, climatic changes must have become a greater problem. There are many records of large agricultural areas that were abandoned because increasing aridity made the yield too low to support the population. Even those nomads who were herders and shepherds and depended almost entirely on their flocks for sustenance were more affected by drought than were primitive men, because they had to move their households and flocks to areas where there was enough forage for their animals. As man became more directly dependent on a particular area for food, a change in climate affected his way of living more and more severely.

Known Climatic Changes of the Past

Although Justice William O. Douglas attributes the demise of the ancient cedar forests of Lebanon to goats, it would appear that man and nature were as much if not more responsible than the animals. Many years ago the Phoenicians cut down the forests to use the timber for shipbuilding and for the lumber trade with Egypt. They also traded oil of cedar, which was used for embalming in the Egyptian cult of the dead. And the increasingly arid climate of the Middle East restricted the growth of young cedars and thus also helped to prevent the regrowth of what had once been a climax forest.

In northern Mexico and adjacent Trans-Pecos, Texas, stone basins and other equipment for producing an Agave pulque (raw beer made from the century plant) have been found in areas that have been without sufficient rainfall for so long that Agave no longer grow there. In northern Chihuahua there are abandoned small settlements that once thrived on irrigation, attested to by dry irrigation ditches and abandoned fields. These places now have neither water nor people, though the condition of the adobe in the buildings indicates that the settlements are less than 60 years old.

Along the eastern ranges of the Rocky Mountains, springs that watered a hundred head of cattle a day 50 years ago have been dry for three or four decades. Many of these springs are in areas where no groundwater is withdrawn, so their disappearance cannot be attributed to the depletion of groundwater supplies. Other writers have cited evidence of the gradual change to a drier climate in South Texas.

Another extensive change in climate has taken place in North Africa and along the Mediterranean coast. North Africa was well populated by Roman and Carthaginian cities and military posts 2500 years ago, when the rainfall was substantially greater than it is today. Today there is not nearly enough rain to feed the ancient Roman aqueducts that 2000 years ago carried water to these defunct cities. In the year 258 A.D. Cyprian, Bishop of Carthage, complained about the decrease "in winter rains that give nourishment to the seeds in the earth."

Climate and Population

When the climate changes, people usually suffer; therefore, the need arises to acquire enough knowledge to predict and plan for climatic change.

American land speculators of the last half of the nineteenth century convinced poor eastern farmers that if they moved west, settled, and planted their crops, their actions would induce greater rainfall because "rain follows the plow." Under the influence of this theory, great numbers of people did move west of the 100th meridian during wetter climatic cycles, but, just as often, they were driven east by increasingly arid periods.

There was a dry period in the West in the 1830s and 1840s, followed by a wetter period in the 1850s and 1860s. After the Civil War many settlers moved west, only to be "drouthed out" in the severe dry spell of the 1880s and 1890s. Another wet cycle began before World War One and lasted through most of the 1920s, only to be followed by the severe drought of the 1930s, which accompanied the Great Depression. During each arid cycle many of the farmers west of the 100th meridian were forced to sell out and return east or move farther west to California, Oregon, and Washington. Such climatic changes have seriously affected the lives of several generations of American farmers.

In the more arid portions of the Southwest, these climatic cycles are superimposed on a still larger cycle. The Southwest seems to have been drying out for the last 500 years, so that each recurring wet spell has less precipitation than the one before. Since the southwestern United States and adjacent northern Mexico are still drying out, one

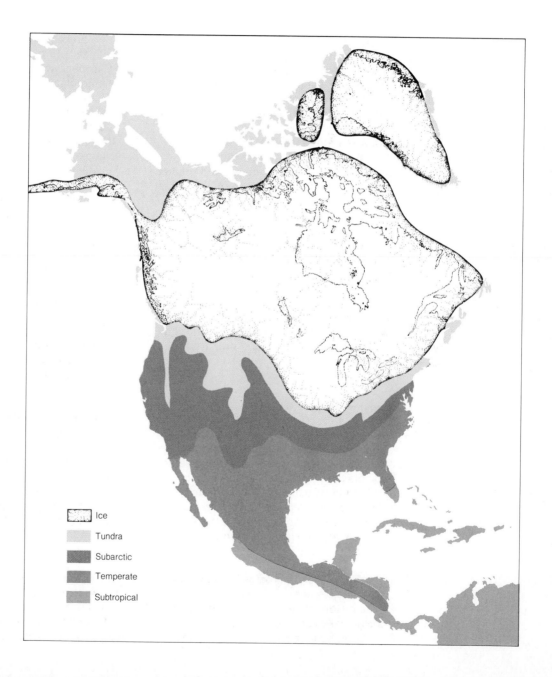

Figure 13–1

Climatic belts during the maximum extension of the Wisconsin ice sheet. (Compare with Figure 13–2.)

Ice

Tundra

Subarctic

Temperate

Subtropical

may ask whether it is wise planning to let them continue to grow and develop at the present exorbitant rates.

The entire world has recently emerged from a colder period, called an ice age. Although the earth's climate is warmer now than it was 14,000 years ago, there is no evidence that the ice age is really over. We may be living in a warm interval between periods of advancing ice and greater cold. Study of the ice age and the last great advance of ice (Figure 13-1) has provided us with data about extremes of cold and warmth on the earth, and rates for some of the more pronounced climatic changes. Climatic changes of lesser magnitude have been recorded for almost the last thousand years in Europe, and for much shorter times in other parts of the world. All these studies have proved that climatic changes are a natural part of the earth's history, and that we should try to predict and prepare for them.

Studying Climatic Change

There are many approaches to the study of climatic change, some more accurate than others. Data from one method alone seldom presents the whole picture, and may be suspect unless supported by confirming data. Researchers usually prefer to use data developed from several methods. Information about climatic changes has been obtained by studying:

1. The history of an area as it recovered from the ice age.
2. The history of sediments deposited in former lakes and lake basins.
3. Drainage changes during the Quaternary.

4. The present and past distribution of animals and plants.
5. European wine harvests in relation to Alpine glaciers.
6. The variations in solar energy received by the earth through time.
7. Temperature records, accurate when they have been maintained as a part of climatic history, and less accurate but still useful in the written records of other cultures.
8. Changes in temperature-sensitive isotope ratios through time.

Although some of these methods overlap, they will be taken up separately for reasons of clarity.

Data from the ice age

Figures 13-1 and 13-2 show the dramatic changes in the climate of North America over the past 15,000 years—before the ice withdrew and the zones moved northward, and then as the zones moved further north after all the ice had melted. Even as late as 10,000 years ago, when there was still ice on the Michigan Peninsula, on Labrador, and around Hudson Bay, the climatic requirements for human life in Wisconsin were very different from what they are at present. As the climate changed over the past 10,000 years, either cold-adapted cultures moved north with their climate, or the men in a certain area had to adapt to the changing climate by altering their habits of eating, housing, and transportation—in other words, changing their entire culture. The effect of the ice age on climatic belts was to restrict the width of the subarctic and tundra zones greatly, and of the temperate and subtropical zones perhaps

Figure 13-2

Distribution of present climatic belts.

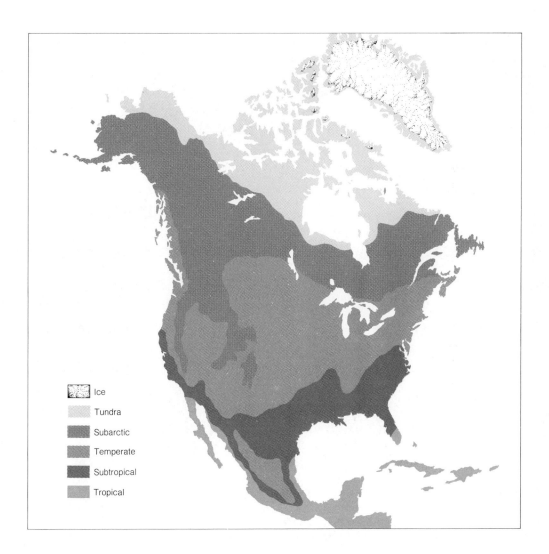

Ice

Tundra

Subarctic

Temperate

Subtropical

Tropical

somewhat less. Without concerning ourselves with real temperatures and the actual widths of the climatic belts, it can be seen that great temperature changes have taken place in both America and Europe during the past 20,000 years.

Even before the Pleistocene, the earth had been subjected to tremendous, though gradual, climatic changes. Figure 13-3 illustrates the northern limit of subtropical flora in North America for about the past 60 million years, during successive epochs

Figure 13-3

Southward migration of colder climatic belts during the Tertiary.

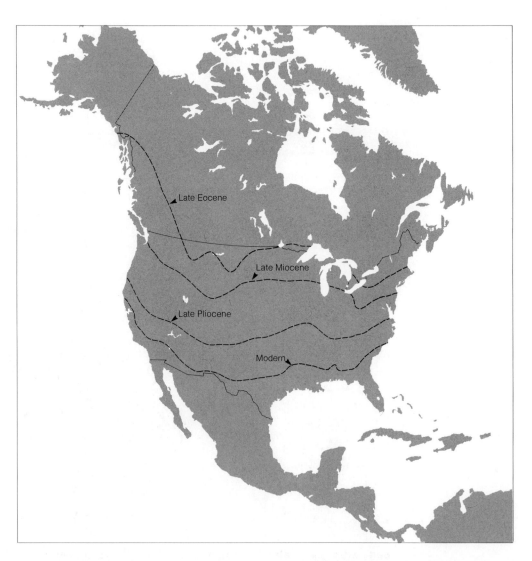

of the Cenozoic Era. Figure 13-4 shows the decline in temperature during the Cenozoic in the lower Rhine River valley in Europe. During the Pleistocene glacial stages, the average temperature was a full 20° Centigrade (36° Fahrenheit) colder than it had been in the early Eocene.

Further climatic effects accompanied the disappearance of the ice. As the ice melted from the land and joined the oceans, there

Lake bed in sand dune area, Death Valley, California

Figure 13-4

Decline of mean annual temperature for the lower Rhine River Valley during the Tertiary and Quaternary.

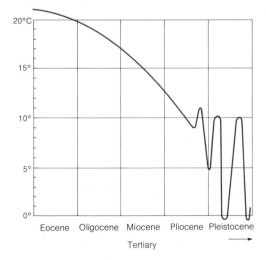

was gradual inundation of the continent (see Figure 11-18). And, as we have seen, if all the ice in Antarctica and Greenland were to melt, most of the major cities of the Atlantic and Gulf Coasts of the United States would be under water.

As the ice melted from the continents there was isostatic readjustment, and the land that had been loaded with ice rose. Resulting from this, central Sweden has risen almost 250 meters since the ice age, with the greatest rise where the ice was thickest. As the ocean encroached more on the land, the remaining land was moderated by warmer air masses, and the average temperature rose. An increase in ocean area warms the atmospheric temperature.

Figure 13-5

Distribution of glacial lakes in the southwestern United States.

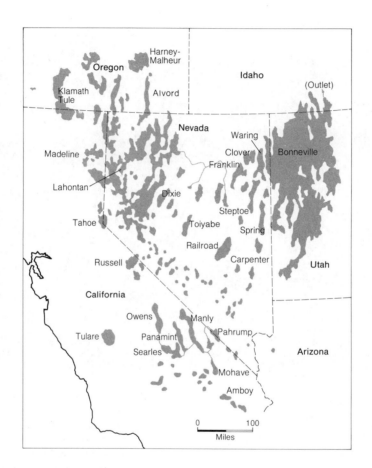

Evidence from ancient lake beds

That the western United States received greater precipitation at some times since the ice age than it does now is suggested by the number and extent of glacial and postglacial lakes (Figure 13-5). One of the most appropriately named is Lake Lahontan, for in the days of European exploration in search of a passage to India, Lahontan was the last European geographer to draw an inland sea on a map of North America. Had Lahontan lived a few thousand years earlier, there would have been an inland sea composed of glacial lakes, and his map would have been closer to reality.

The bottom of Lake Searles, California, consists of many layers of mud deposited under different climatic conditions during the late Pleistocene, and gives an excellent record of variations in climate. Only twice during the Wisconsin—about 24,000 years ago and again about 11,000 years ago (Table 13-1)—was rainfall sufficient for Lake Searles to overflow (Figure 13-6). In the early Recent Epoch the lake was completely dry, but about 6000 years ago rainfall began to increase during a wetter period called the climatic optimum, and continued to increase until about 3700 years ago, when it began to decrease again. Now the lake is almost dry once more.

Table 13-1

North American and European glacial and interglacial classification*

Glacial or interglacial	North America	Europe	Approximate starting date (years B.P.)
	Recent	Recent	10,000
Glacial	Wisconsin	Würm	100,000
Interglacial	Sangamon	3rd IG	300,000
Glacial	Illinoian	Riss	450,000
Interglacial	Yarmouthian	2nd (great) IG	1,100,000
Glacial	Kansan	Mindel	1,300,000
Interglacial	Aftonian	1st IG	1,750,000
Glacial	Nebraskan	Günz	2,000,000
Interglacial		IG	2,200,000
Glacial		Donau	2,500,000

*By convention the 1st interglacial (IG) has been the post-Günz interglacial. Pre-Günz interglacials and glacials were identified later and are not numbered. The 10,000 B.P. date for the end of the Wisconsin is quite arbitrary, since ice began retreating several thousand years earlier in some areas and did not all melt until three or four thousand years later. Some writers have suggested that we are now in another interglacial, and that what we call the Recent should be called the 4th IG.

Figure 13-6

Inferred fluctuations of lake levels in Searles Valley, California (see Figure 13-5) during the last 44,000 years.

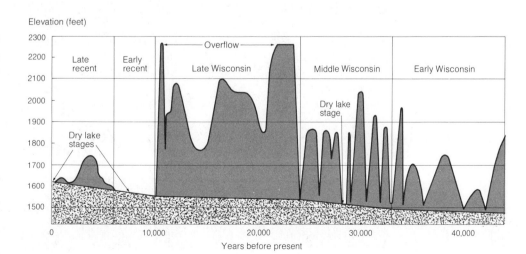

Ancient lake bed, Capitol
Reef National Monument,
Utah

Figure 13-7

Glacial Lake Agassiz on the border of central United States and Canada.

This is a fine record of changes in precipitation in the Southwest over the past 40,000 years.

Another glacial lake, centered on the site of the present Lake Winnipeg, was Lake Agassiz (Figure 13-7). The sediments of this lake form the excellent soils of much of the best wheat land in northwestern Minnesota, eastern North Dakota, Manitoba, and Saskatchewan. The fine chernozem and prairie soils developed on the western part of these lake deposits are among the great assets of southern Canada. This lake demonstrates that there was much more water in the north-central United States and south-central Canada immediately after the melting of the ice than there is now. During part of postglacial time, Lake Agassiz represented impounded ice meltwater accompanying and following the melting of the continental glacier. Greater precipitation later in the Recent, during the climatic optimum, may explain the large size of the lake long after the ice had melted.

Quaternary Drainage Changes

Many river systems were drowned by the submergence of land as the result of the rise in sea level that accompanied the melting of the ice toward the end of the Wisconsin. These systems also give evidence of greatly increased ocean area as a result of melting, which in turn produced still further moderation of climate as more and larger oceanic air masses more frequently covered parts of continents.

The Sunda River system

One of the most dramatic of these drowned river systems is the ancient Sunda River, which presumably headed into the Sunda Strait between Java and Sumatra. This large river flowed northward out of the north end of the Karismata Strait and into the China

Figure 13-8

The ancient Sunda River, Indonesia.

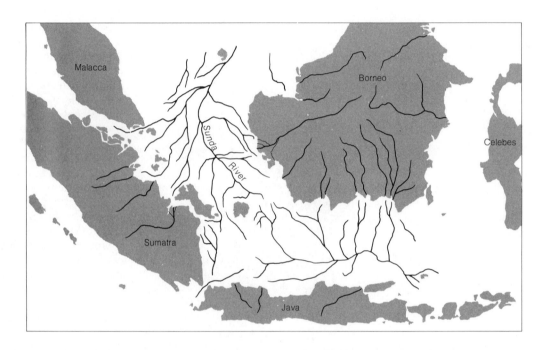

Sea during the Wisconsin ice age. This river had tributaries that are still represented by rivers draining west out of Borneo and east out of Sumatra (Figure 13-8). A second river system, consisting of rivers draining South Borneo and north Java, flowed eastward through the Java Sea. Both these rivers were drowned as the sea level rose with the melting of Wisconsin ice; former tributaries now flow directly into the ocean.

The Thames, Rhine, and Seine Rivers

Another such drowned river system lies beneath the North Sea and the English Channel (Figure 13-9). Near the end of the Wisconsin, the North Sea was mostly dry land, as was the English Channel. The Severn River flowed out of England, westward through the Bristol Channel, and joined a river that drained the area that is now the Irish Sea. The Seine flowed northwest out of France into what is now the English Channel and thence west into the Atlantic Ocean. The Thames, flowing eastward, turned northward around England, and joined the Rhine to become a large river that drained into an early arm of the North Sea that filled the Norwegian Deep west of Norway. The drowning of such lands suggests the environmental problems

Figure 13-9

Drainages of the Severn, Thames, Rhine, and Seine Rivers at the end of the Wisconsin.

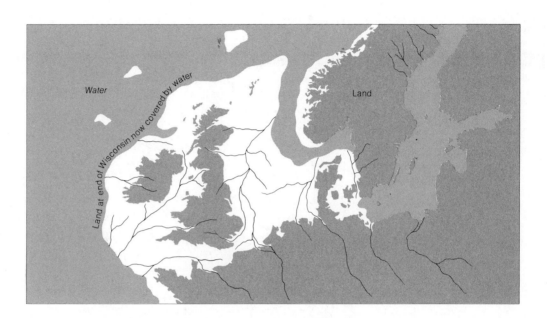

that would besiege a much more populous world if the remaining ice sheets were to melt. Most of the Atlantic coastal rivers of the United States, like the Hudson and the Susquehanna, have also been drowned by post-Wisconsin changes in sea level (see Figure 11-18).

The Missouri River

A more spectacular and somewhat different change occurred within North America during the Wisconsin. Before the Wisconsin, the Missouri River, joined by the Yellowstone and its tributaries in western North Dakota, flowed northeast out of present North Dakota into Canada. There it joined the Red River north of the Minnesota-North Dakota state line and then flowed into Hudson Bay (Figure 13-10). For a while, then, before the Recent, the great mud deposits eroding out of the west drained into Hudson Bay. But as a result of the Wisconsin ice sheet, glacial sediments were piled up along the boundary of the lower Missouri River extending north into western Iowa, to just north of the upper Iowa River, and then on north through South Dakota and into central North Dakota. This pile of rock and debris formed a barrier that

Figure 13–10

Preglacial drainage of the
Missouri and Ohio Rivers.
(© 1956 by The University
of Chicago.)

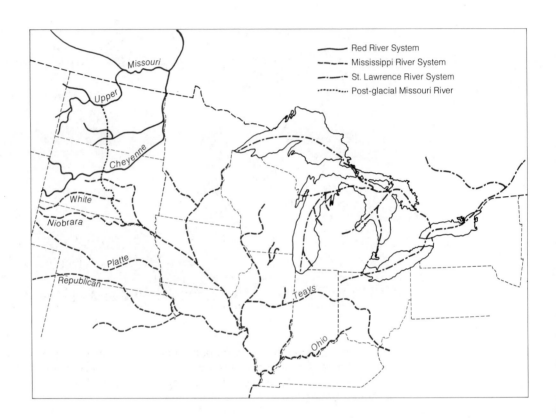

diverted the Missouri River southward along
the dotted line in Figure 13-10 to form its
present path. This circumstance diverted

the great quantity of mud eroded from the
western states away from the Red River
into the Mississippi River drainage system.

Sources of Evidence of Climatic Change

Distribution of plants and animals

Animal Distribution.   Figure 13-11 illustrates
the former distribution of the Asiatic lion,
an animal now almost extinct. There is

more than one way to interpret the change in
the distribution of the Asiatic lion. It can
be said that the lion was killed off by man,
or that the lion died off because of the
disappearance of its main food source,
which in turn had been killed off by man.
Or the lion's demise can be interpreted as

an effect of climate. Much of the decrease in its numbers seems to have occurred during climatic changes that resulted in a continual drying out of certain parts of the Middle East; this process in turn produced a decline in vegetation, and hence in the herbivores on which the lion fed—and so finally in the predator, the lion.

The presence of the wooly mammoth, *Mammuthus primigenius*—a cold-climate animal—in the northern part of the United States in the last stage of the Wisconsin

ice age is also attributable to climatic change. Similarly, the former distribution of the cave bear indicates the warming of Europe since that time. Another piece of evidence that colder weather extended farther south during the Wisconsin is the presence of the pike, a cold-water fish, in southern areas where it is not now found. Detailed analysis of many such changes in animal distribution helps identify climatic changes that have altered the habits and cultures of man since the ice age.

Figure 13–11

Distribution of the Asiatic lion in 1800 B.C. (From "Crises in the History of Life" by Norman D. Newell. Copyright © 1963 by Scientific American, Inc. All rights reserved.)

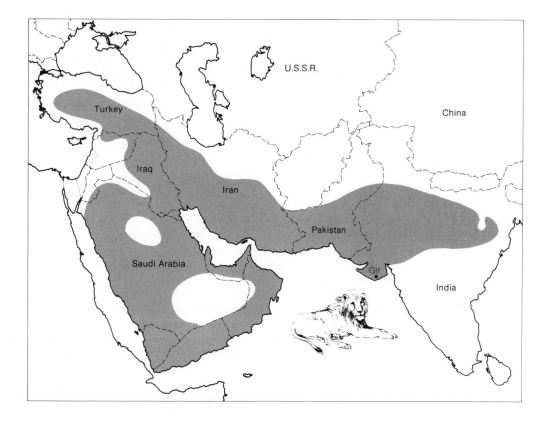

Figure 13-12

Present polar timberline and the maximum southward displacement of polar timberline during the Pleistocene.

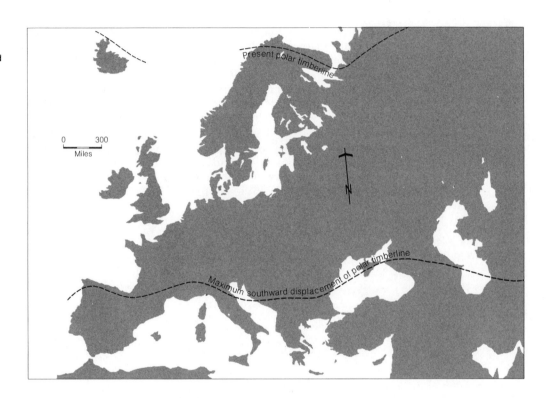

Plant Distribution. Plants usually provide better evidence than do animals about the details of climatic variations, because they reveal more about ecosystems and community conditions than do the animals that feed upon them (see page 9). During the maximum cold of the Wisconsin glacial stage, the polar timberline moved many hundreds of miles south of its present position (Figure 13-12). This in itself demonstrates the change in climate between the maximum cold of the Wisconsin and the climate of the present.

The study of plants, and particularly of pollen, records the northward retreat of the polar timberline over the 20,000 years since the maximum advance of the Wisconsin glaciers. As the ice retreated all over North America and Europe, water filled the basins and depressions that were left on the earth's surface, producing lakes. The lakes were filled with sediments containing pollen from trees growing on the shores, or brought down by streams from higher lands, or even from a more distant source. Sometimes the lake sediments are **varves**— alternating light and dark layers of a sediment, a single light-dark couplet representing a single year's deposition—which make it possible to count the number of years

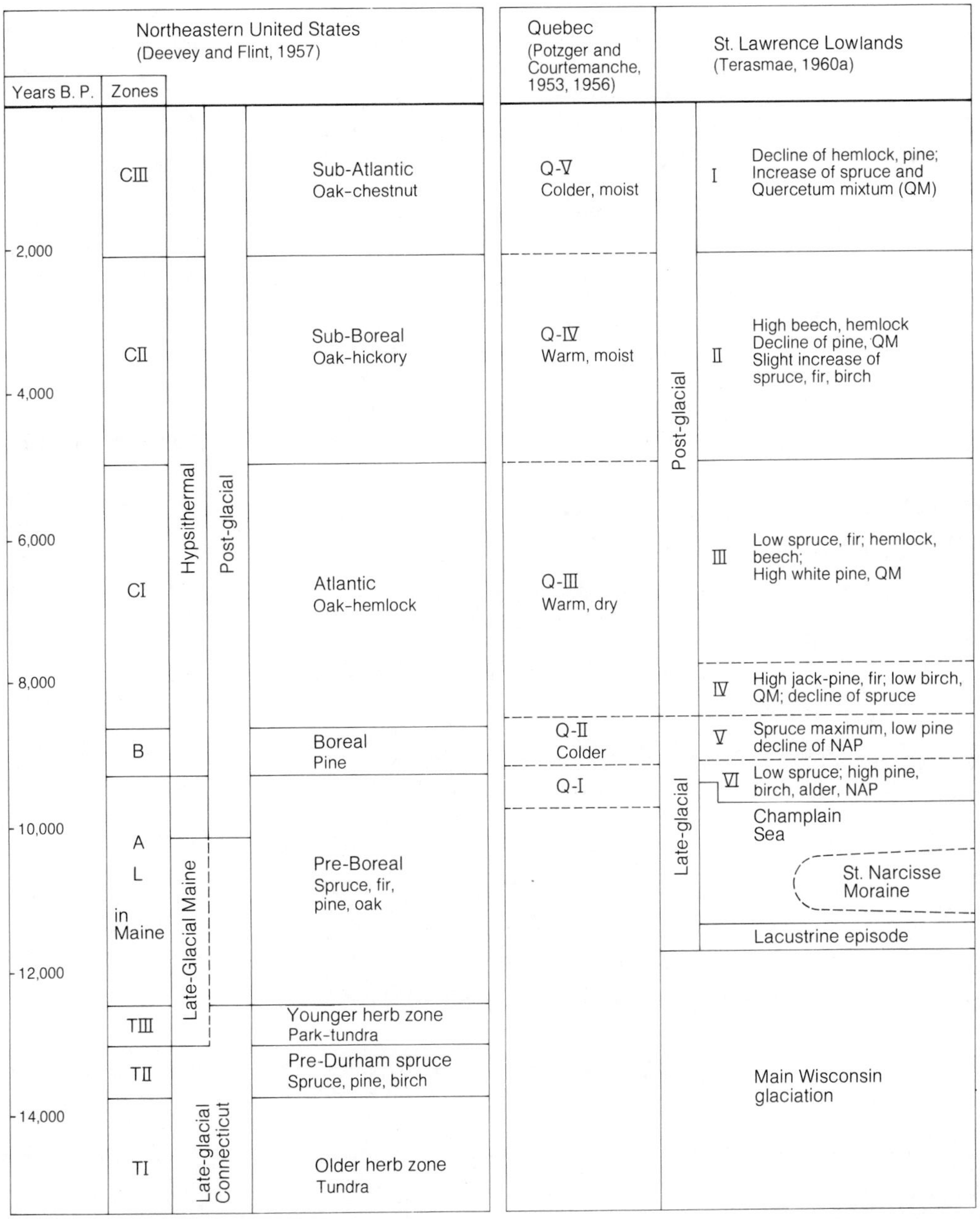

Figure 13-13

Fossil pollen and climatic changes.

Years B. P.	Zones	Northeastern United States (Deevey and Flint, 1957)			Quebec (Potzger and Courtemanche, 1953, 1956)	St. Lawrence Lowlands (Terasmae, 1960a)		
	CIII			Sub-Atlantic Oak-chestnut	Q-V Colder, moist	Post-glacial	I	Decline of hemlock, pine; Increase of spruce and Quercetum mixtum (QM)
2,000								
	CII			Sub-Boreal Oak-hickory	Q-IV Warm, moist		II	High beech, hemlock Decline of pine, QM Slight increase of spruce, fir, birch
4,000								
	CI	Hypsithermal	Post-glacial	Atlantic Oak-hemlock	Q-III Warm, dry		III	Low spruce, fir; hemlock, beech; High white pine, QM
6,000								
8,000							IV	High jack-pine, fir; low birch, QM; decline of spruce
	B			Boreal Pine	Q-II Colder		V	Spruce maximum, low pine decline of NAP
	A L in Maine		Late-Glacial Maine	Pre-Boreal Spruce, fir, pine, oak	Q-I	Late-glacial	VI	Low spruce; high pine, birch, alder, NAP
10,000								Champlain Sea
								St. Narcisse Moraine
12,000								Lacustrine episode
	TIII			Younger herb zone Park-tundra				
	TII		Late-glacial Connecticut	Pre-Durham spruce Spruce, pine, birch				Main Wisconsin glaciation
14,000	TI			Older herb zone Tundra				

over which a particular lake received sediments. Some sequences of varves are thicker or thinner depending on climatic cycles. Before the advent of carbon-14 radiometry, a complex system of geochronology had been developed by counting varves, comparing their thicknesses, and correlating those from one lake with those of another. The varve system has now been outdated, and postglacial sediments are usually dated by the use of carbon-14.

A palynologist—a student of pollen—can now go to a swamp representing a lake that formed after the ice age and continued receiving sediments almost until the present.

He can take a core of muds deposited in that lake, analyze the pollen at different levels, plot the percentages for the different levels, date the levels by carbon-14, and produce a diagram similar to Figure 13-13. Very high percentages of spruce and pine with a minimal amount of fir represent either cold or cool climates. Dominant oak indicates a warmer climate. When climate is warm and moist or warm and dry, birch is at a minimum; when cool and moist but still not cold enough for a tremendous development of pine and fir, the percentage of birch increases.

Foraminifera and Ocean Temperature.
Foraminifera are small marine organisms, some species of which are restricted to warm waters, and others to cold waters. Hence an examination of sediments can tell us whether they were deposited in a warm or a cold Atlantic Ocean, or even in a warm or a cold part of the Atlantic Ocean. Other species have a different coiling pattern in cold water than in warm water. *Globigerina pachyderma,* for example, coils to the left in colder water and to the right in warmer water. Thus a line can be plotted across the North Atlantic between samples containing right-coiled and left-coiled members of this species, and this line will in general separate colder from warmer waters (Figure 13-14). When cores taken from the bottom of the Atlantic are analyzed for a preponderance of right- or left-coiled Foraminifera, one can plot a curve like that in Figure 13-15. The top 10 centimeters of core (the most recent) represent a time of warming (right-coiling). That part of the core from 10 to 40 centimeters in depth represents a cold period (left-coiling).

Figure 13-14

Present distribution of warmer and colder microfossil assemblages in the North Atlantic Ocean. Dots represent sites at which cores have been taken (see Figure 13-15). (From ''Micropaleontology'' by David B. Ericson and Goesta Wollin. Copyright © 1962 by Scientific American, Inc. All rights reserved.)

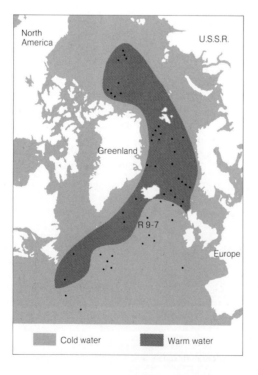

North America

U.S.S.R.

Greenland

R 9-7

Europe

Cold water　　Warm water

Depth in core (Centimeters)

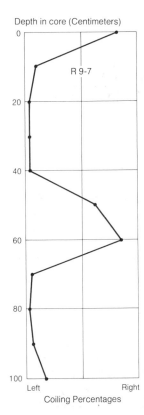

Figure 13-15

Distribution of warm and cold water assemblages from core R9-7 in the North Atlantic Ocean. (From "Micropaleontology" by David B. Ericson and Goesta Wollin. Copyright © 1962 by Scientific American, Inc. All rights reserved.)

From about 40 to about 70 centimeters the core represents a warm period, preceded by one of gradual cooling. That part of the core deeper than 70 centimeters represents a cold Atlantic Ocean. The two cold periods in this core represent two of the glacial substages of the Wisconsin.

Wine harvests and Alpine glaciers

Emmanuel LeRoy Ladurie, a French historian, recently reviewed the records of wine harvests, which have been kept carefully since about 1350 A.D., and spottily back to about 1000 A.D. To produce vintage wine, vineyards need moisture in the spring and early summer, and a strong sun in the late summer. When the harvest is late (the end of October or November) it means that the summer has been too damp and/or cool to produce sweet grapes and give good wine. If the harvest is early the summer has been warm, and if there was enough rain early in the summer followed by a warm sun, the wine will be good. Many monasteries kept meticulous records of the dates of harvest and the taste of the wine. The average difference in temperature between good and bad wine years may be no more than one degree Centigrade. Yet, interestingly enough, the temperature data from wine harvests correlate with the advances and retreats of the Alpine glaciers. It has been observed that as few as six cool summers in succession will cause the Alpine glaciers to advance. Broadly speaking, the evidence from the glaciers and the wine harvests agrees that a colder period occurred between 1500 and 1700, which has been inferred from other evidence; and there is similar agreement on

the warming trend over the past hundred years.

Variations in solar energy

Many years ago a Croatian mathematician and astronomer named Milankovitch studied the aberrations of the sun and earth (mostly cyclical changes in their relative positions) and computed the effect of these aberrations on the amount of solar energy the earth receives (Figure 13-16). Since he lived before computers were invented, his calculations took most of his lifetime. The curve he arrived at is so constructed that it can also be used to predict future climatic changes, insofar as aberrations in the solar system have been accurately identified.

Many scientists discarded the Milankovitch hypothesis because it did not seem to fit facts about the earth's history derived from other evidence. But in the last decade or so, with new and better radiometric dating of Pleistocene sediments, and new and better interpretations of the climatic cycles, there is increasing reason to think that the Milankovitch curve may be valid. According to this curve there have been a number of colder periods through time, including three drops in temperature in the last 150,000 years, one on either side of the 200,000-year mark, two between 400,000 and 500,000 years ago, two more than 600,000 years ago, and severe drops about 830,000 and 930,000 years ago.

There has been other speculation about the amount of solar energy the earth receives. It is known that an increase in sunspot activity marks an increase in solar activity, and that the regular sunspot cycle provides us with an 11-year climatic cycle.

Figure 13–16

The Milankovitch curve, using the 65-degree parallel latitude. (© 1969 by The University of Chicago.

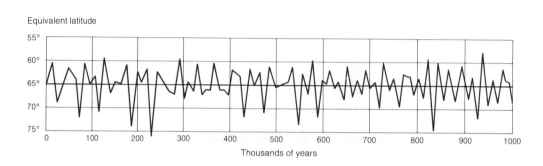

Equivalent latitude

Thousands of years

Are there also longer sunspot cycles? And are there other cycles of solar energy with which we are unfamiliar? Some writers have proposed that cosmic dust clouds, shrouding the earth from the sun and decreasing the solar energy it receives, have produced climatic changes and even glacial stages. Astronomers state, however, that no such dust cloud was close enough to have come between the sun and the earth at any time in the Pleistocene. Decreases in the earth's supply of solar energy can come about through other causes, such as an increase in the earth's albedo—that is, the total reflection of energy from the earth system back into space. The albedo can be increased by an increase in ice on the earth's surface, or by an increase in cloud cover. One interesting theory proposed that an increase in temperature of 4° Centigrade would increase the cloud cover and the albedo, thereby increase snowfall, and bring on another ice age.

Loss of heat from the earth can result from a decrease in retained solar energy. It can also be caused by a decrease in carbon dioxide in the atmosphere or a decrease in the cloud cover, all of which allow more radiant energy to escape into outer space. An increase in the amount of solar energy retained can come about by opposite means. That is, an increase of CO_2 in the atmosphere or an increase of cloud cover will hold in more radiant solar energy.

Isotopic variation

Other important evidence of changes in climate can be derived from studying the ratio of two oxygen isotopes (^{16}O and ^{18}O). The oxygen-16/oxygen-18 ratio is temperature-selective at the time of precipitation of some minerals and of ice. Decreasing temperature of precipitation leads to decreasing content of oxygen-18. In the late 1960s the ice at Camp Century, Greenland, was cored and studied, and a temperature curve was drawn. It is thought that this core represents over 100,000 years of deposition of ice. Studies of oxygen isotope ratios in this core, pollen-derived temperature curves from a section in the Netherlands dated by carbon-14, pollen-derived temperature curves from carbon-14-dated deposits in the Ontario and Erie Basins, and an oxygen isotope study of deep sea cores show a general agreement of temperature curves. Each method gener-

Figure 13-17

Comparison of climatic curves derived by various methods: (a) Oxygen isotope study of Camp Century, Greenland ice core; (b) carbon-14 dated pollen from the Netherlands; (c) carbon-14 dating studies in the Erie and Ontario Basins; and (d) oxygen isotope studies of deep sea cores.

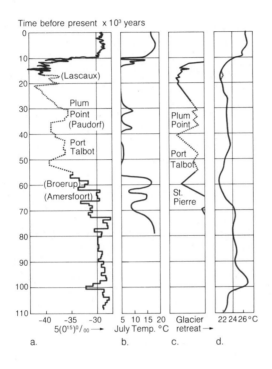

Time before present x 10³ years

a.   5(0¹⁵)⁰/₀₀ →

b.   July Temp. °C

c.   Glacier retreat →

d.

ally supports the findings of the others concerning the times of climatic change, and any disagreement may result from inaccuracies in radiometric dating, rather than temperature interpretation, since the curves are nearly identical though the dates are in slight disagreement (Figure 13-17).

Oxygen-16/oxygen-18 isotope ratios have also been used to measure the temperature of formation of the Orgnon stalagmite (from a cave in France) and to draw a temperature curve for the years from 90,000 to 130,000 B.P. These samples were dated by a thorium-uranium method. More information is needed before an exact correlation of the temperature curve from the Orgnon stalagmite with the Milankovitch curve can be achieved. Figure 13-18 shows a comparison of oxygen isotope measurements by Emiliani with the effective solar radiation curve of Milankovitch; the general correspondence is obvious.

Interpretations

Many writers have pointed to the gradual decline in temperature since the Cretaceous Period (see Figure 13-4) as being of general importance to the evolution of mammals. Furthermore, marked changes of temperature that occurred during the ice ages have seemed to some scientists to correlate with the solar radiation received by the earth and with sea-level changes during glacial ages (Figure 13-19).

The snowline is the lowest elevation at which snow is permanent in any area. Toward the tropics the snowline is at higher elevations than toward the poles. A rise in elevation of the snowline in any particular area indicates a warming climate, and a

Figure 13-18

Oxygen isotope temperature curves for the last 200,000 years compared with the 65-degree north latitude, solar radiation curve of Milankovitch.

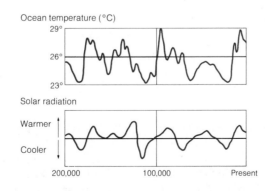

Ocean temperature (°C)

Solar radiation

Figure 13-19

Comparison of changes in
sea level, tropical seawater
temperature, and solar
radiation received by the
Earth. Dates given at base
of figure refer to
years B.P. (before present).
(From "The Changing
Level of the Sea" by Rhodes
W. Fairbridge. Copyright ©
1960 by Scientific American,
Inc. All rights reserved.)

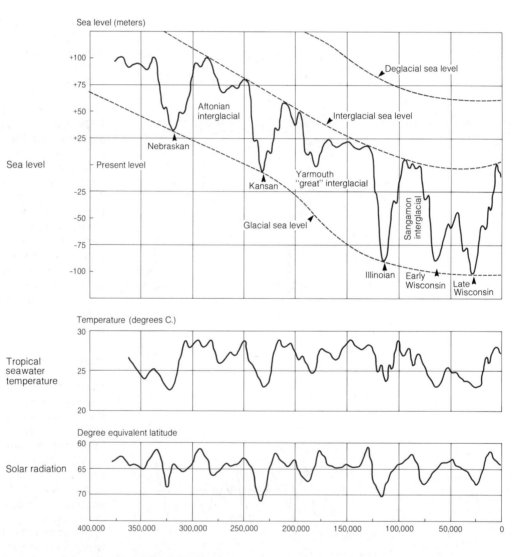

lowering of the snowline indicates a cooling climate. Even in the past few thousand
years there have been some very sharp
changes in climate, as is demonstrated by
the snowline migrations in Norway (Figure
13-20). These temperature changes are also
associated with eustatic curves. In other
words, the association of coastal drowning
with climatic curves seems to be proved.
Since climates can change rapidly enough

Figure 13-20

Climatic changes in the last few thousand years, depicted by the changing snowline in Norway.

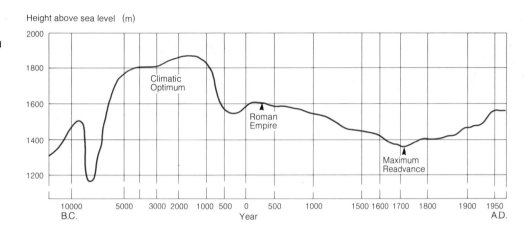

Height above sea level (m)

Figure 13-21

Climatic "warming," indicated by the retreat of sea ice in the Greenland-Iceland areas since 1770. (From "Neoglaciation" by George H. Denton and Stephen C. Porter. Copyright © 1970 by Scientific American, Inc. All rights reserved.)

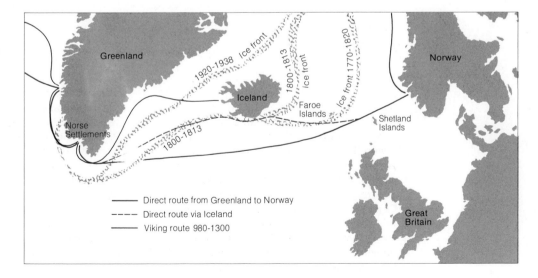

to affect man's environment, the study of future climates is essential if we are to plan intelligently to save such cities as New Orleans and Miami from the transgressing oceans. Marked climatic changes (Figure 13-21) have caused a retreat of sea ice in the Greenland-Iceland area since about 1770. The Norwegian sailing routes from Norway to Iceland to southern Greenland that were used 1000 years ago were im-

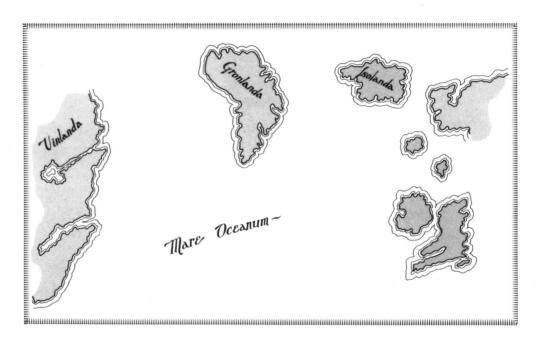

practical from the early eighteenth century (and probably from the seventeenth century) to the middle of the nineteenth century. That Norwegians sailed these routes between the years 980 and 1450 indicates that the area was free of ice during that period. In the early 1960s a map was discovered in Europe that depicts the Atlantic Ocean in a manner that strongly supports the claim that Norsemen discovered America before Columbus. This map has been declared a fraud by some historians. Yet the only accurate part of it is the north coast of Greenland, and one may doubt the possibility of anyone who had never seen it drawing the north coast of Greenland correctly. The map indicates that during the Norse occupation of Greenland,

the north coast was so free of ice that somebody sailed around it (Figure 13-22).

Another indication of changing climate is the dramatic decrease in size of the Caspian Sea between 1930 and 1970, which came about because of a decrease in rainfall in southern Eurasia (see Figure 6-19). The decrease in rainfall in the Sahara Desert over the last 5000 years (Figure 13-23) has severely affected man. Man was driven from cities in parts of southern Arabia about 4000 years ago by what now appears to be a general decrease in rainfall in that region.

Table 13-2 summarizes the general climatic history of the past 10,000 years. In developing theories about causes of climatic change in the last 6000 years, one should consider the probable general relationship between sea level change in Scandinavia, gradual decline in temperature in northwestern Europe, and decline in the ratio of summer to winter solar radiation (Figure 13-24). Furthermore, for the past 5000 years there seems to have been a general relationship between the "tide-generating force" and general rainfall. Also, local peaks in rainfall seem to have followed declines in solar activity, although the process in South Russia may act in just the reverse way.

From about 1750 to about 1950, temperatures gradually increased over such widely separated areas as the Netherlands, Stockholm, cities in Great Britain, and Vienna (see Figure 3-5). Much of this increase in temperature had been attributed to the greenhouse effect of the Industrial Revolution, with its explosive use of fossil fuels, and the consequent release of carbon dioxide into the atmosphere. But the rise in temperature began well before the extensive use of coal (see Figure 13-20).

From 1930 to 1964 there was a gradual increase in rainfall and decrease in temperature in the midwestern United States

Table 13-2

Solar activity and climatic change for the last 10,000 years. From J. R. Bray, Glaciation and solar activity since the fifth century B.C. and the solar cycle. *Nature* 220:674, Table 4, 1968.

Year B.P.	Solar activity L = low H = high	Number of events of temperature increase and deglaciation	Number of events of temperature decrease and glaciation	General climate
0–700	L	0	4	Little ice age, sea level minimum, increased mountain glaciation
700–2000	H	8	4	Climatic recovery culminating in Little Climatic Optimum; greatly reduced mountain glaciation
2000–3300		3	13	Temperature decline; increased mountain glaciation; bog regeneration
3300–4600	H	4	6	Near-maximum post-Wisconsin temperature; maximum sea level; reduced mountain glaciation
4800–5900	L	1	10	Brief temperature decline; minor sea level fall; increased mountain glaciation
5900–7200	H	4	4	Peak post-Wisconsin temperature maximum; near disappearance of continental glaciers; reduced mountain glaciation
7200–8500	L	2	5	Very minor continental glacier readvance; brief temperature decline; minor sea-level decline; increased mountain glaciation
8500–9800	H	4	0	Temperature increase; major continental glacier recession; Lake Agassiz withdrew

Figure 13-23

Areas of the eastern Sahara Desert that receive fifty centimeters of rainfall per year, or less.

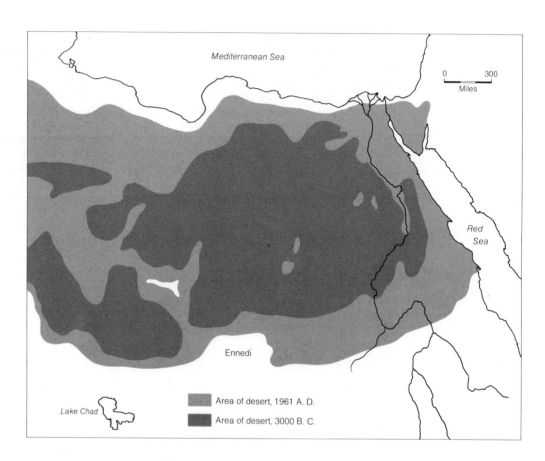

Area of desert, 1961 A. D.

Area of desert, 3000 B. C.

(Figure 13-25). These phenomena have been attributed to the increase of particulate matter in the air, which would provide nuclei for precipitation and increase the earth's albedo. But if the residence time for particulate matter in the atmosphere is only one year, it seems impossible for such matter to affect climate very greatly. The increase in rainfall has also been attributed to irrigation and water conservation in the West, which has released more moisture into the atmosphere. This is the current version of the notion that "rain follows the plow," a doctrine already disastrous to several generations of American dry farmers. To offset these arguments, extrapolations from the oxygen isotope curve of the Camp Century, Greenland, ice core show

Figure 13-24

Comparison of various climatic factors and modifiers for the last 8000 years.

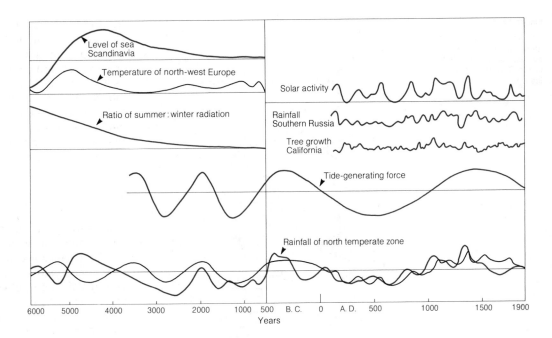

Level of sea
Scandinavia

Temperature of north-west Europe

Solar activity

Ratio of summer : winter radiation

Rainfall
Southern Russia

Tree growth
California

Tide-generating force

Rainfall of north temperate zone

6000   5000   4000   3000   2000   1000   500   B.C.   0   A.D.   500   1000   1500   1900
Years

Figure 13-25

Rainfall and temperature gradients from 1930 to 1962 for Illinois, Nebraska, and Kansas.

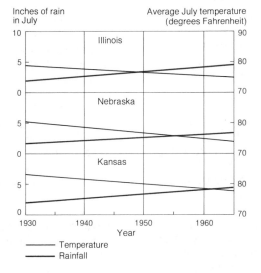

Inches of rain
in July

Average July temperature
(degrees Fahrenheit)

Illinois

Nebraska

Kansas

1930   1940   1950   1960
Year

——— Temperature
▬▬▬ Rainfall

that there would have been a warming trend from about 1820 into the 1930s, followed by a cooling trend until the present (Figure 13-26). This cooling trend is expected to continue until about 1980, when a slightly warmer period should begin. This means that everything that has happened to our atmosphere since the beginning of the Industrial Revolution can be explained just as easily by natural cycles as by changes in the atmosphere caused by man. If man had greatly affected the atmosphere, the cycles would not have continued to the present as repetitions of past cycles. So we are back where we started. Is man altering the climate? The answer is that we do not know.

Figure 13-26

The oxygen isotope temperature curve and climatic cycles, compiled from the Camp Century, Greenland ice core.

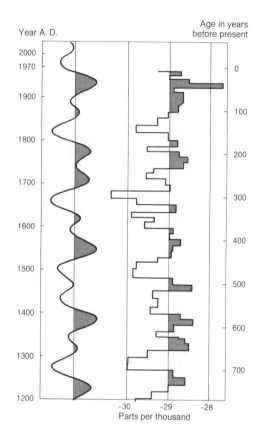

from 1895 to around 1905 there was an eastward migration of the rainfall distribution shown in Figure 13-27, and much more severe eastward migration occurred in the 1930s (Figure 13-28)—a climatic effect also indicated by the Greenland ice cores. In short, as data accumulate, the picture begins to materialize, and patterns emerge that may make long-range prediction in the future more dependable.

Summary

It is clearly not desirable to pollute the atmosphere, but we do not know yet whether man's activities are greatly affecting the total atmosphere from either a geological or a climatic point of view. Yet air pollution creates such serious local health hazards as to completely override the long-term atmospheric changes that pollution may or may not bring about. Much more time and study will be required before we can know whether or not we are changing the earth's climate as a result of our tremendous consumption of fossil fuels. In the meantime climate will change, whether naturally or by man's intervention. Every time an overpopulated nation claims victory over starvation, a new visitation of drought or another climatic change or cycle proves the claim to be erroneous. To prevent starvation in the future, society must be physically and mentally prepared for climatic change. Whether climate warms or cools, becomes wetter or drier, the change will bring benefits to some areas and disaster to others. In short, prediction may be possible, but even limited control is still decades away.

Smaller shifts in climate are apparent from a study of the agricultural history of the Great Plains for the past century (Figures 13-27, 28, 29). If Figure 13-27 is taken as a normal rainfall map of the United States, then such a map for the 1870s and 1880s, and again for the period from 1912 to about 1927, would look like Figure 13-29. During such wet periods dry-land farming was a very successful venture over the Great Plains. But in the years

Figure 13–27

Normal rainfall map for the
United States.

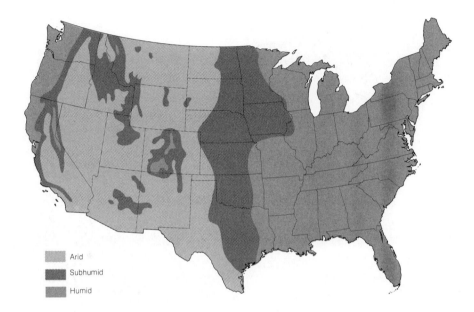

Figure 13–28

Rainfall map for the mid-
1930s (the dust bowl years)
in the United States, showing
the 20-inch rainfall contour
east of its normal position
(see Figure 13–27).

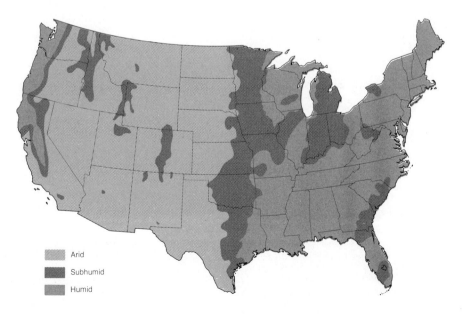

Figure 13-29

Rainfall map for the United States for the early 1920s, with the 30-inch rainfall line displaced to the west.

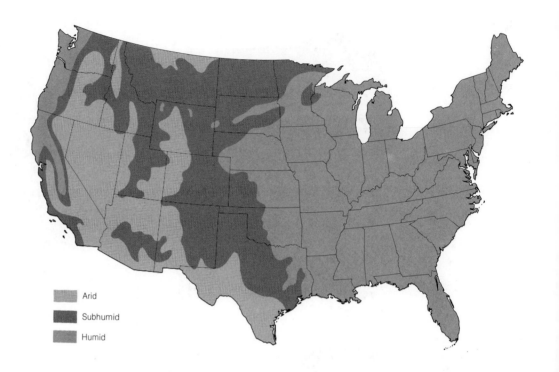

Arid

Subhumid

Humid

Selected Readings

Bibby, Geoffrey. 1969. *Looking for Dilmun.* New York: Alfred A. Knopf.

Bray, J. R. 1968. Glaciation and solar activity since the fifth century B.C. and the solar cycle. *Nature* 220: 672–674.

Budyko, M. I. 1972. The future climate. *Eos* 53:868–874.

Butzer, K. W. 1961. Climatic change in arid regions since the Pliocene. In *A history of land use in arid regions,* ed. L. Dudley Stamp, pp. 31–56. UNESCO, Arid Zone Research, vol. XII.

Dansgaard, W., et al. 1972. Speculations about the next glaciation. *Quaternary Research* 2:396–398.

Emiliani, Cesare. 1972. Quaternary paleotemperatures and the duration of the high-temperature intervals. *Science* 178:398–400.

Engel, A. E. J. 1969. Time and the earth. *American Scientist* 57:458–483.

Johnsen, S. J.; Dansgaard, W.; and Clausen, H. B. 1970. Climatic oscillations 1200–2000 A.D. *Nature* 227: 482–483.

Kurtén, Bjorn. 1968. *Pleistocene mammals of Europe.* London: Weidenfeld and Nicholson.

LeRoy Ladurie, Emmanuel. 1971. *Times of feast, times of famine: A history of climate since the year 1000.* Garden City, N.Y.: Doubleday.

Liestöl, Olav. 1960. Glaciers of the present day. In *Geology of Norway,* ed. Olaf Holtedahl, pp. 482–490. *Norges Geol. Undersökelse,* no. 208.

Mörner, Nils-Axel. 1972. When will the present inter-glacial end? *Quaternary Research* 2:341–349.

Skelton, R. A. 1965. The Vinland map. In *The Vinland map and the Tartar relation,* ed. R. A. Skelton, Thomas E.

Marston, and George D. Painter, pp. 109–240. New Haven: Yale University Press.

Smith, G. I. 1968. Late Quaternary geologic and climatic history of Searles Lake, southeastern California. In *Means of correlation of Quaternary successions,* Roger B. Morrison and Herbert E. Wright, Jr., eds., pp. 293–310. Proceedings of the 7th Congress of the International Association for Quaternary Research. Salt Lake City: University of Utah Press.

Smith, Henry Nash. 1947. Rain follows the plow: the notion of increased rainfall for the Great Plains. *Huntington Library Quarterly* 10:159–194.

Thompson, Louis M. 1964. Our recent high yields—how much due to weather? In *Research on water,* ed. T. J. Army *et al.,* pp. 74–84. American Society of Agronomy special publication 4.

Walcott, R. I. 1972. Past sea levels, eustasy and deformation of the earth. *Quaternary Research* 2:1–14.

Zeuner, Friedrich E. 1959. *The Pleistocene Period; its climate, chronology, and faunal successions.* London: Hutchinson Scientific and Technical.

Chapter 14

Geology and Land Use in the Future

The geologist is not a public health officer—his concern should be. What effect will the accelerating use of metals and fossil fuels have on the composition of the part of the earth and atmosphere inhabited by man? He should be concerned with the environment and what is happening to it; he should be concerned with providing data to those responsible for the health of the species.
Peter T. Flawn, 1968

Thus far, the causes of environmental problems have been considered much more extensively than the cures. It requires careful planning to cure problems; land use planning can aid in such cures. But, more importantly, good land use planning is a method of preventing environmental problems before they develop.

Most problem-solving is undertaken only when the problems have become severe; it is *ex post facto* (i.e., after the fact). Most environmental problems arise from the improper use of land. Thus the urban development of floodplains results in tragedies like those of the Mississippi River Valley in the spring of 1973, and construction on unstable lands results in economic loss like that in the Hollywood Hills. Construction near known and active faults and on unstable ground produces damage from earthquakes, as in San Francisco and Los Angeles. Improper planning for withdrawal of liquids promotes damage from subsidence as in Houston. Flood runoff through storm drains connected to the sewage disposal system compounds the sewage problem of the Milwaukee River. The destruction of pike fishing in the Great Lakes results from lack of coordination of land and water use, as does eutrophication of lakes everywhere. And so on *ad infinitum*. Not all of these problems could easily have been prevented, but death and damage could have been greatly reduced with more forethought about the best use of land. Without local and regional planning, similar events will have even more serious consequences in the future. This danger is becoming so obvious that government and concerned citizens are demanding more and better land use planning.

Table 14-1

Classification of land use. From Sterling B. Hendricks, ''Food from the Land.'' In Preston E. Cloud, ed., *Resources and Man: A Study and Recommendations* by the Committee on Resources and Man of the Division of Earth Sciences, National Academy of Sciences-National Research Council, with the cooperation of the Division of Biology and Agriculture. San Francisco: W. H. Freeman and Company. Copyright © 1969. Table 4-2, p. 67.

	Land use capability class		Best use*
Arable lands	I	Few limitations on use	Cultivation
	II	Some limitations on use, reducing the choice of crops or requiring some conservation practices	Cultivation
Grazing lands	III	Severe limitations on cultivation	Pasture, forest
	IV	Severe limitations on cultivation, reducing the choice of crops or requiring special practices	Pasture, forest
Limited-use lands	V, VI, VII	Soils severely limited—generally unsuited for cultivation	Pasture, range, or forest
	VIII	Soils severely limited and do not give satisfactory returns for management input	

*Forests now occupy some lands of all classes.

What Is Land Use Planning?

Lands are usually classified according to their possible uses—classically, according to agronomic potential. Of the world's 32 billion acres, only about one-fourth is suitable for cultivation (Table 14-1), and only a little over half of this fraction, or about 4.5 billion acres, is actually under cultivation. Some of these 8 billion acres have been irretrievably lost to urbanization. Even to the casual observer, it is apparent that large tracts of productive land are taken up by cemeteries and urban sprawl, particularly in Europe. Worldwide, developers make no attempt to conserve arable land, even when nonarable lands are available for development.

Arable lands do not include all tillable lands. The term arable has come to refer only to those tillable lands that will also satisfactorily produce cultivated crops. In other words, land will not produce just because it can be plowed and planted. The long-used and simple method of land classification outlined in Table 14-1 has not always served well because it emphasizes food production and it uses only a few of the many variables in good land use. Although land use planning is not a new concept, until the last decade most land use studies dealt with actual rather than potential use. Some land use planning in the last decade has provided for the best potential use of land, including aesthetic considerations if possible. Some land is unsuited to the purpose assigned it if aesthetics alone are considered. For example, a county planning commission in Indiana for aesthetic reasons assigned the sand-and-gravel works to an area in a distant corner of the county, out of sight of the city. The drawback to this site was the lack of sand and gravel. In addition to advising the best potential use, the latest studies provide for total land use planning, and are designed to answer such specific questions as, ''Is a particular area suitable for solid waste disposal?'' or ''Is good agricultural land being developed for residences, factories, cemeteries, or other nonagricultural uses when we may, in as little as two decades, need all of it to feed a rapidly increasing population?''

The purpose of land use planning should be to assure the best potential use of land,

Arid land near Ephrata, Washington *before* land reclamation

Same area *after* the Columbia Basin Project has employed a canal to irrigate the land

so that man may derive maximum benefit from ecosystems without destroying them.

Some Examples of Land Use Planning

By the middle 1960s the need for comprehensive land use planning was recognized by a number of states and municipalities, and several studies made increasing use of available knowledge and techniques. While none of these was completely comprehensive, new techniques were developed and, most important, a cadre of planners was trained in new methods. They accomplished a good deal locally, and their shortcomings helped point the way for more comprehensive programs.

The San Antonio survey

In 1966 San Antonio entered into a land-use mapping program with the United States Soil Conservation Service that produced a *Soil Handbook for Soil Survey, Metropolitan Area, San Antonio, Texas.*

This program developed soil maps and a soils handbook, in which the soils were described in terms of strength, stability, suitability for foundations, moisture content, electrical resistivity, shrink-swell potential, and drainage differences. On the basis of these land characteristics, various sites were classified as suitable for recreational, residential, greenbelt, and other uses. Other data included corrosiveness to metal, depth to bedrock, and kind of bedrock to a depth of ten feet. The most consistent users and beneficiaries of the San Antonio project were the City Public Service Board, the City Water Board, the Park and Recreations Department, and the Special Projects Engineers.

The project provided a good deal of information, useful for planning in San Antonio, that had not been previously available. But it was deficient in data on hydrologic factors, the subsurface below a depth of ten feet, geologic factors, resources, erosion rates, flood conditions, detailed topography, and several other useful types of information.

The greater San Francisco Bay project

Several years ago the United States Geological Survey, in cooperation with the Department of Housing and Urban Development, began an analysis of urban geology in San Francisco, called the *San Francisco Bay Region Environment and Resources Planning Study.* A survey of seismic risk areas was followed by studies of surficial deposits, since seismic risk in any area is closely related to the thickness, area, and strength of these deposits. Over 40 maps and reports have since been published. Maps have recently appeared showing subsidence in the Santa Clara Valley and potential landslide areas, and listing by county the annual cost of landslides. The latest contributions include studies of the lead and copper content of surface sedi-

Landfill operation, San
Francisco Bay

ments in the estuaries, and of hydrologic
and storm drainage facilities. The *San
Francisco Bay Study* is producing valuable
data on the most critical problems of greater

San Francisco, and is in many ways the
finest ongoing program of the federal
government to date. But it is not a coordi-
nated land-use planning effort.

Some state programs

*A Pilot Study of Land Use Planning and
Environmental Geology,* published by the
Kansas Department of Economic Development in 1968, illustrated what could be done
to provide geological data for land use
planning, but was a model pilot study and
not an actual land use project. The Alabama

Geological Survey also pioneered in research on problems of environmental geology, concentrating on sources of pollution
of ground and surface water, waste disposal, water supply, and collapse in limestone terrains.

At the same time that Kansas and Alabama
undertook their programs, the Illinois
Geological Survey began to concentrate on

the geology and hydrogeology of sanitary landfills. Several studies were devised to help small communities dispose of waste without polluting their shallow underground water supplies. These projects are also useful, but are directed at specific problems approaching near-crisis proportions rather than overall land use planning.

Land Use Planning and the Illinois Geological Survey. Land use planning is undertaken by the Illinois Geological Survey for limited areas, often the size of a county, and produces maps detailed enough to serve as a basis for complex decisions. The Illinois Geological Survey concentrates on such practical matters as the engineering characteristics of the rocks, the availability and quality of surface and ground water, and local resources such as sand, gravel for roads and other construction, peat, and stone. The survey also stresses waste disposal and describes terrain, so that rock stability and slope can be integrated into planning for construction. Suitability of terrain, lithology, and hydrogeology for various uses is described in tables. In addition to a simplified geologic map, maps showing the following features are prepared if applicable to the needs of the area:
1. geologic conditions affecting waste disposal
2. water resources
3. topography of bedrock
4. geologic conditions affecting construction
5. principal terrains
6. surficial deposits
7. sand and gravel resources.

While these include basic geologic data of interest to local officials trying to solve immediate problems, they do not constitute a complete data bank for land use planning.

Toward a Technique of Land Use Planning

If the aims of land use planning are to alter the ecosystems as little as possible and still provide for human needs, how is it to be accomplished? Many interests and disciplines have been involved in this process. Real estate developers look upon themselves as experts. For many decades geographers were considered the primary land use specialists. Architectural planners, landscape architects, ecologists, engineers, sociologists, psychologists, economists, agronomists—all of these and others—each in his own way—feel that they know how land should be used.

A classic in land use planning is the book *Design with Nature* by Ian McHarg, a landscape architect who integrates geological and hydrological concepts with cultural, public health, and aesthetic aspects of planning. McHarg has the ideas and imagination to take an overall view of the land, but his emphasis on aesthetics tends to obscure a lack of fundamental information. In this weakness he is but one of many. In spite of broad knowledge and experience from many fields, most plans have foundered, or at least fallen short of their goals. The lack of something vital always seemed to decrease the authenticity and utility of even the best of plans. The last few years, however, have produced land use mapping programs, which tell us that the missing ingredient was data. There have long been plenty of good ideas, but a dearth of information on all subjects relevant to good planning.

Accumulation, Storage, and Retrieval of Environmental Data

It is necessary not only to collect enough data, but also to systematize and store it for ready use. The extent of the necessary data is shown in Table 14-2, which lists most of the geologic and biologic factors of potential utility in land use planning. The table also lists various ways of presenting data on each of those factors, including maps. The main problem in preparing and storing data is the production of a retrieval system that provides the most data in the simplest form. With thousands of data points for each category of map or chart, it is categorically impossible to use computers initially. The printout of data would far exceed the reading capabilities of the small local planning staffs that must use and interpret it. But a map can serve as a data storage and retrieval system. On many maps a contour line or rock boundary, or the boundary of a plant community, represents a continuum of thousands of data points. The same data fed into a computer would print out in hundreds of pages. Data for maps can be fed into a computer digitally, but the computer then produces an inadequate caricature of the original. Analog systems can be programmed to produce maps more accurately than do digital systems, but neither produces maps as good as the originals. Generally, then, feeding data from maps into a computer is an expensive way of producing a facsimile, or something less than a facsimile. But the computer does have one advantage over maps: a map contains only those properties that were printed on it. If different sets of data from several maps are fed into a computer, it can be asked to draw a single map comparing two properties from different maps or from a single map that also illustrated several other properties.

Thus the map is the most economical data system for handling this kind of information. The staff of the Bureau of Economic Geology at the University of Texas, in its *Environmental Geologic Atlas of the Texas Coastal Zone,* uses maps as an information storage and retrieval system that comes closer to meeting the standards suggested in Ian McHarg's *Design with Nature* than anything previously produced.

Map units called resource capability units allow one to determine the environmental carrying capacity of any area for different environmental needs or activities. A resource capability unit is an area on the ground that reacts uniquely to a combination of human activities, and that can be mapped for geological and biological characteristics. The *Houston-Galveston Environmental Geology Map* has about 80 such units, each of which can be described in terms of a large number of suitable, limited, or undesirable land uses (human or natural). Resource capability units may be either geological or biological, depending on which category is dominant in restricting the use of the land. No unit is completely ageological or abiological, but human use of some units may depend only on geological or biological criteria.

Since many resource capability units have the same relationship to a single environmental factor (see Table 14-2), but different relationships to other environmental factors, derivative maps can be drawn that illustrate land use suitabilities or limitations for several factors. Derivative

Table 14–2

Environmental factors useful in land use planning.*

Environmental factor	Source map	
Aesthetics	Greenbelt maps	3
	Recreation area and facility maps	3
	Historic sites maps	3
	Scenic value maps	2
Animal life	Animal community maps (Chapter 1)	2
	Biotic provinces (Chapter 1)	2
Bedrock character	Lithologic maps	1
	Geologic maps	
Current land use	Current land use maps	3
	Residential	3, 6
	Industrial	3, 6
	Institutional	3
	Recreational	3
	Agricultural	3, 2
	Forest	3, 2
	Wildlife preservation	3, 2
Deposition	Active processes maps	5
Depth to bedrock	Depth-to-bedrock maps	
	Surficial geology isopachous (thickness) maps	
	Surficial materials maps	4
Drainage basins	Drainage basin maps	1, 2, 6
	Surface drainage maps	5, 6
Earthquakes	Seismic risk maps (Chapter 9, Figures 9-5 to 9-7)	
Erosion	Solid sediment load maps (Chapter 6, Figure 6-8)	
	Dissolved sediment load maps	
	Active processes maps	5
	Susceptibility-to-erosion maps	1, 5
Excavation ease	Lithologic maps	1
	Geologic maps	
	Substrate maps	1
	Depth-to-bedrock maps	
Flooding	Flood frequency charts (Chapter 12, Figure 12-4)	
	Flood maps (Chapter 12, Figure 12-9, 12-10, 12-25)	5, 7
	Flood prediction maps (Chapter 12, Figure 12-14)	
	Active processes maps	5
	Tidal inundation maps	5
	Storm surge inundation maps	5

Table 14-2 *(cont.)*

Environmental factor	Source map	
Landslides	Potential landslide maps	1
	Percent slope maps	8
	Soil and rock stability maps	1
	Soil drainage maps	1
Population	Population density maps	3
	Land value maps	
Precipitation	Isohyetal (rainfall) maps (Chapter 5, Figure 5-3)	7
Resources	Potential resources maps (Chapter 2)	2, 4
	Resource reserve maps (known or projected)	4
	Resource installation maps	4
	Resource-oriented land use maps	3, 4
Soils	Soil maps (Chapter 6, Figure 7-3)	
	Soil drainage maps	1
	Soil stability maps	1
Subsidence	Subsidence maps (Chapter 11, Figure 11-19-11-21)	
Surface water maps	Evapotranspiration maps (Chapter 5)	
	Water surplus and deficiency maps (Chapter 5, Figure 5-4)	
	Runoff maps (Chapter 11)	7
	Streamflow maps and charts (Chapter 11)	7
	Low water flow charts	7
	Riparian water maps	6
	Marshes, freshwater	2, 3, 6
	Marshes, salt	2, 3, 6
	Channelization (potential) maps	1, 6
	Salinity maps	7
Topography (slope)	Contour topographic maps	8
	Hachured topographic maps	8
	Colored topographic maps	8
	Percent slope maps	8
	Physiographic or terrane maps	1, 8
Trace substances	Trace substance maps (Chapter 8, Figure 8-4 to 8-6)	
Underground water	Depth-to-water maps	6, 7
	Equipotential surface maps	
	Soil drainage maps	1
	Aquifer maps	6, 7
	Aquifer recharge area maps	6, 7

Table 14–2 *(cont.)*

Environmental factor	Source map	
Vegetation	Vegetation distribution maps (Chapter 1)	2, 3
	Plant community maps (Chapter 1)	2
	Biotic provinces maps (Chapter 1)	2
Waste disposal	Rock permeability maps	1
	Potential waste disposal maps (Chapter 7, Figure 7-13)	1
Wind	Wind strength and direction maps	5
	Wind strength and direction charts	
	Active processes maps	5

*Figures in parentheses refer to the chapters in which various topics are discussed. Figures in the right-hand column refer to the eight derivative maps described on pages 373–379.

Table 14–3

Classification of resource capability units. From L. F. Brown, Jr., *et al., Resource Capability Units.* Austin: University of Texas Bureau of Economic Geology, Geological Circular 71-1, Table 3, p. 8, 1972.

ACTIVITIES — RESOURCE CAPABILITY UNITS

Resource Capability Units	Surface Disposal of Untreated Liquid Wastes	Disposal of Untreated Liquid Wastes, Subsurface, Shallow	Maintenance of Feed Lots	Disposal of Solid Waste Materials	Construction of Offshore and Bay Platforms	Construction of Jetties, Groins, Piers	Construction of Storm Barriers and/or Seawalls	Placement of Pipelines and/or Subsurface Cables ▲	Light Construction	Construction of Highways	Heavy Construction	Flooding (through dam construction)	Dredging of Canals and Channels, and Spoil Disposal ★	Excavation (includes extraction of natural materials)	Filling for Development	Draining of Wetlands	Well Development	Devegetation	Transversing with Vehicles (marsh buggies, air boats, dune buggies, motorcycles)	Use of Herbicides, Pesticides, Insecticides
Bays, Estuaries, and Lagoons — River Influenced Bay Areas Including Prodelta and Delta Front	X	X		X	0	0	X	0					X		X		0			
Enclosed Bay Areas	X	X		X	0		X	0					0		0		0			
Living Oyster Reefs and Related Areas	X	X		X	X	X	X	X					X		X		X			
Dead Oyster and Serpulid Reefs and Related Areas	X	X		X	0	0	X	0					X		X		0			
Grassflats	X	X		X	X	X	X	X					X		X		X			
Mobile Bay-Margin Sand Areas	X	X		X	X	X	X	X					X		X		0			
Tidally Influenced Open Bay Areas	X	X		X	0	0	X	0					X		X		0			
Subaqueous Spoil Areas	X	X		X	0	0		0							0		0			
Inlet and Tidal Delta Areas	X	X		X	X	X	X	X					X		X		0			
Tidal Flats	X	X		X			X	X	X	X	X		0	X	X		0			
Coastal Wetlands — Salt-Water Marsh	X			X	X		X	X	X	X	X	X	X	X	X	X	0	X	X	X
Fresh-Water Marsh	X			X	0		X	X	X	X	X	X	X	X	X	X	0	X	X	X
Swamps	X			X		0	X	X	X	X	0	0	X	X	0	X	0	X	0	X

Table 14-3 *(cont.)*

LAND CAPABILITY UNITS	RESOURCE CAPABILITY UNITS	Surface Disposal of Untreated Liquid Wastes	Disposal of Untreated Liquid Wastes, Subsurface, Shallow	Maintenance of Feed Lots	Disposal of Solid Waste Materials	Construction of Offshore and Bay Platforms	Construction of Jetties, Groins, Piers	Construction of Storm Barriers and/or Seawalls	Placement of Pipelines and/or Subsurface Cables ▲	Light Construction	Construction of Highways	Heavy Construction	Flooding (through dam construction)	Dredging of Canals and Channels, and Spoil Disposal ★	Excavation (includes extraction of natural materials)	Filling for Development	Draining of Wetlands	Well Development	Devegetation	Transversing with Vehicles (marsh buggies, air boats, dune buggies, motorcycles)	Use of Herbicides, Pesticides, Insecticides
Coastal Barriers	Beach and Shoreface	X	X		X	X	O			X	X	X		X	X			X		O	
Coastal Barriers	Fore-Island Dunes and Vegetated Barrier Flats	X	X	X	X			X		+	+	+		X	X			O	X	X	X
Coastal Barriers	Washover Areas	X	X	X	X		X	X	X	X	X	X		X		X		X			
Coastal Barriers	Blowouts and Back-Island Dune Fields	X	X		X				O	X	X	X		X				X			
Coastal Barriers	Wind Tidal Flats	X	X		X					X	X	X		O			X	O			
Man-Made	Swales	X	X		X					X	X	X		X	X	X	X	X			X
Man-Made	Made Land and Spoil	X	X	X	X							O						X			
Coastal Plains	Highly Permeable Sands	X	X	X	X									O	X			O	X		X
Coastal Plains	Moderately Permeable Sands	X	X	X	X									O	X			O	X		X
Coastal Plains	Impermeable Muds	O								O	O	O	O					O			O
Coastal Plains	Broad Shallow Depressions •	O								X	O	X						O			O
Coastal Plains	Highly Forested Upland Areas •												X					O	X		X
Coastal Plains	Steep Lands, Locally High Relief	X			X					O		O			X				X		O
Coastal Plains	Stabilized Dunes	X	X	X	X							O		X	X			O	X	X	X
Coastal Plains	Unstabilized, Unvegetated Dunes	X	X		X				O	X	X	X		X				X			
Coastal Plains	Fresh-Water Lakes, Ponds, Sloughs, Playas	X			X									X		X	X	O			X
Coastal Plains	Mainland Beaches	X	X		X		X	X		X	X	X		X	X	X		X	O		
Coastal Plains	Areas of Active Faulting and Subsidence	O	O		O				X	O	O	X						O			
Major Floodplain Systems	Point-Bar Sands	X	X	X	X							O			X			O	X		X
Major Floodplain Systems	Overbank Muds and Silts	X	X	X	X					O	O	O	O	O							O
Major Floodplain Systems	Water	X			X								O	X							X

ACTIVITIES — Liquid Waste Disposal · Solid Waste Disposal · Coastal Construction · Inland Construction

X   Undesirable (will require special planning and engineering)
O   Possible problem(s)
+   Barrier Flat only (no construction on dunes)

•   Substrate variable
▲   Also occurs in Offshore Construction
★   Also occurs in Offshore Canals and Dredging

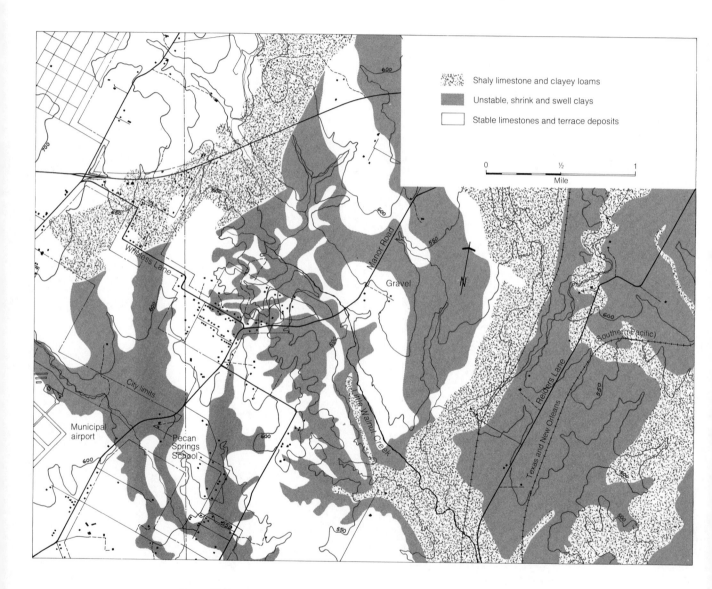

Figure 14-1

Slope stability for part of the East Austin Quadrangle,
Travis County, Texas.

maps are produced by combining under one symbol (or color) all resource capability units that behave alike toward the environmental factors covered on that map. By using the geologic map and eight derivative maps, most of the important data listed in Table 14-1 are covered. The eight derivative maps described on the following pages are (1) physical properties maps, (2) biotic assemblages maps, (3) current land use maps, (4) mineral and energy resources maps, (5) active processes maps, (6) man-made features and water systems maps, (7) maps of rainfall, stream discharge, and surface salinity, and (8) topography and bathymetry maps.

Physical properties maps (1)

On physical properties maps, the map units are groupings of resource capability units that have the same physical properties for a single environmental factor use (Table 14-3). In other words, a physical properties map contains units that combine all the resource capability units from the geologic map that are alike in lithology, susceptibility to erosion, substrate, landslide potential, soil and rock stability, soil drainage, rock and soil permeability, and potential for waste disposal. These maps can be used in solving problems of construction, excavation, drilling, channelization, and waste disposal. Figure 14-1 is a map illustrating rock stability. It combines those formations from the geologic map of the same area (Figure 14-2) that have similar load, flow, and shrink-swell properties. When permeability and soil moisture factors are added to it, the map becomes more

complex (Figure 14-3), but still not as complex as the geologic map. If ease of excavation were to be considered, an additional unit would be added, because one of the units of limestone (represented by coarse diagonal lines on Figure 14-3) is more difficult to excavate than the others.

Environments and biotic assemblages maps (2)

Environments and biotic assemblages maps group together those resource capability units that have similar biologic assemblages (Table 14-4). Underwater units are based mainly on bottom dwelling assemblages, and land units on plant assemblages. The Galveston-Houston sheet contains 31 different biotic assemblages, as mapped for the *Environmental Geologic Atlas of the Texas Coastal Zone* (Figure 14-4). When these units are compared to current land use maps, the extent of man's modification of the total environment is easier to see. To some users these are resource maps: the shell dredger and the oyster fisherman both consider oyster reefs a resource. For different reasons timber lands are resource units to the lumberman, the conservationist, and the sportsman. Salt marshes, bays, and estuaries, as spawning grounds for countless marine organisms, are also resource units. Subaerial biotic assemblage units can tell the planner a good deal about substrate, because different plant communities grow on different soils of different parent rocks. From these maps, local agencies or individuals can prepare inventories of the assemblages. These maps provide information that makes it possible for man to avoid environments where his

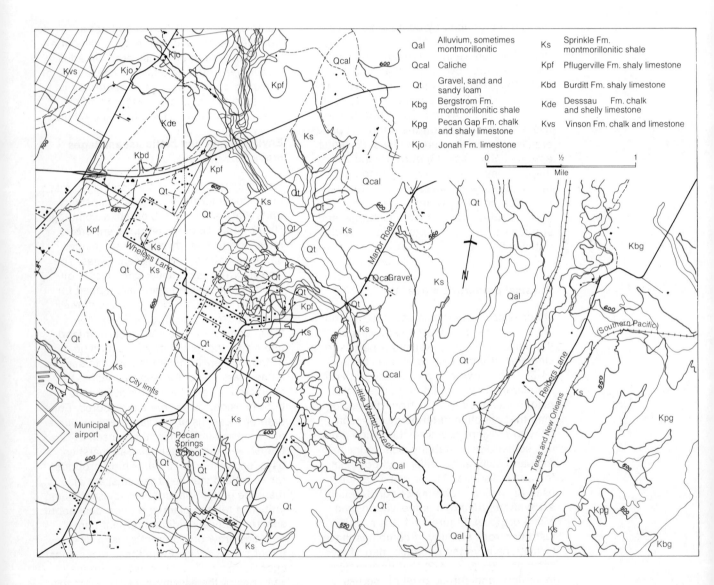

Figure 14-2

Geologic map of part of the East Austin Quadrangle.
(Compare with Figures 14-1 and 14-3.)

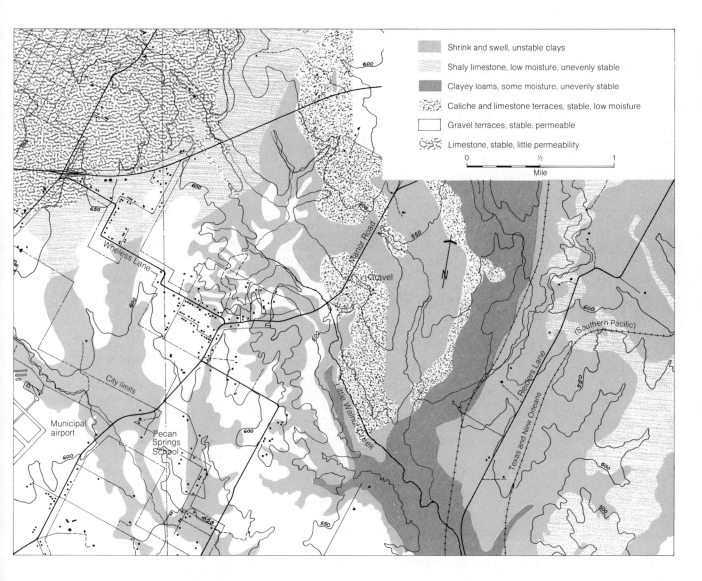

Figure 14-3

Rock stability and soil moisture map, part of the East
Austin Quadrangle.

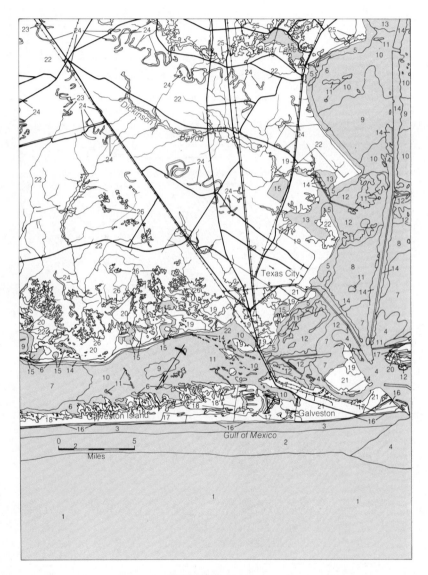

Figure 14-4

Environments and biotic assemblages map for the Houston-Galveston area.

Table 14-4

Data for Environments and Biotic Assemblages Maps

Animal communities
Plant communities
Biotic provinces
Agricultural lands
Forest lands
Wildlife preservation areas
Drainage basins
Areas of scenic value
Potential resources (biologic)
Freshwater marshes
Salt marshes
Oyster reefs

activities would be detrimental to himself or the environment.

Current land use maps (3)

Current land use maps (Table 14-5) record human uses of land and untouched natural lands (where these still exist). Present use often determines planning of adjacent land; for example, one does not build a bakery adjacent to the sewage disposal system. Lands usually remain natural only because (1) they are valuable as resources, or (2) they are unsuitable or uneconomic for development.

Mineral and energy resources maps (4)

Mineral and energy resources maps (except water) show both producing and potential resources (Figure 14-5). Aggregate —whether shell, sand, gravel, or crushed

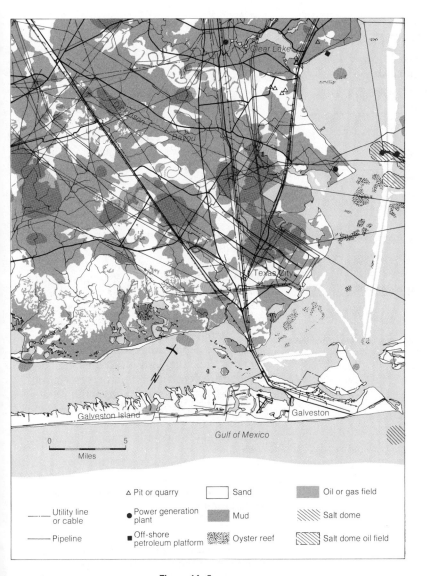

Figure 14–5

Mineral and energy resources map for the Houston-Galveston area.

Table 14–5

Data for Current Land Use Maps

Residential areas

Industrial areas

Institutional areas

Recreational areas and facilities

Agricultural areas

Forested areas

Wildlife preservation areas

Greenbelts

Historic sites

Population density

Resource-oriented land use

Freshwater marshes

Salt marshes

Plant distribution

limestone—is the local resource most important to the developing community or city. Aggregate commodities have such a low cost, however, that if the distance from source to user is over 16 miles, more than half the cost is for transportation. Clay and lime for cement are also necessary for large urban centers. The developing nation of Ghana has no limestone formations within its boundaries. Imagine trying to develop an entire nation without cement, or with cement that is very expensive because the limestone from which it is made must be shipped long distances. In urban development low-cost commodities are more important than high-cost commodities. Mineral and energy resources maps make it possible to plan for the use of low-cost resources before development so raises land prices that it is uneconomical to use commodities at all (Table 14-6).

Table 14–6

Data for Mineral and Energy Resources Maps

Surficial materials of resource value
Potential resources
Resource-producing areas
Resource-oriented land use
Resource installations

Table 14–7

Data for Active Processes Maps

Active processes
Surface drainage
Susceptibility to erosion
Flooding
Tidal inundation
Storm surge inundation
Wind strength and direction
Areas of excessive deposition

Table 14–8

Data for Maps of Manmade Features and Water Systems

Residential areas
Industrial areas
Drainage basins
Surface drainage
Freshwater marshes
Salt marshes
Lakes
Reservoirs
Bays
Estuaries
Manmade land (including spoil)
Channelization
Irrigation installations
Depth to groundwater
Aquifers
Aquifer recharge areas

Active processes maps (5)

Active processes maps (Table 14-7) of the Texas coastal zone illustrate nearshore phenomena, but the same principle can be applied to lakes and large rivers in other kinds of environments. Areas of extreme erosion, rapid deposition, and flooding are the most important features of these maps. The record floods of 1972 give impetus to planning for the better and safer use of floodplains, as do the Mississippi River floods of 1973.

Manmade features and water systems maps (6)

Because many manmade structures outside cities are located on or near water, it is convenient to record both natural and manmade water systems on the same map as other manmade features (Table 14-8). In addition to urban areas, roads, railways, and the like, manmade structures include drainage, irrigation, and barge canals; reservoirs; piers; jetties; groins; and a number of other kinds of water facilities. Knowledge of natural and manmade water systems as they operate together is essential to planning.

Rainfall, stream discharge, and surface salinity maps (7)

Surface salinities are important in all nearshore areas, and their maintenance is necessary to biologic assemblages (e.g., shellfish) that require water of specific salinity (Table 14-9). As freshwater lakes become more saline due to human intervention,

Table 14-9

Data for Rainfall, Stream Discharge, and Surface Salinity Maps

Flooding
Precipitation
Runoff
Salinity
Depth to groundwater
Aquifers
Aquifer recharge areas

Table 14-10

Data for Topography and Bathymetry Maps

Topography
Contours
Hachures
Colors
Physiography
Percent slope

changes in them must also be monitored, for the good both of man and of other animal and plant assemblages. Rainfall and stream discharges are essential to planning by providing information on the amount of water available for human consumption and potential flooding, as many cities in the Southwest will discover in the next two decades when they deplete their available sources of water. Drainage systems and bodies of water, both natural and artificial, add to planning problems, but increase aesthetic value if treated properly.

Topography and bathymetry maps (8)

From topographic and bathymetric maps, percent slope maps can be constructed (Figure 14-6). If the topographic and bathymetric maps are colored, percent slope maps can be overlaid in a pattern; if topographic and bathymetric maps are contoured, percent slope maps can be overlaid in color or in a pattern. Relief is important in all planning (Table 14-10).

Who Should Do Land Use Planning

This chapter has thus far dealt primarily with the collection and storage of data. But the collectors of data seldom make planning decisions. The purpose of derivative maps is to put data into a form that can be used and understood by everyone, not just by specialists. Yet there is still a gap between collecting data and making decisions, and the question that continually arises is: who should do the land use planning? Since land use decisions are usually political, many think that data collectors should not be involved in them. Thus potential political decisions might have less influence on the objectivity of data collection. But this separation of functions usually leaves decision making to people unfamiliar with the data. Perhaps this drawback can best be remedied by training a relatively new category of specialist—a broadly trained individual who combines some knowledge of data-gathering and political disciplines. Only this type of person is likely to make optimal use of planning data.

In most cities and communities of less than 500,000 people, developers control most of the planning and sincerely believe that they are the most capable planners. Although developers are necessary for community development, the contention that they are the best planners has proved fal-

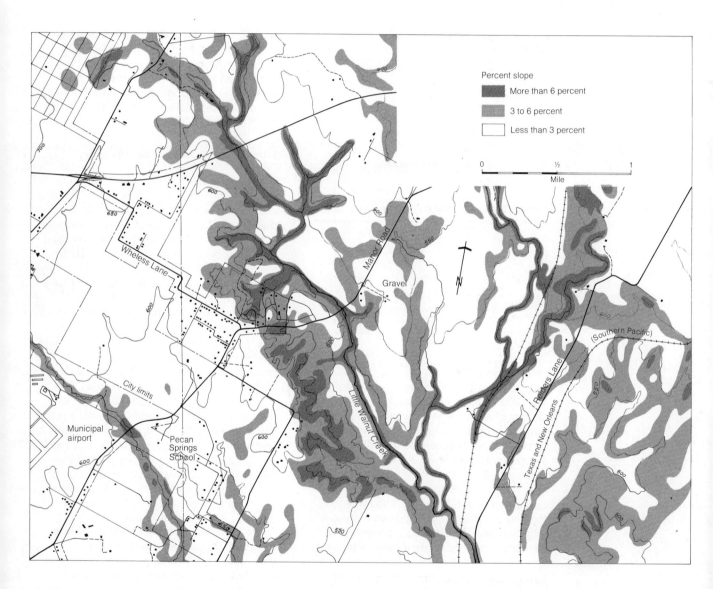

Figure 14-6

Percent slope map of part of the East Austin
Quadrangle.

Man-made lakes constructed by the Santee County Water District, California, to recycle waste water from a nearby sewage treatment plant

lacious in many situations. Cities and communities usually do their own planning, and planning outside of their corporate limits is restricted by law. But the environment is not always best served if the land beyond the city limits is not planned in conjunction with the city. City or municipal planning alone is inadequate. Consequently, area councils of government (COGs) have gradually evolved to coordinate planning of larger areas. Most COGs have not been operating long enough to prove themselves, although a few are showing good results. COGs have some authority, largely because federal funds can be withheld from all counties and communities in a particular COG unless certain planning decisions are made. The withholding of federal monies is sometimes unfair, but is an effective stimulus to decision making.

Planning authorities are needed for areas larger than COGs. In any area, such as a river system, where several COGs have jurisdiction or where part of the system has no COG, it is not always possible to obtain the cooperation needed to coordinate waste disposal (both municipal and industrial), water purification, water planning, power planning, or other types of planning that should be systemwide. This is because COGs do not cooperate or have different priorities, or because there is no planning in that part of the system without a COG. It does a river system little good if one COG plans for environmental protection and the remainder of the system continues to deteriorate.

Therefore it has been proposed that river basin authorities, originally established to undertake water resource planning, also be responsible for land use planning. The proposal has merit, but river system authorities were organized and staffed for water resource control only. Where they have been assigned planning and environmental control, their work has been shoddy. If properly staffed and organized, however, river basin authorities could become viable planning agencies.

Federal legislation

In order to stimulate planning at lower governmental levels, Congress and various federal agencies provide either legal or financial stimuli to plan toward certain goals.

In 1964 Congress established a Public Land Law Review Commission, "to study existing laws and procedures relating to the administration of the public lands of the United States, and for other purposes." The commission was chaired by Representative Wayne Aspinwall (Colorado), and composed of six senators, six members of the House of Representatives, and six members appointed by the President. The commission had a staff of 37 and 45 professional consultants.

The Public Land Law Review Commission published its extensive report, *One Third of the Nation's Land,* in 1970. In Chapter III, "Planning Future Public Land Use," the commission made 15 recommendations for future use of public lands. Of these recom-

mendations, four, with slight modification, could be applied to all land use planning. These are:

Recommendation 1: Goals should be established by statute for a continuing, dynamic program of land use planning. . . .

Recommendation 2: Public land agencies should be required to plan land uses to obtain the greatest net public benefit. Congress should specify the factors to be considered by the agencies in making these determinations, and an analytical system should be developed for their application.

Recommendation 11: Provision should be made for public participation in land use planning, including public hearings on proposed federal land use plans, as an initial step in a regional coordination process.

Recommendation 15: Comprehensive land use planning should be encouraged through regional commissions along the lines of the river basin commissions created under the Water Resources Planning Act of 1965. Such commissions should come into existence only with the consent of the states involved, with regional coordination being initiated when possible within the context of existing state and local political boundaries.

Although municipalities and cities might at first object to still another planning authority, the best approach would be for river basin authorities, with federal legislative backing, to issue broad land use goals toward which local governments could plan. Such a procedure would, for example, prevent the disposal of wastes or the siting of industries at localities or under restrictions that might prove detrimental to some other part of the system. Other legislative proposals delegate land use planning authority to the separate states, with restrictions or directions similar to those discussed above for the river system authorities. In the United States responsibility for decision making is determined by slow legislative and judicial processes.

The Environmental Impact Statement

Further definition and direction was given to the federal sponsorship of land use planning by the Environmental Policy Act of 1969, which required that every subsequent report or recommendation for proposals on legislation or other federal action should contain an **environmental impact statement** (EIS). These statements are to discuss:

1. the environmental impact of the proposed action,
2. any adverse environmental effects which cannot be avoided should the proposal be implemented,
3. alternatives to the proposed action,
4. the relationship between local short-term uses of man's environment and the maintenance and enhancement of long-term productivity, and
5. any irreversible and irretrievable commitments of resources which would be involved in the proposed action should it be implemented.

Executive orders and legislation since 1969 have included not only federal projects, but also all projects totally or partly funded by the federal government or guaranteed by federal funds. Thus housing developments secured by federal loans were included.

An environmental impact statement must be made available to the public at least ninety days before any action is taken on a project. Some such statements, like the EIS for the proposed trans-Alaska pipeline, take years to prepare. Statements on nuclear energy plants fill large volumes, and even a statement on a decision to terminate federal helium purchase contracts runs to 80 pages, with over 300 pages of supplementary evidence and appendices.

Some federal agencies, such as the Tennessee Valley Authority and the Army Corps of Engineers, prepare their own impact statements. Others collaborate on projects of mutual interest. The statement on the Salton Sea Geothermal Test Well contains appendices from the U.S. Forest Service, the U.S. Fish and Wild Life Service, the Bureau of Land Management, the U.S. Soil Conservation Service, and other agencies.

By legislative act, the environmental impact statement must be part of land use planning decisions. The answers to the questions posed by the legislation are provided by data collectors. In this way the EIS bridges the gap between data collection and decision making.

The U.S. Geological Survey, in its Circular 645, lists 85 environmental characteristics that can be disturbed by man in 98 different ways. Coordinating these against potential modifications is one method of preparing an environmental impact statement. If in Table 14-1 we replace the word "activity" with "modification," we have a chart from which to prepare an impact statement, since most human activities modify the ecosystem. Each activity pertinent to any project can be examined in terms of each of the resource capability units within the province of the project. Considering the pertinent activities in relation to the pertinent resource capability units will produce a competent environmental impact statement for the proposed project.

Summary

As population increases and food sources barely hold their own or lose ground, it seems that political units must become more rigid in classifying land. We are no longer in a position to waste arable land for urbanization when there is much nonarable land in the vicinity. That many cities are attacking the special tax laws that protect farm land from high urban taxes is probably a valid indication that such laws are needed. However, the planning problems posed for cities by such laws are real. The only solution would seem to be to give regional planning commissions and cities the authority to plan beyond city limits, and to enforce much more rigid land use than is now practiced, including the restriction of arable land to agriculture. This should not be done unless state or national laws become sufficiently strong to protect different kinds of land from unsuitable uses.

So many geological factors are basic to environmental problems that geologic maps are the best basis for land use planning, provided that derivative maps are produced and provided that the geologic base is overlaid with biological, geographical, hydrological, cultural, aesthetic, and recreational information to complete the necessary data bank. In addition, special maps —on seismic risk, depth to bedrock, soil, subsidence, evapotranspiration, water surplus and deficiency, and trace substances— are needed for those areas that have special problems. Some data change so rapidly that other special maps are necessary. Examples are flood prediction maps and land value maps. Culture and land use also change through time, but not as rapidly as flood predictions and land values. Because of changing culture and land use, land use planning maps should probably be updated about every ten years. As land use planning becomes more critical, basic geologic input will be more and more necessary.

Selected Readings

Brown, L. F., Jr.; Fisher, W. L.; Erxleben, A. W.; and McGowen, J. H. 1971. Resource capability units: their utility in land- and water-use management with examples from the Texas coastal zone. University of Texas Bureau of Economic Geology, Geology Circular 71-1.

Bybee, H. P. 1952. The Balcones fault zone—an influence on human economy. Texas Journal of Science 4:387–392.

Fisher, W. L.; McGowen, J. H.; Brown, L. F., Jr.; and Groat, C. C. 1972. Environmental geologic atlas of the Texas coastal zone—Galveston-Houston area. University of Texas Bureau of Economic Geology.

Flawn, P. T. 1971. Environmental impact statement. Geotimes 16(9):23–24.

Frye, John C. 1967. Geological information for managing the environment. Illinois Geological Survey Environmental Geology Notes no. 18.

Hackett, James E. 1968. Geologic factors in community development at Napierville, Illinois. Illinois Geological Survey Environmental Geology Notes no. 22.

Hilpman, Paul K., et al. 1968. A pilot study of land-use planning and environmental geology. Kansas Department of Economic Development. Report no. 15D.

Lamoreaux, Philip E., et al. 1971. Environmental geology and hydrology, Madison County, Alabama, Meridianville quadrangle. Geological Survey of Alabama, Atlas series 1.

Leopold, Luna B.; Clarke, Frank E.; Hanshaw, Bruce B.; and Balsley, James R. 1971. A procedure for evaluating environmental impact. U.S. Geological Survey Circular 645.

McHarg, Ian L. 1969. Design with nature. Garden City, N.Y.: Natural History Press.

Mohorich, Leroy M. 1972. The role of geologic input in urban planning. Mountain Geologist 8:209–219.

Powell, W. J., and Lamoreaux, P. E. 1969. A problem of subsidence in a limestone terrane at Columbiana, Alabama. Geological Survey of Alabama circular 56.

Taylor, Fred A., and Brabb, Earl E. 1972. Map showing distribution and cost by counties of structurally damaging landslides in the San Francisco Bay Region, California, winter of 1968–69. San Francisco Bay Region Environment and Resources Planning Study, U.S. Geological Survey, Miscellaneous Field Studies Map MF-327.

U.S. Geological Survey. 1967. Engineering geology of the northeast corridor, Washington, D.C., to Boston, Massachusetts: Coastal Plain and surficial deposits. U.S. Geological Survey Miscellaneous Geology Investigations Map I-514-B.

Wright, R. L. 1972. Some perspectives in environmental research for agricultural land-use planning in developing countries. Geoforum 10/72: 15–33.

Yeates, M. H. 1965. The effect of zoning on land values in American cities: A case study. In Essays in Geography for Austin Miller, J. B. Whittow, and P. D. Wood, eds., pp. 317–333. University of Reading.

Chapter 15

Too Many People

Precisely because population is the most difficult and slowest to yield among the components of environmental deterioration, we must start on it at once.
Ehrlich and Holdren, 1971

Man is a part of all ecosystems. He is also a heterotroph—an eater of a great variety of plants and animals. Were he a top carnivore only, and few in number, his position in the ecosystem might not be critical. But his success lies in his ability to adapt to and survive at nearly all levels of the food chain. This success may also be man's downfall. As he increases in number, he adapts his food requirements to those levels of the food chain that are easiest for him to harvest in sufficient quantity (man can also upset all higher levels). By this process man himself becomes more dependent on single foods, and cultivates them in such abundance that they in turn become more subject to disease explosions (e.g., corn blight), drought, or other disasters.

The paradox is that man, as part of the ecosphere, can outproduce his place in it, and destroy either the entire ecosphere or himself. His alternative is as an intelligent social and political animal to apply new controls, or reinstate those old natural controls that his technology and his genius have allowed him to subvert. "A storm of crisis problems" is the term applied by John Rader Platt to man's present predicament. Contributing to this predicament are three phenomena related to population: population increase, the availability of food, and good medicine.

The Malthusian Principle

Glen D. Everett has described the bewilderment of an old man named John Eli Miller, who had experienced his own population explosion: when he celebrated his 95th birthday Miller had a grand total of 410 descendants. At the time of his death he was receiving a letter once every ten days announcing the birth of a new descendant. Fortunately for the United States, few of us live to see 410 living descendants. But the story of John Miller emphasizes both the power of the sexual drive and, when enough food (and medicine) are available, the results of that drive. T. R. Malthus, in his almost notorious *Essay on the Principle of Population,* maintained that human populations have the potential to increase so much more rapidly than their food supply that most of humankind is continually on the verge of starvation.

Several years before Malthus' essay appeared, the Reverend Joseph Townsend applied its thesis to animals. The good reverend described how at one time there had been left on an island in the South Seas one male and one female goat. With no predators and no diseases to contend with, these goats after some generations populated the entire island. British privateers discovered it, and both pirates and British seamen replenished their meat supplies at the island. The Spanish decided to put an end to this British supply station, and introduced to the island one male and one female greyhound. With plenty of goats for food, the greyhounds in turn fulfilled the Malthusian principle, multiplying to the ex-

tent of their food supply. Finally the goats remained only in the high rocky places where the greyhounds found footing difficult. The goats ventured out of the rocks for food, and only the swiftest and most alert returned. In turn, only the swiftest and most alert of the greyhounds could get enough food to survive. A balance was achieved between predators and prey.

A similar balance was maintained by early man until the New Stone Age, when the development of weapons reached the point that man began to destroy the balance of nature (see Chapter 1). The technology with which man can destroy his environment has become ever more efficient, particularly in the past 40,000 years. With improved technology man's numbers have continually expanded to meet his new food-gathering ability. Intentional population control has been practiced by only a few small cultures such as that of Tahiti, where excess children were thrown into the sea.

Population Control in Animals

In nature population control is the rule. Its maintenance in all animal populations is not always understood, but the Reverend Townsend's example illustrates one common mechanism, and there are many others.

The hare-lynx cycle

For example, in a certain area in Canada, the hare population rises and falls in cycles of slightly less than ten years, and over that period the number of hares ranges from 20,000 to 80,000 or more. The lynx cycles follow the hare cycles, because hares are a primary food of the lynx (Figure 15-1).

Figure 15–1

Relationship of hare (herbivore) to lynx (predator) populations in Canada.

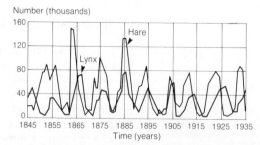

Number (thousands)

Time (years)

Rabbits in Australia

Failure in population control accompanied the introduction of rabbits into Australia. Because there were no natural predators, rabbits reproduced rapidly and became a scourge to the eastern half of the continent. To prevent rabbits from invading the western half of the continent, the government decided to build a rabbitproof fence some 1500 miles long across the country. But it was also decided to import from Europe a rabbit disease, the *Myxomatosis* virus, and to introduce it into the rabbit population. Since the virus did not exist in Australia, freely breeding rabbits had lost their resistance to it. The virus cut the rabbit population so rapidly (Figure 15-2) that the Australian government decided not to complete the fence. But the remaining resistance in the genetic system of Australian rabbits prevented their complete eradication, and they again began to multiply rapidly. The disease did not spread as rapidly as the

Rabbits converge on a waterhole during a drought in Australia

Figure 15-2

Effect of the *Myxomatosis* virus on the rabbit population in Australia.

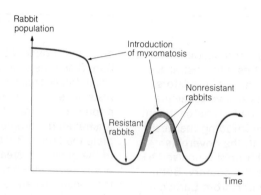

population increased, and soon nonresistant rabbits were plentiful again. Consequently, the disease was introduced a second time, and there was again a rapid decrease in rabbits, with only the resistant strain surviving. These resistant rabbits then fulfilled the Malthusian principle and increased to the extent of their food supply, until eastern Australia was again overrun with them and the Australian government was again building its rabbitproof fence.

Factors Controlling Population

Plants

Plant populations are controlled by nutrients, climate, light, and disease. If some nutrient is at the minimal level for plant growth, the plant population is controlled by the amount of that single nutrient available. If none of the required nutrients is at a minimum, a plant population is controlled by the interaction of nutrients or by some other means. Climate is an effective control on some plants. Minimum or maximum temperatures are effective population controls, especially on the fringe of habitat tolerance. Thus spruce and birch near the tundra can live only in protected areas.

Evapotranspiration has been used to explain the boundaries between forests and grasslands. Forest life extends into the grasslands on damp northern slopes where evapotranspiration is less, and along stream valleys where there is more water available.

Diseases like the Dutch elm disease and the virus destroying so many flamboyant trees in Puerto Rico can also control or destroy a population. The activities of many plants are controlled by light. Some plants, like the succulents in the lower levels of the tropical forest, need little light. Others, including many desert species, require much intense light. Some deciduous trees are so sensitive to light that the appearance of leaves in the spring is not a function of temperature, but of the increasing length of the day.

Animals

Animal populations are controlled by food, space, climate, disease, and stress. Through many generations, populations may grow tolerant or immune to once-destructive diseases, and those controls become ineffective.

Populations of herbivores adapted to specific food plants are controlled by plant distribution. For example, the *Agave* (century plant) moth does not occur where there are no *Agave*. The distribution of food sources organizes and controls the size of different parts of the animal community, and no part of the community can exceed a size commensurate with its food source.

In some animal populations space is a population control mechanism. With plants space is related to light requirements, since too-dense growth restricts light, which in turn restricts plant growth. With animals there seems to be a psychological control. The male African springbok that cannot find and defend a territory attracts no female and sires no offspring. Many birds have territories, and a young male that cannot find and defend a territory attracts no mate. In these compelling behavioral situations, when all the territories in a breeding area are occupied, the population starts to stabilize.

There is evidence that some species, when under the stress of overpopulation, produce fewer offspring. This is not an efficient control mechanism, however, for— as with man—stress begins to operate only when the quality of life has already deteriorated. The ability of a population to grow is called its **biologic potential**. This factor ranges widely among species, and the larger the animal the less the potential. Thus, elephant populations do not have nearly the biologic potential of fruitflies (fortunately); man has less biologic potential than mice or rabbits. Under natural conditions, complex interactions of many phenomena prevent the biologic potential from being fully realized.

Population Control in Man

The human population was once controlled by the same factors that control other animal populations. Technology has quite changed this. For example, it has improved man's food-gathering ability so that he can support much greater numbers. War, disease, famine, sanitation, medicine, and food-producing technologies control the size of human populations.

War

War has always been thought of as a stringent population control, but the facts show otherwise. There may be exceptions, such as the piling of a million skulls in the square of Baghdad by Tamerlane in the fourteenth century. In man, however, as in most animal populations, the birth rate is not reduced by the killing of males only.

Disease

Epidemics have long been thought to exert a severe form of population control, and it is often said that if man does not control his own population, epidemics will do so for him. Malthus reportedly derived some of his ideas on population from studying population increases in Europe following the bubonic plague. Crusaders returning from the Middle East brought with them the Norwegian (black) rat and the virulent relative of typhus called bubonic plague. The Norwegian rat had been wild in the Middle East, probably becoming associated with man only a few centuries before the outbreak of the plague in Europe.

When the Tatars invaded Europe from the East, making their farthest advance into Hungary in 1241, they brought with them the common brown rat, later called the European rat. The brown rat had apparently run wild in northeast Asia until about the

seventh century, when it became associated with man. Although there may have been typhus-like diseases in the Mediterranean area and in Europe before the advent of the rat, modern typhus, like the plague before it, appears to have been brought west from Asia by men and rats.

Deaths from bubonic plague reduced population in Europe a number of times between the fourteenth and the seventeenth centuries. Between 1340 and 1400 more than 25 million people—almost a third of the European population—died of the plague (Figure 15-3). Yet in the next two centuries the population doubled, as Malthus would have predicted. Again in the seventeenth century the plague hit England, where it created havoc. However, lack of immunity or tolerance to bubonic plague barely slowed total population growth in Europe, since recovery after each great

dying-off was much more rapid than the normal curve of increase had been. The effect of the plague was only temporary; population increased rapidly after each epidemic to meet the food-producing capability.

Three million deaths from typhus in eastern Europe around the end of World War One hardly slowed the increase in population. The rapid spread of cholera across Asia, starting in Hong Kong about 1937 and moving westward into Africa and Europe, does not even appear to be a significant point on the total population curve of Asia, Africa, or eastern Europe.

Famine

It is estimated that between 20 and 30 million people starve to death every year. This figure does not include all those who die of disease because malnutrition reduces their resistance. Though much of humanity is very near starvation, population still increases where starvation is most prevalent. In these areas, though drought may kill tens of millions in any year, population continually increases to a number that cannot be provided with more than a minimal diet.

Sanitation and medical science

Improved sanitation in the developed countries was responsible for the major increase in population in the nineteenth century, just as good medicine has been in this century. The rate of population growth is determined by the number of births compared to the number of deaths. True zero population growth is reached when the number of

Figure 15-3

Effects of bubonic plague on European population. (From "The Black Death" by William L. Langer. Copyright © 1964 by Scientific American, Inc. All rights reserved.)

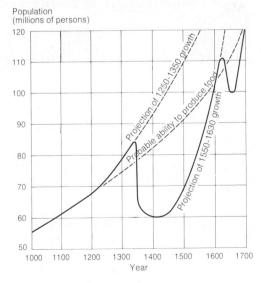

Beggars trying to get food
in Dacca, Bangladesh

deaths equals the number of births. Sanitation has lowered the number of deaths, particularly among children, and good medicine has lengthened the lives of all. The population continues to grow, even with a low birth rate, as long as the birth rate exceeds the death rate. Even with zero population growth, the population does not cease growing until all those born prior to achieving zero population have had their two children.

Food production

Improved food production technology allows us to produce more food so that we can bear more children who in turn improve food technology, *ad infinitum*—until we ruin our soils and deplete our trace substances. And still most of the people are at starvation level! The net effect of all the above is that good medicine, sanitation, and improved food-producing technology

Table 15-1

World population projected to the year 2050 (millions of people). From Nathan Keyfitz, "United States and World Populations." In Preston E. Cloud, ed., *Resources and Man: A Study and Recommendations* by the Committee on Resources and Man of the Division of Earth Sciences, National Academy of Sciences-National Research Council, with the cooperation of the Division of Biology and Agriculture. San Francisco: W. H. Freeman and Company. Copyright © 1969. Table 3.3, p. 57.

World Population

	1970	1985	2000	2050
United States (official estimate)	205	251	304	470
United Nations "medium" variant				
Total developed countries	1,082	1,256	1,441	2,040
Total underdeveloped countries	2,510	3,490	4,688	8,320
World total	3,592	4,746	6,129	10,360

have allowed population to increase far more than war, famine, and pestilence have held it down.

Prospects for World Population

It has been estimated (Figure 15-4) that a million years ago the population of the world was just over 100,000. From then until 10,000 years ago it increased about tenfold, to about 1 million people. Because of the agricultural revolution it then took only 9000 years (that is, to 1000 A.D.) to increase

100 times to about 100 million people. Then, because of steadily improved agriculture and the first use of fossil fuels, the next 900 years (through the Industrial Revolution) saw another tenfold increase. These figures are only rough estimates, but are accurate enough to suggest how dramatic the population growth has been.

Most of the world's people have lived within the last thousand years. Table 15-1 shows population estimates for the United States and the world for 1970, 1985, 2000, and 2050. The apparent steepness of a population curve, which is sometimes very impressive, varies with the method of plotting. If thousands of years are plotted on a line three inches long, the curve will appear much steeper than if only 30 years are plotted. If the units on the vertical scale (*y* axis) are spread out, the curve will appear much steeper than if they are not. Figure 15-5 shows gradually steepening curves for the populations of the world, undeveloped regions, and developed regions. The population of the developed regions is increasing much less rapidly than that of the undeveloped regions. The curves appear much steeper in Figure 15-6, which covers a much longer period of time than does Figure 15-5. Figure 15-6 shows the increase in population accompanying the Industrial

Figure 15-4

Estimated increase in world population for the past million years.

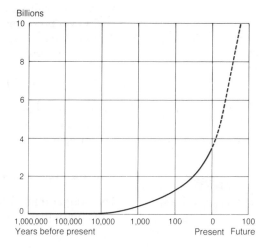

Figure 15–5

Population of undeveloped and developed areas of the world, projected to the year 2000. (From *Resources and Man: A Study and Recommendations* by the Committee on Resources and Man of the Division of Earth Sciences, National Academy of Sciences-National Research Council, with the cooperation of the Division of Biology and Agriculture. W. H. Freeman and Company. Copyright © 1969.)

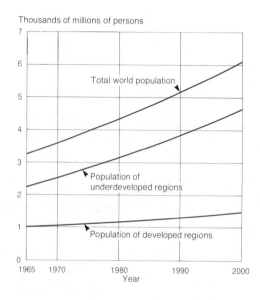

Thousands of millions of persons

Total world population

Population of underdeveloped regions

Population of developed regions

Year

Figure 15–6

Population curves for various regions of the world for the last three and one-half centuries.

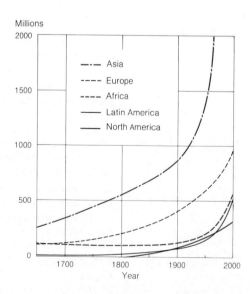

Millions

--- Asia
---- Europe
---- Africa
— Latin America
— North America

Year

Revolution. The greatest population increases are in Asia, Africa, and Latin America, the undeveloped regions. Europe and North America have a much slower rate of increase. The doubling times of the world population for the past 2000 years are given in Table 15-2. That there can be disagreement in projecting populations to the year 2000 is illustrated in Figure 15-7. The lower curve of D. J. Bogue, dating from 1967, projects greater use of birth control than is reflected in the other curves. Notice that the United Nations estimates vary from above to below a 2 percent increase per year.

Starvation and Malnutrition

Food problems have been discussed earlier in this book (pages 177 and 182). Population creates our greatest food problems, since it is the increase in population that has kept most of the world's peoples on the verge of starvation, just as Malthus predicted. The relation of the food problem to population can be discussed in the form of answers to four questions:
1. Can we produce enough food?
2. If not, should we continue to feed those who will produce children destined to starve?
3. If we continue to maximize food production, will we wear out our soils at the expense of future generations?
4. Can new sources of food counteract shortages?

Can We Produce Enough Food?

The hunger map of the world (see Figure 7-8) shows that only Australia, Argentina, the United States, Canada, Iceland, northern Europe, and northern Asia have really adequate diets. Central America, Ecuador, the Caribbean Islands, nearly all of Africa, Madagascar, southern Asia, the Arabian Peninsula, and most of the islands of the East Indies have starvation or malnutrition diets. As Table 15-3 illustrates, protein-deficient countries must depend on plant protein sources whereas protein-sufficient countries enjoy a much larger ration of protein from animal foods. Research in process suggests that mental deterioration occurs more rapidly in the elderly among vegetarians than among meateaters, because vegetarians are likely to experience protein deficiency. This finding should be considered in outlining food programs for the future. Figure 15-8 is an estimate of the increased demand for food over the next three decades in countries with protein and/or caloric deficiencies.

None of the estimates is for less than a 100 percent increase, and for Asia and the Far East estimates range above 150 percent. From these data and others impossible to cover here, the United Nations Food and Agricultural Organization (FAO) has estimated the projected animal protein gap in the developing countries (Figure 15-9). According to the most optimistic projection, the gap between available supply and minimal requirements—much less target requirements—is destined to increase rapidly. Hence, the answer to the question, "Can we produce enough food?" is an emphatic "No!"

Should We Feed People Who Will Produce Children Destined to Starve?

Even if we produce the minimum food requirement, what then? A Thai woman who

Table 15-2

Time required to double world population. From *Biology and the Future of Man* edited by Philip Handler. Copyright © 1970 by Oxford University Press, Inc. Reprinted by permission of Oxford University Press, Inc. Table 20-1, p. 902.

World population	Year	Time required
250,000,000	1	
		1649
500,000,000	1650	
		200
1,000,000,000	1850	
		80
2,000,000,000	1930	
		45*
4,000,000,000	1975*	
		30*
8,000,000,000	2005*	

*Estimate

Figure 15-7

Estimates of world population, projected to the year 2000. (From *Resources and Man: A Study and Recommendations* by the Committee on Resources and Man of the Division of Earth Sciences, National Academy of Sciences-National Research Council, with the cooperation of the Division of Biology and Agriculture. W. H. Freeman and Company. Copyright © 1969.)

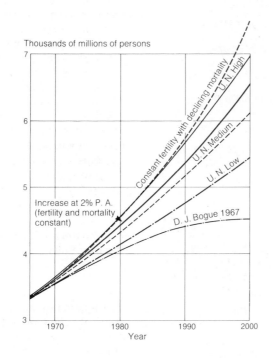

Thousands of millions of persons

Constant fertility with declining mortality

U.N. High

U.N. Medium

U.N. Low

Increase at 2% P. A. (fertility and mortality constant)

D. J. Bogue 1967

Year

Table 15-3

Relative importance of vegetable protein compared to animal protein in different parts of the world. Figures are in grams per day per person. From Food and Agricultural Organization of the United Nations, *The State of Food and Agriculture, 1964*. Rome: United Nations, 1964, Table III-3, p. 108.

Area	Total protein prewar	Total protein 1961–1962	Animal protein prewar	Animal protein 1961–1962
North America	86	93	51	66
Western Europe	85	83	36	39
Mexico	53	68	18	20
South Asia	52	50	8	7
Far East, including mainland China	61	56	7	8

had been brought down out of the hills emaciated and suffering from malnutrition was kept at a clinic for several weeks, given a good diet, shot full of vitamins, and nursed back to robust health. When she returned to her home in the hills, the late Dr. Tom Dooley was asked, "What will she do now?" He replied, "Have another child." It should be asked now, as then, "What is the justification for nursing women back to peak health so that they can bear more children into abject misery?" The quick, unthinking solution to this problem is to say, "Send them our excess food." But all the excess product of North America can give only a small percentage of India's millions a sufficient diet, leaving nothing for the rest of the world. Thus the answer to our second question is another "No!" But when we see the miseries of those now living, we cannot deny them. We face a moral question that will not be answered solely by geologists or agronomists: should we be swayed by the greater hunger of the next generation, or the lesser but immediate hunger of the present one?

Figure 15-8

Percentage increases expected in the demand for food by the year 2000 (top) and percentage increases in grain yield per acre from 1934 to 1960 (bottom).

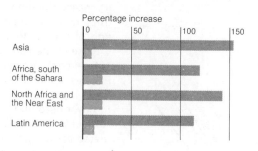

Figure 15-9

Projected gap in animal protein needs for undeveloped countries.

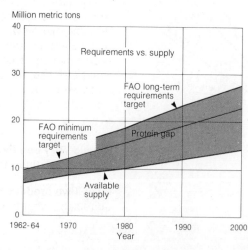

Will Maximum Food Production Eventually Destroy Soils?

If we continue to maximize food production, will we wear out our soils at the expense of future generations? As indicated in our study of soils (pages 180–181), the answer is "Yes." The answer need not always be affirmative but before it can be changed, a much greater knowledge of various

Figure 15-10

Rice production in different countries. (From *Resources and Man*: *A Study and Recommendations* by the Committee on Resources and Man of the Division of Earth Sciences, National Academy of Sciences-National Research Council, with the cooperation of the Division of Biology and Agriculture. W. H. Freeman and Company. Copyright © 1969.)

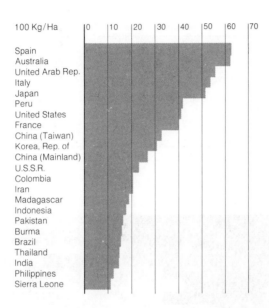

Figure 15-11

World consumption of potash (as fertilizer).

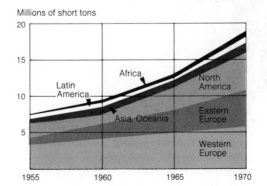

time soils are used contribute to these differences. Many parts of the world have increased their use of fertilizer tremendously in the past 15 years (Figure 15-11). In time Asia, Latin America, and Africa will probably follow the examples of Europe and North America. But the more rapidly we add fertilizers, the more rapidly our soils will wear out. The special disadvantage of Asia, Latin America, and Africa is that their soils are innately poor.

On the other hand, the Dutch have farmed and fertilized some of their soils for so many centuries that it is impossible to determine their original type. Nonetheless, these are still good soils, because they have been enriched with natural fertilizer (animal waste), not the artificial concentrates used in such quantities in most of the rest of the world. The soils in Ireland have also been altered by early man. It may not be necessary for us to wear out our soils, but we must give them thought and care if we are to avoid doing so.

Can New Food Sources Offset the Widening Caloric and Protein Gap?

Culture and food production

Increases in the food supply are attributable to a surprising variety of sources. For example, there is a relationship between yield of wheat per acre and the literacy level of the farm population. It is thought that the farmer who can read the directions on packages of seeds, sacks of fertilizer, and bags of weedkiller and pesticides follows directions more accurately and produces more food.

agricultural methods is needed. Figure 15-10 shows that rice production ranges from about 24 bushels an acre in Sierra Leone, Africa, to about 130 bushels an acre in Spain. Factors such as type of soil, amount of fertilizer, climate, and length of

Table 15–4

World land area, arability, and soil groups (billions of acres). From the President's Science Advisory Committee, *The World Food Problem*, vol. II. Washington, D.C.: U.S. Government Printing Office, 1967, Table 7.1, p. 423.

Soil group	Potentially arable	Grazing	Nonarable	Total
Alluvial	0.79	0.43	0.25	1.47
Tundra	0.00	0.00	1.28	1.28
Desert	1.07	2.26	1.93	5.26
Mountain and others	0.34	1.38	6.91	8.63
Podzols and yellow-red podzols	1.17	1.98	2.91	6.06
Prairie	1.42	1.04	0.29	2.75
Latosols	1.07	2.06	3.92	7.05
Total	5.86	9.15	17.49	32.50

Figure 15–12

Animal protein supply per person in selected countries. (From *Population, Resources, Environment: Issues in Human Ecology*, 2nd ed., by Paul R. Ehrlich and Anne H. Ehrlich. W. H. Freeman and Company. Copyright © 1972.)

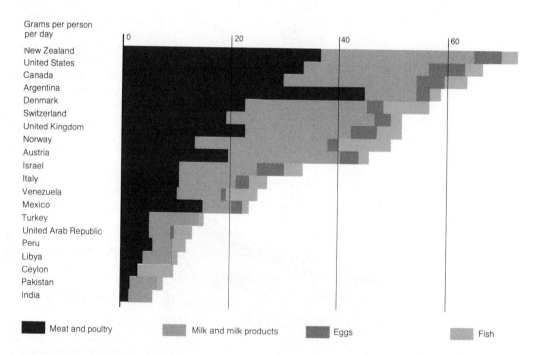

Grams per person per day

New Zealand
United States
Canada
Argentina
Denmark
Switzerland
United Kingdom
Norway
Austria
Israel
Italy
Venezuela
Mexico
Turkey
United Arab Republic
Peru
Libya
Ceylon
Pakistan
India

■ Meat and poultry   Milk and milk products   Eggs   Fish

Increasing agricultural acreage

Geographers have long insisted that much of the world's soil resources are not being used to grow food. Table 15–4 illustrates that most of the soils not now under cultivation are highly weathered soils that would be inefficient food producers without highly sophisticated agricultural methods, including proper fertilizers for leached latosols—fertilizers not now in existence. These highly weathered soils have not been worn out by man, but by the processes of nature—principally leaching.

Ocean farming

Many have pointed to data like those in Figure 15-12 and said, "Look, the world uses very little protein from the oceans." Kleiber has discussed the capacity of the earth to support human beings on different food sources (Table 15-5). It is his opinion that about one-hundredth of an acre devoted to the intensive culture of algae could support one person, and that 1.81 billion acres could support 2.26×10^9 people. He also gives figures for potatoes, grain, prunes, and other foods. But Kleiber has overlooked two considerations: (1) there is no assurance that 1.81 billion acres of continental shelf, or a comparable area in any part of the ocean, can be submitted to algal farming; and (2) it is unlikely that man can live satisfactorily on algal protein alone, considering the variety of plant and animal proteins and the 40 or more trace elements man uses. Besides, a strict diet of algal protein could soon become intolerable. Kleiber recognizes these drawbacks. But his chart is misleading and not useful in estimating the number of people the earth can support. It is interesting to note, however, that eggs have so few calories that, according to Kleiber, if man lived on eggs alone the earth could support only a fraction of its present population!

The contributions of the oceans to the food reserves of the world cannot be overlooked. But we know that overfishing alters ecosystems. The elimination of the whale, for example, is a man-produced disclimax. The whale is at the top of his food cycle and is the defining animal of a climax marine community. As man tries to harvest more food from the ocean, he will upset different oceanic communities, producing more disclimaxes. If a disclimax occurs low in the food cycle it will probably be disastrous, because all marine life above it in the food cycle will be damaged or destroyed, depending on the degree of dependence on the missing part of the food pyramid.

The decline of marine bird communities on the west coast of Peru, concurrent with rapid expansion of Peruvian fisheries, suggests that the capacity of the ocean to produce food for man may be greatly overrated. As fishing tonnages increased, the Peruvian fertilizer industry declined proportionately. The fertilizer was guano, the dung of sea birds (cormorants) that live on fish. In six years so many fish were harvested that the population of cormorants decreased to 20 percent of its original number, and fertilizer production dropped accordingly. If the expansion of Peruvian fisheries can so thoroughly upset the food cycle in the Humboldt Current in as little as six years, it seems inevitable that similar expansion of fisheries in other parts of the world's oceans would have similar effects.

In rebutting the optimistic supporters of the "food-from-the-sea" myth, Paul and Anne Ehrlich use the whale as an example. They argue that the decline in the number of

Table 15-5

Capabilities of food production for man. Modified from Max Kleiber, *The Fire of Life*. New York: John Wiley & Sons, 1961, Table 19.5, p. 341.

Food source	Acres per person	Billions of people*
Algae	0.01	226.25
Potatoes	0.15	12.13
Grain	0.30	7.07
Prunes	0.37	5.07
Milk	0.37	5.07
Pork	1.0	1.81
Eggs	7.4	0.25

*Billions of people that can be fed on 1.81 billion acres of arable land.

Figure 15-13

The killing of whales. (From *Population, Resources, Environment*: *Issues in Human Ecology*, 2nd ed., by Paul R. Ehrlich and Anne H. Ehrlich. W. H. Freeman and Company. Copyright © 1972.)

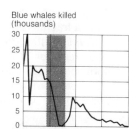

Blue whales killed (thousands)

First, the industry killed off the biggest whales—the blues. Then in the 40's as stocks gave out

Since 1945 more and more whales have been killed to produce

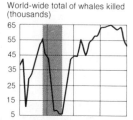

World-wide total of whales killed (thousands)

Figure 15-14

Whale kills versus whale oil production since 1930. (From *Population, Resources, Environment*: *Issues in Human Ecology*, 2nd ed., by Paul R. Ehrlich and Anne H. Ehrlich. W. H. Freeman and Company. Copyright © 1972.)

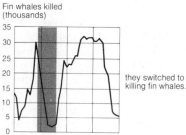

Fin whales killed (thousands)

they switched to killing fin whales.

less and less oil.

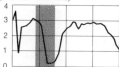

World-wide whale oil production (millions of barrels)

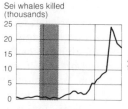

Sei whales killed (thousands)

As fin stocks collapsed, they turned to Seis.

Catcher boats have become bigger

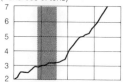

Average gross tonnage of catcher boats (hundreds of tons)

and more powerful,

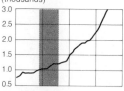

Average horsepower of catcher boats (thousands)

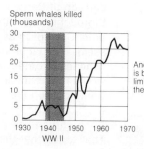

Sperm whales killed (thousands)

And now, the sperm whale is being hunted without limit on numbers—the ultimate folly.

but their efficiency has plummeted.

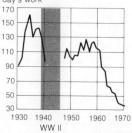

Average production of whale oil (barrels) per catcher boat per day's work

whales shows what will happen when man tries to harvest any trophic level of the sea (Figures 15-13 and 15-14). The best method of producing large quantities of food from the oceans is intensive and controlled farming of bays, estuaries, lagoons, and other shallow nearshore areas, rather than chancing a disastrous disclimax in the open oceans. Intensive farming of ocean margins is impossible under present expanding open-beach and open-water policies. If nearshore areas are to be farmed, people must be kept out of them. In other words, recreation and food requirements are not usually compatible. New legislation will be required to resolve the impasse.

It seems unlikely that the oceans can supply more than a minor part of the food needed for an increasing world population. Although oceans can produce more food than they do now, ocean harvesting is not the whole answer to the problem of feeding billions of people.

Genetics and food production

Geneticists are continually working to develop new hybrids and varieties of plants that will grow more rapidly and produce more calories and proteins, and convert more solar energy to plant food. Genetics, the increased use of fertilizers, and the distribution of more water kept food production even with population growth in the 1950s, but not in the 1960s. N. E. Borlaug, who in 1970 received a Nobel Peace Prize for his work in genetics devoted to greater food production, has said that there is no chance of catching up with the present rate of population growth.

The development of special high-protein hybrids of some cereals has produced plants of less vigor and less resistance to disease. This explains the decreases in rice production in the Philippines and southeast Asia in the early 1970s, and accounts for the lowered resistance of corn to corn blight in the United States. Can we expect better of other crash programs?

More efficient use of solar energy

It may be possible to increase plant production through better use of more of the solar energy that reaches the surface of the earth. According to some ecologists, available carbon dioxide is the limiting factor in plant production in nondesert areas. If so, increasing the carbon dioxide content of the air should increase the growth of plants. At present less than 0.3 percent of the solar energy reaching the earth is converted into plant tissue. If the plant biomass can be increased by more carbon dioxide, as has recently been accomplished in experimental laboratories, or by breeding experiments that yield more efficient plants, more food may be grown.

Man and protein

The solution to the protein problem does not lie entirely in the realm of food crop production. Since man evolved as a primary carnivore, his body either lost or never developed the capacity to manufacture certain substances (including some proteins) necessary for optimal functioning. This means that man cannot function at his best on a diet consisting of a single plant food without extensive supplements of proteins and vitamins. Eventually human beings may subsist optimally on a single food source and many artificial supplements. Too many

food faddists and population experts think only of protein in general, not of the great variety of proteins used in biological systems, the great variety of their sources, or the specificity of their use. Neither now nor in the near future can technology supply all the dietary supplements we require.

So the answer to our fourth question, "Can other food sources offset the widening food gap?" is another resounding "No!"

People Pollution: The Quality of the Environment

Twenty years ago, a geology class traveling through Yellowstone National Park on the Fourth of July found no space at all in the camping areas. Now, though the camping areas are much larger and more numerous, there are also more people, and camping has deteriorated further. To one familiar with nature it is incredible to see a lone bullmoose grazing at Fishing Bridge with two or three hundred people, ignorant of moose behavioral patterns, watching and photographing. But this happens in Yellowstone.

The time may soon arrive when a visitor must write weeks ahead for permission even to enter a national park, and attendance will be not only controlled, but restricted, so that not everyone who wants to visit the park will be given permission. Reservations may go to those who ask first, or be allotted by some ethnic or educational formula. The presence of too many people is ruining our national recreation areas, and their quality must be maintained if they are to go on being recreational. Overcrowding and the consequent decrease in environmental quality are results of people pollution and a threat to the quality of life.

Depreciation of the human species

Many animal species have density control mechanisms. When the population reaches a certain density, either the birth rate drops off or, if new space and food are available, the new space is occupied. Human populations do not seem to operate this way. But even in many animal populations, the quality of the environment seems to drop off before such controls set in. This is true of deer in Texas, where men have decreased the numbers of their natural predators. Perhaps this is because deer evolved to depend on predators, rather than a density control mechanism, for population control.

Even under the worst conditions human birth rates often continue to increase. Human populations on a low protein diet seem to select for the gene suite that can function with minimal needs, rather than for a controlled birth rate and the best possible gene suite. In areas of protein starvation the best brain-and-body combination, which requires more protein and will not function on less, is selected against in favor of a brain and body that will function less well, but at least will function. In such conditions man is not selecting for individuals of high metabolic or intellecutal potential, but for individuals of mediocre potential, because those who have lesser metabolic needs can get by and still function. This is a form of people pollution that can degrade the quality of mankind.

Dominance of the human species

Some say man differs from other animals because of his brain. Others say he is distinguished by his unique self-awareness. Still others maintain that he is different because he has something they call a soul. Some ecologists and ethnologists maintain that man's most distinguishing characteristic is his propensity to dominate ecosystems in ways that no other animal can. Because he has become a heterotroph, and needs a variety of food types for optimal living, his ability to dominate ecosystems and destroy their variety makes him vulnerable to his own activities. As population increases, man's dominance of ecosystems increases, making them more subject to destruction. People, simply by virtue of their numbers and their group behavior, pollute ecosystems.

Resources and population management

For the entire period of his evolution man has exploited the earth at will. Only in the last few decades have there been any serious attempts at resource management, and these have been largely unsuccessful. Not only commodity resources, but also recreational and aesthetic resources, are coming under management. How can we save the mountain lion, or the grizzly bear, or any number of other endangered species, when under the pressure of increasing population men keep moving into the territories these animals occupy? Why develop recreational land only to allow squatters to occupy it, even temporarily, as has been done in parts of the public domain along the Colorado River of Texas, Central Park in New York City, the Boston Common, and islands along all of our coastal lands—or even to allow activists to camp on it in times of dissatisfaction, as on the mall in Washington, D.C., preventing its intended aesthetic or recreational use? People are attracted to aesthetically pleasing areas in such numbers that they destroy the very

values they come to enjoy. Population pressure, furthermore, drives less competitive people into the very areas the majority are trying to save.

Until better population management is possible, legal means are needed to prevent the natural results of population pressures. Strict zoning ordinances must be enacted, not only in cities, to prevent encroachment on lands necessary to aesthetic and mental welfare. Land use planning is one of our chief needs.

What is gained by rationing fuel oil or gas if we let the population increase and every individual becomes a little colder? What is gained by trying to save our water and our soils if the population continues to grow and every individual become a little hungrier, a little thirstier, or a little dirtier and hence more subject to disease? As John Eli Miller asked, when told he had 410 living descendants, "Where will they all find farms?" In the long view, resource management without population management is a game in which the cards are stacked against us.

Summary

The only answer to people pollution is population management. We should calculate the population our system will support at a desired standard of living and restrict it to that level. Such controls are not likely to be instituted in the near future, because we lack data and because people are not yet ready to accept such decisions. Perhaps our greatest fault as a nation and a world is that we are so pragmatic that we can neither think nor plan beyond the next price rise. To counter this tendency Abrahamson (1970) has pulled some figures out of thin air.

He says that if a $12,000 annual income can provide a satisfactory "quality of life," the world as a steady-state system can support only 500 million people, about one-sixth of its present population. His 500 million is an educated guess, but he must be applauded for supporting population reduction and a nongrowth economic system. He is also thinking of the earth as a spaceship. Men have to live on an earth that can feed them and absorb their wastes. Our earth cannot continue to provide these services to an ever-increasing population. It appears that the earth cannot cure the waste of as many people as it can feed. Increasing population threatens survival, and the controls that would be needed to manage great numbers of people quickly threaten such values as individual liberty, aesthetic freedom, freedom of choice, free will, and all those other abstractions that make man an individual and a human.

Every effort to prevent pollution and produce more food and other resources is bound to be short-lived under present world population policies. But such temporary procedures can provide lead time so that people can be educated to the need for limiting population to that number for which the world can provide. If this education is unsuccessful, all other measures are in vain.

Selected Readings

Abrahamson, Dean E. 1970 Effects of pollution on population growth. In *Population growth: Crisis and challenge,* John R. Beaton, and Alexander R. Doberenz, eds., pp. 43–50 and panel discussion, pp. 51–76. Proceedings of the 1st Population Symposium, University of Wisconsin at Green Bay.

Cloud, Preston, et al. 1969. *Resources and man.* San Francisco: W. H. Freeman.

Cook, Robert C. 1962. How many people have ever lived on earth? *Population Bulletin* 18:1–19.

De Camp, L. Sprague. 1972. *Great cities of the ancient world.* Garden City, N.Y.: Doubleday.

Deevey, Edward S., Jr. 1960. The human population. *Scientific American,* September: 194–203.

Ehrlich, Paul R., and Ehrlich, Anne H. 1970a. *Population, resources, environment: Issues in human ecology.* San Francisco: W. H. Freeman.

———. 1970b. The food-from-the-sea myth. *Saturday Review* 4 April: 53–55, 64–65.

Ehrlich, Paul R., and Holdren, John P. 1971. Impact of population growth. *Science* 171:1212–1217.

Everett, Glen D. 1961. One man's family. In *Population, evolution, birth control,* ed. Garrett Hardin, pp. 47–51. W. H. Freeman: San Francisco.

Handler, Philip. 1970. *Biology and the future of man.* New York and London: Oxford University Press.

Hardin, Garrett. 1970. Nobody ever dies of overpopulation. *Science* 171: 527 [editorial].

Holt, S. J. 1969. The food resources of the ocean. *Scientific American* 221 (2):178–194.

Kleiber, M. 1961. *The fire of life.* New York: John Wiley & Sons.

Langer, William L. 1972. Checks on population growth: 1750–1850. *Scientific American* 226 (2):92–99.

Maddox, John. 1972. Problems of predicting population. *Nature* 236:267–272.

Malthus, T. R. 1798. An essay on the principle of population. In *Population, evolution, birth control,* ed. Garrett Hardin, pp. 2–20. San Francisco: W. H. Freeman.

Mayer, Jean. 1972. Toward a national nutrition policy. *Science* 176:237–241.

Platt, John. 1969. What we must do. *Science* 166: 1115–1121.

President's Science Advisory Committee. 1967. *The World Food Problem,* 3 vols. Washington: Government Printing Office.

Ridker, Ronald G. 1972. Population and pollution in the United States. *Science* 176:1085–1090.

Townsend, Joseph. 1786. A dissertation on the poor laws. In *Population, evolution, birth control,* ed. Garrett Hardin, pp. 29–33. San Francisco: W. H. Freeman.

Wynne-Edwards, V. C. 1964. Population control in animals. *Scientific American,* August.

Zinsser, Hans. 1935. *Rats, lice, and history.* Boston: Little, Brown.

Chapter 16

Evolution and Behavior

Man's great struggle today is to find the correct and most constructive sublimation for his aggressive drives.

Konrad Lorenz, *On Aggression*

In recent decades there have been two polar views of the origins of human behavior. One is that human aggression is mostly acquired. The other holds that most aggressive and predatory behavior in man is inherited, and has roots that go back in time to our Pliocene ancestors. Today it is generally accepted that primates can modify inherited behavior by experience more readily than can other animals, and that man can do so more readily than other primates. Yet the last two decades have turned up more than a little evidence that many of man's behavioral problems are basically hereditary. Because of the genetic base of many primate behavior patterns, some socially acquired behaviors often fail under sufficient stress.

Charles Scott Sherrington in 1941 was one of the first to emphasize that body and mind are interdependent, and that neither could have evolved independently of the other. Since then, the discovery of the close association of inherited behavioral traits with physical traits at a single chromosome site in experimental animals has borne out this view.

Man's Ancestors

Man is a primate. He belongs to a large group of animals that includes lemurs and tarsiers, many varieties of monkeys, gibbons, and the great apes (chimpanzees, gorillas, and orangutans). Some biologists also include tree shrews, small insectivores (insect feeders) about the size of mice. Lemurs and tarsiers are small tree-dwelling quadrupeds now mainly restricted to Madagascar and South Africa; they include primitive types that go back as far as the Paleocene and the early Eocene.

The primate lineage is probably more than 70 million years old, and mammals similar to tree shrews are known to have existed from the Cretaceous. Lemuroids and tarsioids are known from the Eocene (Figure 16-1), and during the Eocene primitive monkeys divided into two different groups: the Cebioidea, or new-world monkeys, and the Cercopithecoidea, or old-world monkeys. By the late Oligocene a

branch of the old-world monkeys had given rise to the Dryopithecidae, a group of apelike animals from which evolved two important primate groups, the Homininae (men and great apes) and Ponginae (orangutans) (Figure 16-2). These two groups make up the family to which man belongs, the Hominidae. This family, a second that contains the modern gibbons, and a family, of manlike primates now extinct, called the Oreopithecidae, make up the Hominoidea (Figure 16-2). All of these

Figure 16 – 1

One interpretation of the evolution of primates.

Evolution of primates

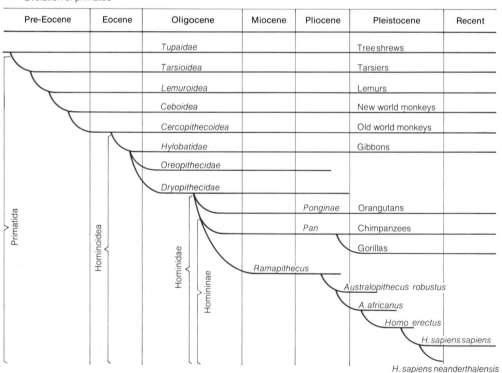

Pre-Eocene	Eocene	Oligocene	Miocene	Pliocene	Pleistocene	Recent
		Tupaidae			Tree shrews	
		Tarsioidea			Tarsiers	
		Lemuroidea			Lemurs	
		Ceboidea			New world monkeys	
		Cercopithecoidea			Old world monkeys	
		Hylobatidae			Gibbons	
		Oreopithecidae				
		Dryopithecidae				
				Ponginae	Orangutans	
				Pan	Chimpanzees	
					Gorillas	
			Ramapithecus			
					Australopithecus robustus	
					A. africanus	
					Homo erectus	
					H. sapiens sapiens	
					H. sapiens neanderthalensis	

Primatida

Hominoidea

Hominidae

Homininae

Figure 16-2

Evolution of apes and men from the Dryopithecidae, a group of Miocene primates.

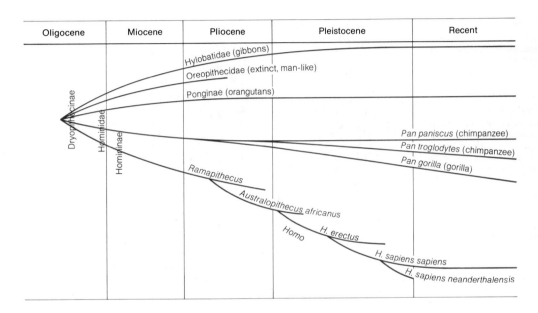

Oligocene	Miocene	Pliocene	Pleistocene	Recent

families date at least from the Miocene and probably from the late Oligocene.

The subfamily Homininae includes the genus *Pan,* comprised of the gorillas, chimpanzees, and their ancestors on one hand, and a lineage of men leading to *Homo* (modern and mid-Pleistocene men) through, *Australopithecus* (primitive men from the late Pliocene and early Pleistocene, mostly from south Africa and southeast Asia), from *Ramapithecus,* the Pliocene preman that connects *Australopithecus* to the Dryopithecidae on the other (Figures 16-1 and 16-2).

It is often said by apologists for the theory of evolution that man is not descended from apes, but that both are descended from a common ancestor. If one thinks only of modern men and modern apes, this statement is correct. But it is misleading. If we could be transported back 20 million years to meet that common ancestor, we would call it an ape.

Skeletal Evidence

Some evidence for the evolution of man comes from fossils. Paleontologists—geologists who study fossils—cannot classify fossil primates on the basis of color, flesh, skin, and hair, for little or no trace of the soft parts survive fossilization. Instead, paleontologists of man study comparable features of the skeletons of both fossil and modern men and apes. Some of the more important of these features are discussed below.

Figure 16-3

Basal views of skulls of (a) dog, (b) chimpanzee, and (c) man.

Figure 16-4

Rear view of skulls of (a) chimpanzee, (b) *Australopithecus*, (c) *Homo erectus*, a mid-Pleistocene man, and (d) *Homo sapiens*.

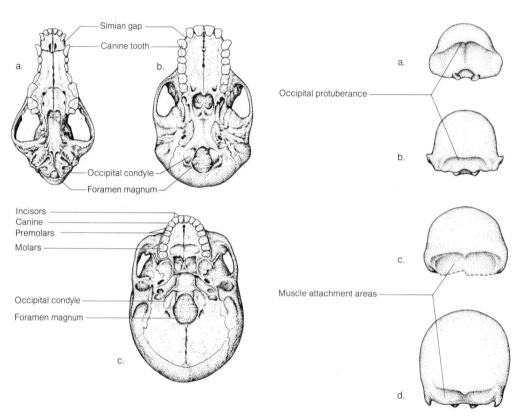

The skull

The foramen magnum (Figure 16-3) is the hole in the base of the brain case through which the brain stem connects to the spinal cord; it moved forward as man's head became more nearly balanced on his spinal column, instead of extending forward and even downward like the head of a dog. In the chimpanzee the foramen magnum occupies an intermediate position, though still well to the rear. During this evolution the occipital condyles (Figure 16-3), two projections on either side of the foramen

magnum that articulate with the first vertebra of the neck, also moved well forward of the position they still occupy in modern apes, maintaining their position on either side of the foramen magnum.

The occipital protuberance is a flange or protruding attachment surface for the neck muscles present in animals that hold their heads up. During the evolution of man this feature became less pronounced (Figure 16-4) until in modern man, who has little need for strong neck muscles since

Figure 16-5a

Side views of the skulls of (a) dog, (b) chimpanzee, and (c) man.

Figure 16-5b

Side views of skulls of (a) female gorilla and (b) *Australopithecus.* (© 1955 by The University of Chicago.)

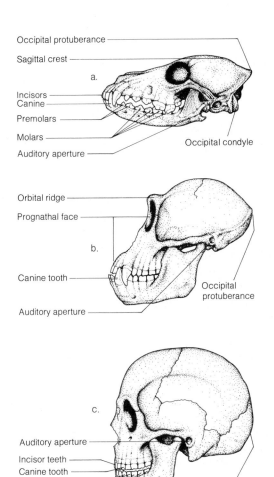

Occipital protuberance
Sagittal crest
a.
Incisors
Canine
Premolars
Molars
Auditory aperture
Occipital condyle

Orbital ridge
Prognathal face
b.
Canine tooth
Auditory aperture
Occipital protuberance

c.
Auditory aperture
Incisor teeth
Canine tooth
Premolars
Molars
No occipital protuberance

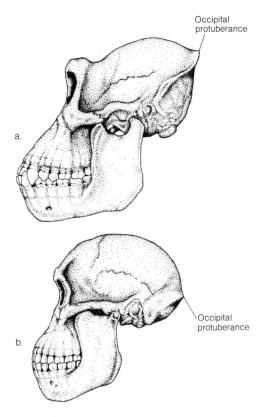

Occipital protuberance
a.
Occipital protuberance
b.

his head is upright, there is only slight occipital protuberance above the attachment area for his weaker neck muscles. This protuberance is relatively pronounced in modern apes, in which the occipital condyles are rearward and the head is not

balanced in an erect position on the spinal column.

The auditory aperture (Figure 16-5) is a hole in the side of the skull through which the ear connects to the brain. This aperture is lateral to the foramen magnum in primates, also moved forward during the evolution of man, and can be used to locate the approximate position of the foramen magnum and occipital condyles when those features are not preserved. Prognathism,

Table 16-1

Average cranial capacities of modern and fossil men compared to that of gorilla. Data from David Pilbeam, "Early hominids and cranial capacities." *Nature* 227:747-748, 1970.

Species	Cranial capacity (cc)
Pan gorilla	500
Australopithecus (early)	475-500*
Australopithecus (late)	650
Homo erectus (Java)	750-1,000*
Homo erectus (Peking)	850-1,300*
Homo sapiens	1,350

*Ranges are given because specimens differ in age; some evolution may be included in the figures.

the forward extension of the jaws beyond the upper part of the face (Figure 16-5), decreased during the evolution of man.

Perhaps modern man differs most from the great apes in the size of his brain. In modern men brain size and intelligence have not been successfully correlated. But among the primates as a group, the more intelligent species have greater cranial capacities (Table 16-1).

The orbital or eyebrow ridge is a thickening and elevation of the bone that protects the eyes (Figure 16-5). Any blow from above would kill modern man, with his high forehead, through brain injury before damaging the eye. Thus orbital ridges in modern man have no function and are slight or absent. Orbital ridges are pronounced in great apes and in those fossil men whose braincase was to the rear of the eyes.

Some animals with particularly strong jaw muscles have a median or sagittal crest (Figure 16-5), a bony ridge extending along the top of the skull from just above the forehead to the occipital protuberance at the rear. To this large surface the strong jaw mucles are attached. The weak jaw muscles of modern man find sufficient attachment on a smooth, ridgeless skull.

Teeth

The higher primates (Figures 16-3 and 16-6) have 32 teeth, eight on each side of each jaw: two incisors; one canine, so-called because it is a fang in dogs; two premolars; and three molars, the rear one being the wisdom tooth. The wisdom tooth erupts last and is absent in some modern men with small jaws.

If the canine tooth is a fang, there is a gap between it and each incisor in the upper jaw, and between it and each premolar in the lower jaw. This simian gap accomodates the projecting tip of the canine in the other jaw when the jaws are closed. It is present in all modern great apes, but absent in all men, fossil and modern. All modern great apes have long canines, but man's are more like his incisors, usually chisel-shaped.

The dental arcade of the great apes has straight sides; that is, the molars and canines are aligned in straight rows that are nearly parallel (Figure 16-6a and 16-6e). This is not true of man and his immediate ancestors, in whom the entire dental arcade is more arcuate—more rounded or convex forward.

Pelvic structure

Apes do not have the spinal curvature (Figure 16-7) that enables modern man to hold his head erect easily. But in fossils the degree of spinal curvature is difficult to determine because vertebrae are small and easily scattered and lost or gnawed by

Figure 16-6

Dentition of upper jaw in (a) male gorilla, (b) *Rama-pithecus*, (c) *Australopithe-cus*, (d) an Australoid example of *Homo sapiens*, and (e) *Pongo*, the orangutan.

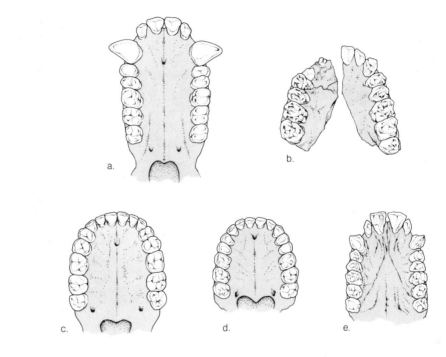

Figure 16-7

Skull, spinal column, and pelvis of a chimpanzee compared with those of modern man. (From *Man-apes or Ape-men? The Story of Discoveries in Africa*, by Wilfrid E. Le Gros Clark. Copyright © 1967 by Holt, Rinehart, and Winston, Inc. Reproduced by permission of Holt, Rinehart, and Winston, Inc.)

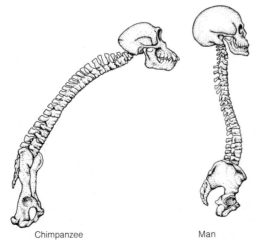

Chimpanzee   Man

small animals. A complete vertebral column is rare except in relatively recent burial sites. Pelvic bones are more commonly preserved, and those of man are quite distinct from those of modern apes. Modern apes do not carry themselves erectly and have long and relatively narrow pelvic bones (Figure 16-8a-c). Those of man are shorter, and the broad upper end allows for the attachment in the buttocks of bunched muscles that help to maintain balance and erect posture under a variety of conditions (Figure 16-8d-h). Modern apes do not have such bunched muscles in the buttocks.

Figure 16–8

Lateral views of pelvic bones of (a) gorilla, (b) chimpanzee, (c) orangutan, (d) man, (e) *Homo erectus*, and (f, g, and h) *Australopithecus*. (Copyright 1955 by The University of Chicago.)

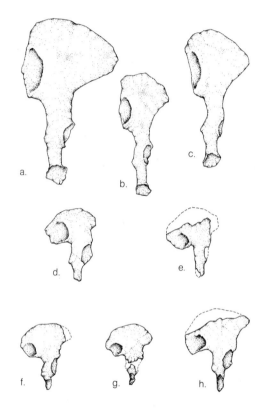

except for a few minor details; all have an opposable thumb capable of grasping. The greater dexterity in the hands of man is an adaptation accompanying the gradual evolution of the use of tools.

Summary

In summary, fossil and modern men differ from modern apes in the following ways (these differences converge to a common ancestor toward the late Oligocene):

Men:	Modern apes:
no simian gap	simian gap
canines are similar to the incisors	canines are fangs
arcuate dental arcade	cheek teeth arranged in parallel rows
short wide pelvis	long narrow pelvis

In addition, the following changes gradually occurred during the evolution of man, but not during that of modern apes.

1. The foramen magnum and occipital condyles moved from near the rear to near the center of the base of the skull.
2. The auditory aperture gradually moved from near the rear to near the middle of the side of the skull.
3. These changes accompanied forward and rearward enlargement of the brain case as the head became more erect on the spinal column.
4. The erect balance of the head was accompanied by the gradual decrease in size and eventual disappearance of the occipital protuberance.
5. As the braincase enlarged and the originally massive jaw decreased in size,

Hands and feet

The feet of modern men differ greatly from those of modern apes, which generally retain a grasping great toe (Figure 16-9). The evolution of the foot of modern man must be closely associated with his evolution as a carnivore capable of running great distances in an upright posture, a feat impossible for apes.

Unlike his feet, the hands of modern man are remarkably like those of modern apes

Figure 16-9

Bottom views of the left foot of (a) chimpanzee, (b) forest gorilla, (c) mountain gorilla and (d) *Homo sapiens*.

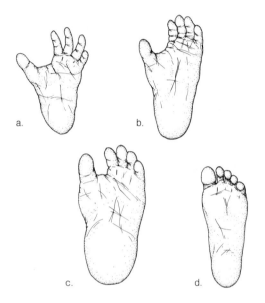

a.

b.

c.

d.

there was a gradual decrease in prognathism.

6. There was also a gradual but three- to fourfold enlargement of the braincase as greater vocal and muscular proficiency required much more accurate correlation of these with visual, auditory, and tactile stimuli.

7. The orbital ridge declined in size and prominence as its protective importance decreased.

8. The sagittal crest gradually disappeared as the need for strong jaw muscles declined.

9. As jaws became smaller the eighth or wisdom tooth began to disappear in adults, and as jaws decreased in size all teeth decreased in size.

10. Spinal curvature gradually developed with more erect posture.

11. To maintain balance more easily by bunching muscles in the buttocks, the lineage of man developed the broad, short pelvis before the late Pliocene.

12. The foot of man gradually changed to accommodate the more erect posture. The big toe moved in parallel to the other toes to allow for greater speed as the ability to hunt, or to evade the hunter, increased.

These and other features led those paleontologists who study fossil men to conclude that *Australopithecus* is the ancestor of *Homo,* and that *Ramapithecus* is not only ancestral to *Australopithecus* but in turn is descended from a group of apelike primates of the Miocene collectively called Dryopithecidae.

The Major Races of Man

The diversity of man has always been a puzzle. Why have certain races been dominant on certain continents? Why is human diversity so great if gene flow has been as continuous as some have maintained? Anthropologist Carleton Coon has modified a theory of the races of man first proposed by Franz Weidenreich, the great student of fossil man, and enlarged upon by William Howells. Coon's theory is developed here because it agrees with available evidence and is the only one that tells a consistent story. This concept divides late Pleistocene and modern men into six major races distinct enough to be called **subspecies**: Australoid, Capoid, Mongoloid, Congoid, Caucasoid, and the extinct Neanderthaloid (Figure 16-10). Race, a looser term than subspecies, is also applied to some lesser

Figure 16-10

Pleistocene evolution of the major races of man.

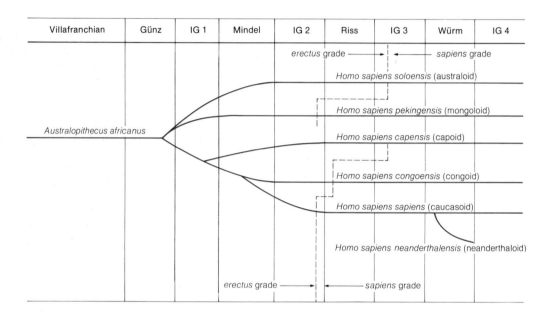

Villafranchian	Günz	IG 1	Mindel	IG 2	Riss	IG 3	Würm	IG 4

erectus grade ——— sapiens grade

Homo sapiens soloensis (australoid)

Homo sapiens pekingensis (mongoloid)

Australopithecus africanus

Homo sapiens capensis (capoid)

Homo sapiens congoensis (congoid)

Homo sapiens sapiens (caucasoid)

Homo sapiens neanderthalensis (neanderthaloid)

erectus grade ——— sapiens grade

categories of man to which subspecies would not be appropriate.

The Australoid race includes natives of Australia and New Zealand, and small groups in the mountains of northern India, the highlands of Thailand, the Philippines, and New Guinea, who have managed to avoid competition with more numerous and aggressive peoples. The Capoid race in its pure form is probably limited to a few thousand Bush people of the Kalahari Desert of Southwest Africa, although many Bush traits appear in the intermixing zone between Bush and Congoids along a line running diagonally northwest from southeastern Africa (Figure 16-11). The Congoid race is comprised of many of the so-called Negroes of Africa, a diverse group ranging from the pygmies of the Ituri forest of

Zaire to the tall Ghinkos of the upper Nile. The different tribes frequently referred to as Bantu are Congoids with some infusions of Capoid and Caucasoid along their boundaries.

Caucasoids are various white, brown, and black peoples who in the past 50,000 years seem to have migrated out of the Indo-Iranian Plateau in waves of conquest. They include Europeans, Arabs, many Asian Indians, and several tribes of black Africans.

Mongoloids include the Chinese, the Japanese, most of the American Indians, and mixed groups in Southeast and Northern Asia. The Neanderthaloids were a race especially well adapted to the glacial environment of northern Europe who became extinct before the end of the last ice age. Thus there remain five major races of men.

Figure 16-11

Distribution of the five major races of man before the 15th century B.C.

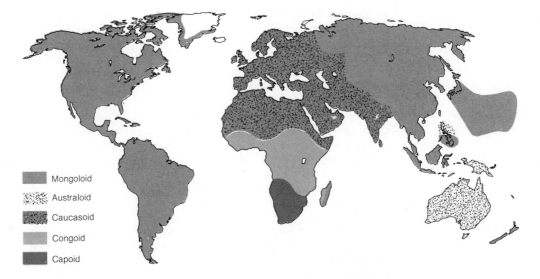

- Mongoloid
- Australoid
- Caucasoid
- Congoid
- Capoid

Ages of the races of men

The Pleistocene ice age is now thought to have lasted from two and one-half to four million years. Table 13-1 gives an average estimate of its length in the light of radiometric dating and new interpretations. Because these estimates have changed, the races of man are now thought to be somewhat older than was previously believed. Until the last decade or two the Pleistocene was thought to be only a million years long, and modern races were thought to be only a few tens of thousands of years old. With the advent of accurate radiometric dating, this has all been changed.

The oldest identifiable Australoids (Java man) are pre-Riss, or more than 500,000 years old (see Table 13-1). Likewise, Peking man, the first identifiable Mongoloid, is from the great (second) interglacial and is also 500,000 or more years old. Heidelberg man, the earliest known Caucasoid, is Mindel and therefore over a million years old. If Steinheim and Swanscombe fossils are Caucasoid, as Carleton Coon believes, they are 500,000 or more years old, since they are pre-Riss.

The relationship of the Capoid and Congoid races to others is based on indirect evidence, because there is no well-preserved fossil record of them. As will be shown later, they appear to be more closely related, morphologically, to Caucasoids than to Mongoloids or Australoids.

Neanderthal man is well-dated, appearing less than 100,000 years ago and becoming extinct about 30,000 years ago.

Racial Differences

The races of man differ in a remarkable number of ways, major and minor, some quite distinct and some not. Many differences are genetic. Many have to do with soft parts of the body that are not preserved in fossils.

Dentition

The ancestor of all modern men, *Australopithecus*, had a large jaw. Although the incisors were not especially large, the cheek teeth (molars and premolars) were so large that isolated finds of teeth in southeast Asia were originally ascribed to a race of giant men, though their owners were actually less than five feet tall. Australoids (see Figure 16-7) have larger cheek teeth than other extant races, and fossil Australoids are thereby distinguishable from fossils of other races. Mongoloids differ from other races in possessing comparatively larger incisors and canines than molars. Almost 70 percent of Mongoloids have no third molars (wisdom teeth), but it is not known how far back in their lineage this statistic is valid.

Dental differences between Caucasoids, Capoids, and Congoids are less distinct, although Coon states that Capoids and Congoids are more alike in minor tooth features. Central incisors larger than lateral incisors are a typical Caucasoid feature, and "buck" teeth (extremely large upper median incisors) seem to appear in Caucasoids only.

Blood flow

Mongoloids who live in cold climates have larger distal blood vessels, permitting greater blood flow, than do other races. In addition, as temperature decreases, metabolism rises and more blood is pumped to the extremities. Thus the Mongoloid is better able to withstand cold than are other races. As one goes south in Asia these traits of Mongoloids are still present genetically. The important fact, however, is that adaptation to cold by increased flow of blood in terminal blood vessels and increased basal metabolism is restricted to the Mongoloid race.

In Australoids who live in cold climates, the arteries of the legs and arms are intertwined with the veins, so that outgoing blood warms incoming blood, conserving heat energy in the torso. During the long nights of sleeping naked in cool to cold weather, the feet and hands grow much colder than does the torso. This adaptation to cold is typical of Australoids native to cold climates, and elsewhere is found only in the Lapps of northern Scandinavia, a small group of probable Caucasoids who also live in a cold climate. This method of adaptation to cold also occurs in many other mammals living in cold climates.

It was originally expected that Bush people (Capoids), who also sleep outdoors almost naked in near-freezing weather, would exhibit some special adaptation to cold. They do not, and thereby differ from Australoids and Mongoloids. Capoids, Caucasoids (with the exception of the Lapps), and Congoids have no known special adaptations to cold involving the circulatory system.

One might speculate that the evolution of special mechanisms in the circulatory system would take a very long time, and that this would constitute firm evidence that races had been separate and distinct for hundreds of thousands of years (see Figure 16-10). On the contrary, if the survival value of a trait is very high (like the prevention of death or maiming by freezing), the adaptation could evolve rapidly—whatever "rapidly" means in the context of evolution.

Skeletal differences

Neanderthal man had an orbital ridge, less curvature of the spine than other races, and curved thigh bones. The combination of these traits is not found in modern races of men, though orbital ridges still appear in Australoids and in early fossils of all modern races. It has been thought for some time that Neanderthal man was an adaptation to the cold glacial climate of the Würm, and it has recently been suggested that he might have had an inherited vitamin D deficiency, producing ricketsial bone structure.

Some races have greater mineralization (density) of bones than others. Congoids have the densest bones, followed by Caucasoids; Mongoloids have the least dense bones of the three most numerous races. As a result hospital records in the United States show a much higher incidence of hip fracture from falls for elderly white women than for elderly black women. This degree of difference in the mineralization of bones also produces differences of behavior in teenaged girls. For example, in cafeterias in integrated high schools in New York City, black girls habitually select a diet higher in calcium than do white girls. This is behavior resulting from an inherited need for calcium.

A skeletal feature occurring recently in Caucasoid and Mongoloid races is the shortening of the skull. Longer, narrower skulls are known as dolichocephalic; shorter, broader skulls are brachycephalic. The Mongoloids can be divided into two subraces: (1) primitive Mongoloids are largely dolichocephalic, but (2) advanced Mongoloids have a higher percentage of brachycephalic individuals. In the last few millennia brachycephalism has increased in Caucasoids. Australoids and Congoids are almost devoid of it.

Among subraces and smaller groupings in the hierarchy of men, two very interesting "rules" apply. Allen's rule states that extremities are shorter in colder climates; Bergman's rule states that bodies are bulkier in colder climates. These observations apply to all animals, and the conditions result in the conservation of body heat. Heavy, compact torsos with short appendages (as in the Eskimo) have less surface skin area per unit of volume and radiate less heat. Long, thin bodies (like those of the Masai and Watusi in tropical Africa) radiate much more body heat and are adapted to climates in which the torso needs to rid itself of heat.

Miscellaneous Inherited Characteristics

Many other inherited characteristics are racial. One partial survey lists 66 inherited human diseases that result from either genetic malfunction or the genetic omission of an enzyme. Over 10 percent of deaths in the United States are from inherited diseases.

Cholesterol levels are partly inherited. Diabetes is an inherited disease that occurs only under stress on the pancreas by over-indulgence in carbohydrates. Blond hair and blue eyes are completely absent in all races except the Caucasoid. Capoids alone are characterized by steatopygia (the deposition of fat reserves in the buttocks), hair growing in isolated spirals, and a number of other features, including heart-shaped faces. Indeed, it is this complex of characters that sets off the Capoids as a distinct race.

American Jews differ markedly from American Protestants and Catholics, not only in ABO blood group combinations but also in dermatoglyphics (fingerprint types). Although all these are Caucasoids, the distinct dermatoglyphics of Jews would lead one to suspect that there has been far less intermarriage of Jews and non-Jews than has generally been supposed.

Blood group factors

Blood factors are beginning to show significant ratios in many ethnic groups, especially now that nearly 50 factors are being studied in addition to the ABO and Rh positive and negative factors. The relation of the incidence of disease to blood grouping is statistically significant. The frequency of stomach cancer in the A group is higher than the average, as is the frequency of peptic ulcer in the O group. American Indians had been isolated from smallpox for such a long time that by genetic change they lost all resistance to the disease; it wiped them out far faster than did their own wars or massacres by Whites.

Diseases

To discuss the inheritance of a disease it is necessary to know something about its history, and we have this kind of knowledge for only a few diseases of the bone. Arthritis can be identified in fossil men, fossil elephants, and other fossil animals. Dinosaurs, even 150 million years ago, appear to have suffered from tuberculosis of the bone. Some diseases may have begun in the primates long before the lineage of man, and evolved with man. Chimpanzees, for example, suffer polio epidemics. Did poliomyelitis afflict the ancestors of great apes and men before the differentiation of men and chimpanzees, or are men and chimpanzees so closely related that the polio virus which evolved with one can be transmitted to the other?

For diseases to be racially associated (1) they must have evolved after racial separation, or (2) partial immunities must have been developed or lost by one or more groups after racial separation. Either way, the cause is genetic. Either the immunity of the American Indians to smallpox was lost because there was no smallpox in their environment for the thousands of years they lived in North America, or smallpox developed after they migrated to America.

Protein and protein requirements

Races appear to differ in the protein content of their blood sera and in their protein requirements. The potassium content of blood sera is racially significant in man and other animals. Since potassium in the blood is partly responsible for regulating the heart, potassium blood sera differences probably indicate basic physiological differences.

Protein differences lead us to the problems of protein intake and natural selection. It appears that man's brain may not have approached its present development until he became primarily a carnivore, and even then not until he was able to kill enough meat to consume large quantities of proteins. As was implied in Chapter 15, meat contains some proteins that man needs but does not manufacture, and that are not available from plants (see page 403). Furthermore, meat contains a larger variety and amount of protein per pound than do other foods. Therefore, meat eaters have more proteins available to them than do vegetarians. Primitive man, primarily a carnivore, ate a great variety of proteins—from the pancreas, kidneys, liver, guts, brain, thyroid, and all of the soft parts—and gnawed the bones. Modern selective meat-eaters—steak-eaters for example—ignore up to 90 percent of the more valuable proteins.

Food and the brain

The chemist Linus Pauling has given considerable attention to the food requirements of the brain, some but not all of which are protein. The brain's requirements for different metabolites (either food for or catalysts for enzymes of the brain) often vary over twentyfold from one person to another for genetic reasons, and may vary as much as one hundredfold. For example, one individual may require less than 75 milligrams of vitamin C a day for optimum efficiency, whereas another may require as much as 1000 milligrams a day. Such differences are inherited. If the best brains are those that use a quantity and variety of metabolites, then the races of man are in trouble. With most of the world on diets deficient in protein and brain metabolites, nature is not now selecting for the potentially best brain, but for the ordinary brain that can still function on a minimum of metabolites. We are not selecting for quality but for efficient mediocrity. This bodes ill for the future of man, and is the reason that some of our estimates of food requirements for undeveloped nations are suspect.

Other distinct genetic characteristics that are or may be racially significant include the following:
1. The viscosity of ear wax is unique in some races.
2. The composition of esters (odor producing fats) secreted with perspiration differs from race to race.
3. Polydactylism (more than five fingers or toes) is inherited, but racial incidence has not been ascertained.
4. Curly hair differs among races, and does not exist in Mongoloids.
5. The eyefold in advanced Mongoloids is absent in other races.
6. The vitamin C requirement is an inherited characteristic, but racial incidence of inherited requirements is unknown.

7. The incidence of various perceptual illusions varies with race.
8. The pepsinogen content of the blood varies with race.

Many other inherited characters may or may not be racially distinctive.

Skin Pigmentation

Because it has played such an inordinate part in ancient and modern racial and ethnic problems, skin color deserves separate discussion. Most racial or ethnic problems are a complex of cultural phenomena that have little to do with race. Skin color is something the racist can see and dwell on, though it is not the real cause of anything. Why is there so much variety in the color of our skins? The answer lies in the vitamin D requirements and toxicities of the body, and is also related to the incidence of skin cancer.

Vitamin D is manufactured by the deeper layers of the skin in response to the triggering mechanism of ultraviolet light, invisible to man but necessary to the proper functioning of the body in several ways. In large amounts vitamin D is toxic, and the optimal range is very narrow. This is why light-skinned persons sunburn easily. Sunburn results from the rapid development of pigment between the source of ultraviolet light and the vitamin D-manufacturing layer of the skin, in order to reduce the production and toxicity of vitamin D. In darker-skinned peoples protection by pigment is inherited. The adaptation to ultraviolet light associated with climatic belts produces the phenomenon of lighter-skinned peoples in polar climates grading to darkest-skinned

peoples in tropical areas (Figure 16-12). This phenomenon is less pronounced in Mongoloids. Mongoloids produce keratin as a skin pigment, whereas other races produce melanin. This fundamental difference may also reflect an early separation of the Mongoloid race, and probably accounts for the lesser skin-color differentiation in Mongoloids. Generally, however, the greater the amount of ultraviolet radiation received by the body as a whole, the darker the skin.

Skin cancer has recently been induced experimentally by ultraviolet radiation. It is well known that albinos in African Congoid populations suffer severely from skin cancer, presumably produced by ultraviolet radiation, whereas their normal relatives do not. Skin cancer is prevalent in the desert of the southwestern United States, and the Mexican sombrero was developed for a purpose. Blue-eyed people are much more subject to skin cancer than are brown-eyed people, so skin cancer is predominantly a disease of Caucasoids, and is associated with the inheritance of skin color and eye color. The darker the skin the greater the resistance to skin cancer.

If a light-skinned Caucasoid race were to migrate to the tropics, under conditions in which protection from the sun was difficult, there would be rapid selection for darker skins to prevent vitamin D toxicity and skin cancer. This has happened many times in the past, producing the dark-skinned Caucasoid Indians of India, the Berbers and Tuaregs of North Africa, the Coptic peoples of Ethiopia, and the Masai of Kenya. The Caucasoid Arabs of intermediate latitudes are intermediate in skin pigmentation. Skin color has little value in defining major races,

Figure 16-12

Skin pigmentation and latitude. (From ''Rickets'' by W. F. Loomis. Copyright © 1970 by Scientific American, Inc. All rights reserved.)

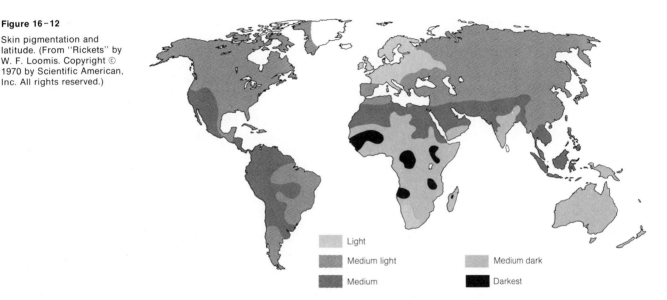

Light

Medium light

Medium

Medium dark

Darkest

because some of the lightest peoples on earth are Caucasoids—as are some of the darkest.

All in all, skin color is probably the most unstable and thus the poorest of all racial criteria, except in Mongoloids, whose primary skin pigment is keratin instead of melanin. Yet the presence of keratin is not easily observed with the unaided eye and must be tested for chemically. Since most primates related to man are dark-skinned, and since man evolved in the tropics or near-tropics, presumably without clothing, it is probable that all men were originally black, or at least very dark. Light skins probably did not evolve until changes in way of life subjected men to less ultra-violet light, so that they required less pigment to keep the production of vitamin D below the harmful level. Such changes

could have included living in caves or other dark dwellings, or migrating into colder climates where sunlight was less intense, making it necessary to wear clothing.

Isolating Mechanisms and the Origin of Race

Barriers to widespread gene flow in man before modern transportation included oceans, deserts, glacial epochs, mountains and jungles, and certain cultural mechanisms, mostly behavioral.

Oceans

Until the invention of the boat, large bodies of water were effective isolating agents. Although there may have been a number of accidental, or even purposeful, Atlantic crossings before Columbus, there is considerable doubt that they produced enough gene exchange to affect the genetic makeup of any population. The groups that migrated into Polynesia were not large, and the migrations probably occurred only in the last 2000 years. Wallace's line, a deep oceanic barrier through the seaways between East Indies islands, successfully protected the Australoids in Australia from intermixture for many thousands of years.

A few hunters crossing the Bering Strait could not produce enough gene exchange to alter populations. For this to occur migrations of whole populations are necessary, whether by boat, by land during periods of lowered sea level, or by ice during a glacial stage. Until the last four or five thousand years, transportation was not good enough for people to cross oceans in large enough numbers to alter the genetic makeup of populations.

Deserts

Deserts are effective barriers to gene flow. After the disappearance of its water about 200 A.D., the Takla Makan Desert in Sinkiang, north of the Himalaya Mountains, reduced the contact of eastern and western cultures for centuries. In earlier times, until the camel and the horse became conventional modes of transportation, deserts were formidable barriers. The Sahara Desert, enlarging as the water supply decreased, finally defeated the camel, but only a few decades before the airplane. As a result of the separation enforced by the growing desert, contact between Libyan and Nubian peoples may be less frequent now than it was three centuries ago.

Ice ages

The alternation of glacial and interglacial stages produced contrasting phenomena. During the ice ages the deserts were reduced, and populations that had been separated by deserts were free to make contact. During interglacial stages, east-west contacts were made easy in temperate latitudes, but north-south migrations across the deserts then in the northern horse latitudes were impeded. The desert of the Asian horse latitudes probably originated in the Miocene and was active during all subsequent nonglacial periods. During those periods when much ocean water was immobilized in ice, land bridges were wider and straits narrower. In these times it was easier to migrate across the Bering Strait, along the Isthmus of Panama, or between the islands of Indonesia and Malaysia.

Mountains and jungles

Mountains and jungles are also effective barriers to migration. If it were not for boats and airplanes there would be little gene flow across the Andes today. This is especially true during interglacial periods, when the jungles of the Amazon are most effective as a barrier. The combination of Pakistani-Iranian-Afghanistan-Sinkiang deserts, Himalyan-Hindu Kush Mountain ranges, and the jungle east of the Himalayan Mountains have created an exceptional and continuing east-west geographical barrier across Asia.

When we add to all these physical barriers those cultural and behavioral characteristics that make people suspicious of each other, it is remarkable that gene flow

Picture of man's aggression

no doubt support some current interpretations and modify others.

The Sources of Man's Behavior

If the threat of overpopulation is our greatest environmental problem, primate behavior itself runs a close second. Indeed, this behavior is at the root of the population explosion. It is responsible for economic growth, discrimination of all kinds, littering and unsatisfactory disposal of wastes, crime, war, drug addiction, and man's inhumanity to man. In short, human behavior is the common denominator of most modern problems. For the past several decades it has been assumed that most of man's behavior is culturally acquired and modified by his environment, but recent evidence has led some to believe that inheritance plays a larger part in it than had long been supposed. The argument continues in our scientific journals, and probably itself reflects the drive called "territoriality," which causes extremists on both sides to make extravagant claims for total possession of the truth.

If, indeed, primate adaptability is largely responsible for man's successes, what then is the problem? It seems to have three components: (1) Under stress, acquired behavior may be submerged by the sudden reemergence of inherited behavior patterns. (2) Moreover, it appears that some genes function only when certain chemical conditions exist. It is thus possible that adrenalin generated by excitement can trigger an enzyme or other chemical to produce behavior previously not experienced. (3) Even if it were possible it is

has occurred at all. The races of man have been evolving for a long time. When certain isolating barriers broke down, some mixing took place. The appearance of new barriers created further isolation. The story, which can only be briefly sketched here, is complex, and is only now being clarified. Evidence turned up in future decades will

questionable whether it would be ethical to condition individuals to perform like social robots, as in George Orwell's *1984* and Aldous Huxley's *Brave New World.*

Inherited Behavior

Some behaviors have always been thought to be inherited. There has never been any doubt that a mother's love for her infant is inherited. This behavior has great survival value to the species because it greatly reduces the dangers of infant mortality. Likewise, the love of an infant for its mother has always been assumed to be inherited, though there is now some question whether this trait may have been misnamed.* In many animals the type of play even the very young engage in is inherited.

The sex drive is also generally thought of as instinctive; however, young male baboons are essentially taught the sex act by their female elders, and primates that have been socially deprived are frequently incapable (perhaps emotionally) of reproduction. So even behaviors thought to be basic to existence have an overprint of experience or the lack of it.

Man and the social contract

Nearly all primates punish infractions of social discipline, though the punishment sometimes appears to human observers to be unwarranted. The most severe punishment, ostracism, is not often imposed. Each baboon instinctively obeys a system in which individual isolation almost always means death. It is the complete dependence of the individual on the group for survival that has allowed baboons to create such a strong social contract.

Modern man is also subject to a social contract that he as an individual did not sign. His social contract is not so firm as that of baboons and may be broken more easily without fatal consequences to the individual or the society. But as population increases and society insists on protection from the aberrant behavior of individuals and groups, man often finds that legal institutions control him, perhaps more tightly than the natural social contract. As population continues to increase, legal institutions will restrict individual freedom more and more.

Destruction of social discipline arises from several social pathologies. Many who have studied population pressure in rats, rabbits, and other animals wonder if it will produce similar effects on man. John B. Calhoun's rats developed a behavioral sink, an area of aberrant social behavior, in the overpopulated group. The sinks included sexual and social aberrance, in which rats could not stay away from crowds of other rats. Females were more prone to this behavior than were males. Perhaps more distressing were plummeting birth rates and increased death among the young due to lack of maternal care. Population studies of other groups of rats and mice found less aggression and a great determination to do nothing but satisfy curiosity. Ambition and purpose, in terms of normal rat behavior, were lost.

Have we bred and are we now breeding behavioral sinks in our own ghettos, condominiums, densely populated apartment complexes, and other areas of overcrowding, including camping areas for rebellious dissidents? In other animals the excitement

*Instead of "love," the infant may inherit a basic need for security, which only its mother is able to give it in full measure. Infant chimpanzees left orphans at an early age frequently become psychotic or die.

of the behavioral sink is contagious and attractive to individuals, though harmful to the population. Is this also true of man?

Social Discipline.   According to Robert Ardrey's concept of the "social contract," each individual born into a primate society inherits an obligation to perform a function that helps to secure the welfare of the society. Ardrey does not limit the application of his concept to primates, though they are our main concern here. In primates below the agricultural level the performance of this obligation lengthens the survival time of the individual. Any individual may exceed or fall short of the average, but the probability of a longer lifespan is increased. The study of inheritance involves the basic question whether any individual primate is born with a need for social discipline after 50 million years of evolution under a strict social regimen. This seems true of some primate societies, but modern societies of men are so complex that answers to this question are difficult. There are numerous examples of different groups of primates that did not survive when social discipline was disrupted. Much the same thing has happened to rats and mice under population pressure.

All primate groups have rigorous social disciplines, but that of baboons is perhaps the most tightly controlled. A hierarchy of males, usually three or four, control and protect the group. There is a job for everyone, except the very young. Young adult males and young maturing males are outriders, who give warnings of danger. The male leaders replace the young males when danger threatens. Females care for the young. Living on the savannah near or in trees, and foraging on the ground most of the time, baboons are thought to occupy much the same environment as did the progenitor of man, *Ramapithecus,* and early man, *Australopithecus.* Because rigid social discipline is necessary for survival in this environment, it is surmised by many students of behavior that man probably evolved in such a system.

Antiwar and Symbolic War.   Vilhjalmur Stefansson tells of traveling with a group of Eskimos in the Canadian Arctic Islands. During the journey another Eskimo group was encountered. As they passed, the two groups circled each other, carefully remaining about a half-mile apart. There was no contact, physical or verbal. Stefansson interpreted this behavior as an antiwar precaution, since every Eskimo group, under natural conditions, depended on a single prime hunter for food. This hunter would also have been the prime warrior had trouble started. Group survival demanded that the prime hunter not be killed, and therefore that there be no fighting. Peace was assured by complete avoidance.

This no-war rule is characteristic of most animals, particularly primates. Monkeys defending a territory are all noise and no fight. So are jackals and, to a lesser extent, hyenas. In groups of primitive men symbolic war is more common than real war, and even among many of our own Plains Indians it was less honorable to slay an enemy than to count coup (touch an enemy without harming him). Only two mammals wage war on their own kind—rats and men. Somehow in these two species the antiwar rule has been subverted. But not entirely. The American writer Mary McCarthy described the London demonstration of 1967 as "monkey warfare."

Gibbon family

Dominance and status

All primates have pecking orders except orangutans and gibbons. Gibbons and orangutans feed and travel in family groups. Consequently, their environment does not call for a strict social order outside the family group. The order of dominance within a primate group operates as if the rewards of the social contract—position, attention, and food—are given to those who bear the greatest responsibility for the survival and welfare of the group. These rewards generally do not go to females, because they use all their energy caring for the young. If mother-child attachments were not strong, the young would not survive long enough to insure the survival of the species. Among chimpanzees orphans six years old or younger, even if cared for by an older sister, may die or become socially maladjusted. Furthermore, except among orangutans and some gibbons, sharing responsibility with the father is impossible, for no mother knows which male sired any particular infant. Dominance and status are a way of life among nearly all primates, including man. True of all species, sex is not an important aspect of dominance or status. Any good baboon will give up the "light of his life" if his position is threatened.

Drives for dominance and status in man often take the form of a pursuit of power: marrying the prettiest woman, driving the largest car, living in the largest mansion on the highest hill, writing the most popular books, and so on. Unfortunately, dominance and status as goals in human society are not likely to diminish rapidly. They are too deeply entrenched in the primate biogram.

Discrimination Against Females.   All the higher primates except orangutans and gibbons practice discrimination against females, as do most of the old-world monkeys. Although this fact is not proof that discrimination against females is inherited, it constitutes circumstantial evidence. If there is an inheritance factor in discrimination against females, it will take longer for learned cultural behavior to override it.

Territorialism and aggression

Territorialism is the occupation, and usually the defense, of an area. It has been described in fish, birds, reptiles, herbivores, carnivores, and nearly all primates. In fish it is most often the defense by a male of the nest and surrounding territory of a mating pair. In birds the territory may be the nest area, which both male and female defend, or simply a space that the male defends from his own kind. Most territorialism operates within species, though some species defend territories from other species. In herbivores a territory most often belongs to a male, and females come and go as they choose, though most female herbivores also have a "home range" for grazing that may not be defended. When herbivore males defend territories, the females usually choose the male and the territory they desire, although females of some herbivores belong to harems of dominant males.

Among carnivores the territory is usually

Cattle have a grazing
territory

a hunting area protected from others of the same species, whether it is wolves, lions, hyenas, or jackals. Territorialism in most primates is group-against-group, in defense of feeding areas. Is man territorial? Yes, although the instinct is largely redirected. Professors defend their research areas with the fervor of stickleback fish chasing away intruding males. A typical example is James D. Watson's description of the competition among scientists on two continents to discover the structure of the DNA molecule.

Although there have been many attempts to interpret the behavior of man in terms of animal behavior, none is completely satis-factory. Yet the sum of them makes it plain that aggression in man deserves careful scrutiny. Man's aggression has been blamed on poverty, population pressure, lack of property, territoriality, desire for dominance and status, and the legacy of a carnivorous ancestry. Konrad Lorenz, one of the great students of animal behavior, has suggested that nations promote spectator sports to their fullest in order to redirect aggressive tendencies. But he specifically states that such measures are only temporary stop-gaps until enough can be learned about the causes and cures of behavior to lead us to more permanent and meaningful solutions.

African bull elephants
defending their territory

Racism

Nearly all animals are endowed with curiosity, suspicion, and fear. Most racism is founded on suspicion and fear of the unknown, the different, the exotic. As educational, cultural, and economic differences appear, and competition develops, breeding its own forms of suspicion and fear, these feelings may turn into dislike and even hatred.

Fear and suspicion of phenomena outside the experience of the individual and the group generate feelings much like those of Stefansson's Eskimos. Color is the most visible group character, and is con-

sequently at the root of much racism. The East Indian caste system was originally based on color, with the light-skinned at the top and the dark-skinned at the bottom. Racism is as old as man, although it has not always had the connotation it has in the United States today.

In the Bible racial hostilities begin with the three sons of Noah, each of whom is later described as the progenitor of a major segment of the world's people. Ham was given the land of Kush, which later scholars mistakenly thought to be Ethiopia. According to Genesis, Ham's race was also to serve the other peoples of the world for eternity.

The prophesied battles between the forces of light and darkness have been commonly misinterpreted as battles between light-skinned and dark-skinned peoples, rather than as symbolic warfare between enlightenment and ignorance or good and evil.

Month by month new racial problems are reported in the world press. Anthropologists fear that the Ainu, the natives of the north island of Japan, will disappear because they will fail to meet the population and cultural pressure of the Japanese. In Yugoslavia the Macedonians, a distinct ethnic entity, maintain that they are discriminated against. According to most American Plains Indians the Crow Indians were the filth of the earth. In the early decades of this century it was the practice of some newspapers to publish cartoons of Orientals with large fangs (enlarged canines). Discrimination against Chinese and Japanese was rampant in the western states during the last century and culminated in the Chinese Massacre at Rock Springs, Wyoming, 1885. Since the Chinese miners worked for lower wages than Americans, this racial discrimination had an economic basis. The treatment of the Irish in the northeastern United States in the last century was callous and indifferent. But man never learns, and now the Irish are discriminating against Poles, Italians, and Blacks.

The independent nations of Africa are no less troubled with racism than the few remaining colonies. In colonial times everyone could dislike the discriminating imperialist. When the major imperialists departed, older hatreds flared up again among members of the same race as well as different races (Figure 16-13).

Most migrations have been caused by racial and national hostilities. In Mali and the Sudan, Congoids are leaving Caucasoid-dominated countries, and Caucasoids are leaving Congoid-dominated Zaire for the Sudan. The war in Nigeria in the 1960s was between the economically superior Ibos and other Congoids. Congoids are leaving Portuguese-dominated Angola to escape discrimination by Caucasoids. Caucasoids are leaving, or being driven out of, Kenya and Uganda because of Congoid dominance. The many migrations between Uganda, Burundi, Rwanda, and Tanzania are largely caused by Blacks discriminating against Blacks as a result of hatreds extending back to precolonial times. The discrimination of North Vietnamese against South Vietnamese has gone on for more than 600 years. Such feelings are not likely to be abolished by the withdrawal of American and Russian support and influence. Still, the recent war in this area could be interpreted as "monkey warfare" between two societies that do not want a war of their own—the United States and the Soviet Union.

Learning and the Use of Tools

The use of tools is inherited in some species and acquired in others. Inherited use of tools is exemplified by the Galapagos finch, a small bird that uses a cactus spine or twig to pry worms out of holes in trees or logs. Likewise, the Egyptian vulture throws rocks at eggs. When unbreakable plastic eggs were substituted for real ones, two vultures exhausted themselves trying to get a meal. Rock-throwing is also inherited behavior in the banded mongoose. The mongoose throws rocks between its hind legs at an egg in much the same way a football center passes the ball to the punter. Birds inherit nest-building behaviors. A

robin raised in an incubator or a bluebird's nest, in such a manner that it never sees a robin's nest, will build a robin's nest.

Other animals learn certain behaviors, though the means of communication by which they do so may not be clear to us. No cowboy has ever been able to explain how a few cows in a herd of longhorns with calves knew that it was their turn to "babysit" the calves on a certain day while the other cows walked off to graze.

In England there is a bird popularly called a blue tit. A few years ago a London woman discovered one morning that the cap had been pried out of the milk bottle on her front step. This happened two or three mornings in a row, until one day she saw a blue tit standing on the rim of the milk bottle prying the cap out with its bill. This was indeed remarkable, but what followed was even more so. After three months blue tits all over London were opening milk bottles. They had learned from each other.

In primates it is often hard to tell whether behavior is learned or inherited, but certain behaviors seem to be learned. Each troop of chimpanzees can be identified by certain unique patterns of behavior. The chimpanzees of the Gombe Reserve, Kenya, have two behavioral patterns not known in others. They love the taste of termites and to catch them find a grass stem or strip the leaves from a twig or a small branch. They then push the twig down a hole in a termite nest, and when the soldier termites grasp the twig the chimpanzees remove them, smacking their lips while eating these delicious morsels—a source of proteins unavailable to other troops of chimpanzees in other areas. Since the small twigs or grass stems break easily and do not last long, smart chimpanzees prepare several at a time before approaching the termite nest. In other words, they anticipate the need for additional tools. The young must learn how to catch termites by watching their mothers, and a mother not adept at termite-baiting produces an inept offspring.

A second type of behavior unique to the Gombe chimps is a method of getting water. In the dry season water can be found in hollow logs. When the water can be reached only through a small hole in a log, the chimpanzees in this group have learned to chew leaves, to make them absorbent like a sponge, and then stick them into a hollow log to sop up the water in it. The leaves are then withdrawn and the water is squeezed into the mouth. This behavior is also learned by young chimps from watching their mothers. Other chimpanzee groups have their own unique patterns of behavior. As with baboons, the father of a baby chimpanzee is not known. This means that the mother is the chief conveyor of survival tactics and social procedures to the young.

It had been speculated that bipedalism (walking on two feet) originated from a need to carry food or nest-building materials to a desired locality. A few years ago the Japanese macaque, an old-world monkey, outpopulated its food supply. The macaques were then fed a new food, sweet potatoes, which they quickly learned to carry to a preferred eating site by walking on their hind legs. This behavior, invented to fill a need, reinforced such speculation.

Summary

Man inherits a gene suite from which his environment may select a number of behavioral options. His behavioral pattern can be modified by experience or by cultural indoctrination from options offered by the genetic system. Not every person inherits the same gene suite and, consequently, experience has a different set of behavioral options to modify in each individual.

Perhaps a major error in Western culture has been faith in the innate goodness or innocence of man—the "romantic fallacy," as Robert Ardrey calls it. This idea has had a long evolution. According to the pastoral tradition of the Greeks, man once lived in a Golden Age of perfection. In the Old Testament, the early part of which was borrowed from the Accadian peoples further east and south, the first man lived in a perfect Eden, and his innocence was destroyed by the serpent with the connivance of his own mate, Eve. In the late eighteenth century Jean Jacques Rousseau argued that primitive man was innocent, a noble savage who was corrupted by acculturation (though Rousseau did not use that word). With Karl Marx capital replaced the snake, and with Prince Kropotkin and the anarchists the source of evil became property.

Modern biology takes the opposite view—that man has never been innocent, and that in a harsh world he competed with other animals for survival long before and long after he became human. If there is a primate biogram that dictates basic behaviors and attitudes through inheritance, the cure for human weakness and folly is harder to achieve than if behavior can be modified by positive reinforcement.

Anthropology and biology tell us that primitive man was a primary carnivore, and that he was never pure, noble, or innocent. He had to fight for survival in a rugged and often cruel environment. Most of his behaviors were responses selected either for easier survival in that rigorous environment, or because they added greater security and pleasure to living.

Man no longer lives in a natural world, but in a brain-made world—a world that may be changing more rapidly than evolution, working through his genetic system, can select viable biologic alternatives. Behavioral modification, no matter what its source, is today meeting its severest test.

Selected Readings

Ardrey, Robert. 1970. The social contract. New York: Atheneum.

Bastide, Roger. 1967. Color, racism, and Christianity. Daedalus Spring 1967: 312–327.

Béteille, André. 1967. Race and descent as social categories in India. Daedalus Spring 1967: 444–463.

Blum, Harold F. 1968. Does the melanin pigment of human skin have adaptive value? In Man in adaptation, ed. Yehudi A. Cohen, pp. 143–156. Chicago: Aldine.

Calhoun, John B. 1967. Population density and social pathology. Scientific American 215 (2):139–148.

Cattell, Raymond B. 1971. The structure of intelligence in relation to the nature-nurture controversy. In Intelligence: Genetic and environmental influences, ed. Robert Cancro, pp. 3–30. New York: Grune & Stratton.

Clark, W. E. LeGros. 1967. Man-apes or ape-men? New York: Holt, Rinehart & Winston.

Coon, Carleton S. 1962. The origin of races. New York: Alfred A. Knopf.

Eisenberg, J. F.; Muckenhirn, N. A.; and Rudran, R. 1972. The relation between ecology and social structure in primates. Science 176:863–873.

Galle, Omer R.; Grove, Walter R.; and McPherson, J. Miller. 1972. Population and pathology: What are the relations for Man? Science 172:23–30.

Harlow, H. F.; Harlow, M. K.; and Suomi, S. J. 1971. From thought to therapy: Lessons from a primate laboratory. American Scientist 59:538–548.

Kuttner, Robert E., ed. 1967. Race and modern science. New York: Social Science Press.

Lorenz, Konrad. 1966. On aggression, trans. Marjorie K. Wilson. New York: Harcourt, Brace.

Mowat, Farley. 1963. Never cry wolf. New York: Dell.

Pauling, Linus. 1968. Orthomolecular psychiatry. Science 160:265–271.

Rife, David C. 1967. Race and heredity. In Race and modern science, Robert E. Kuttner, ed., pp. 141–168. New York: Social Science Press.

Simons, Elwyn L. 1969. The origin and radiation of the primates. Annals of the New York Academy of Sciences 167:319–331.

Simpson, George Gaylord. 1966. The biological nature of man. Science 152:472–478.

Tinbergen, Niko. 1968. On war and peace in animals and man. Science 160 (3835):1411–1418.

Washburn, S. L. 1970. Comment on "a possible evolutionary basis for aesthetic appreciation in men and apes." Evolution 24:824.

Washburn, S. L., and Devore, Irven. 1961. The social life of baboons. Scientific American June:62–71.

Figure A–1

Major features of the continents and oceans.

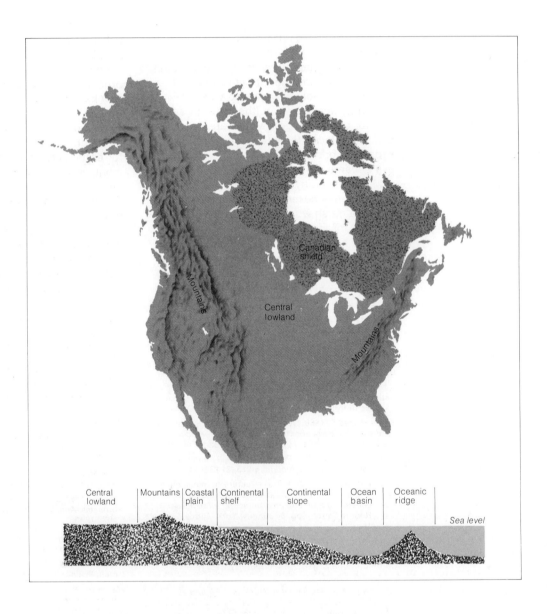

Appendix A

Physical Geography of the Earth

The largest physical features of the earth are the continents and the ocean basins. Major features of the continents are mountains, lowlands, continental shelves, continental slopes, and shields (Figure A-1). The shields are exposed areas of very old (Precambrian) rock.

The major features of the ocean basins are the oceanic plains, the midocean ridges and the island arc systems. Typical island arcs are the Japanese Islands and the Aleutian Islands. Parts of the midocean ridges and the island arcs may rise above sea level.

Earth features undergo slow but perpetual change. Mountains, eroded away by wind and water, are renewed by the internal forces of the earth only to be worn down again. Basins and geosynclines—great troughs in the earth's crust that have received sediments for hundreds of millions of years—are uplifted, faulted, and folded to become great, new mountains. Then new streams cut new valleys and pour new sediments into the seas, and a new geosyncline is initiated to start the process over again.

Figure B–1

(a) One of several concepts of the relationships of the continents about 200 million years ago—prior to the start of the last great period of continental drift. Subsequent stages shown are in (b) the Triassic, (c) the Jurassic, (d) the late Cretaceous, and (e) the late Cenozoic.

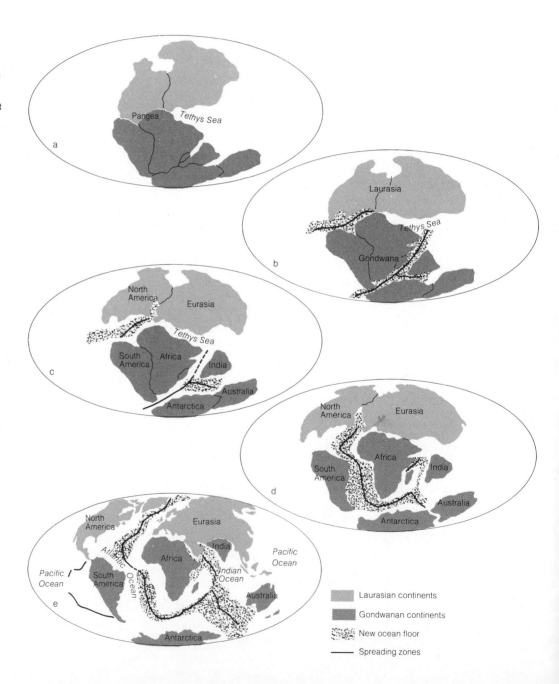

Even the continents and the ocean basins are no longer thought to be permanent or stable, but to move over the earth and take on new configurations and relations to each other. A theory of continental drift was first stated over a hundred years ago, and in the ensuing decades many mechanisms and configurations have been proposed.

Figure B–1 shows several stages in the fragmenting of a world continent (called Pangea) and the drifting of its parts—our continents—to their present positions. They are thought to be still drifting at rates of from one to five centimeters per year. According to this theory, continents have been drifting apart and back together throughout geologic time, and this latest drift is thought to have begun about 200 million years ago.

The concept of continents and adjacent seafloors spreading from midocean ridges provided a viable theory (for skeptical American geologists), if not a mechanism, for continental drift. This theory of seafloor spreading—the forerunner of modern plate tectonics—was initiated by the late Harry H. Hess of Princeton University.

The process of continental drift produces new crust at the midocean ridges by the upwelling of magma from the earth's mantle which in turn reacts with the oceans, adding new mineral materials to both the rock and the hydrosphere cycles (see Chapter 1). According to this concept, the upwelling mantle creates new crust at the boundaries of rigid plates—great segments of the earth's crust. At other boundaries plates collide, as along the Pacific coasts of North and South America, and one plate usually slides beneath another in a process known as subduction. At these subduction sites, great oceanic deeps and continental mountains may form as the margin of one plate sinks and the other rises. Beneath the continent, simatic material may be added to the sial in part of the rock cycle.

Although some geologists do not accept the theory of continental drift, it has become a working hypothesis to explain many phenomena, including the distribution of fossils, soils, and unique rocks, as well as continental distribution, continental shapes, and the addition of mantle materials to the ecosystems.

Figure C–1

Divisions of geologic time. Figures for the geologic periods are given in millions of years.

Era	Period	Epoch	Age to beginning of period 10^6 years	Duration of period 10^6 years	History of life	Dominant animals	Other events
Cenozoic		Pleistocene				Man	
Cenozoic		Pliocene	67	67			Mountain building
Cenozoic		Miocene	67	67		Mammals	
Cenozoic		Oligocene					
Cenozoic		Eocene					
Cenozoic		Paleocene					Coal
Mesozoic	Cretaceous		138	71			Mountain building
Mesozoic	Jurassic		195	59		Reptiles	
Mesozoic	Triassic		225	30		Reptiles	Deserts
Paleozoic	Permian		280	55			Mountain building
Paleozoic	Pennsylvanian		315	35			Coal
Paleozoic	Mississippian		345	30		Amphibians	
Paleozoic	Devonian		395	50			First large trees
Paleozoic	Silurian		440	45		Fishes	
Paleozoic	Ordovician		500	60		Invertebrates	Mountain building
Paleozoic	Cambrian		570	70		Invertebrates	
Precambrian			4,600	89% of earth history		Protista	Several periods of mountain building

History of life (columns): Protista — Algae — Invertebrates — Fishes — Terrestrial plants — Amphibians — Reptiles — Mammals — Flowering plants

Appendix C

The Ordering of Geologic Time

The latest events of continental drift that we have just considered occupied the last 200 million years of the history of the earth out of a total of some 4 1/2 billion years, that is, within the last 5 percent of the earth's existence. In order to describe the history of the earth, geologists have constructed the Geologic Time Scale (Figure C–1), comprising—from the beginning of earth history to the present—four eras of unequal length. The earliest of these, known as the Precambrian, is over 600 times as long as the Cenozoic (which means new life) and represents some 85 percent of the earth's history. More recent geologic processes have so obscured events in the distant past that a geologist could write many more pages on the last 60 million years (the length of the Cenozoic) than on the 4 billion or more years that preceded it.

To further clarify the description of the earth's history, the Paleozoic, Mesozoic, and Cenozoic Eras are divided into periods and these are further divided into epochs The only epochs listed here are those for the Cenozoic. Some geologists separate Pleistocene and Recent from the other Cenozoic epochs, calling the pre-Pleistocene epochs the Tertiary Period, and the two most recent epochs (comprising about the last 2 1/2 to 3 1/2 million years of geologic time) the Quaternary Period. The Pleistocene Epoch represents the time of the most recent ice ages, and the scheme given here, without a Recent Epoch, emphasizes the probability that the latest ice age has not yet ended.

Appendix D

Length, Area, and

Volume Conversions

Metric

1 kilometer	= 0.6213 mile = 3,281 feet = 1,000 meters
1 centimeter	= 0.3937 inch
1 meter	= 39.37 inches = 3.2808 feet = 1.9036 yards
1 hectare	= 0.01 square kilometer = 10,000 square meters = 2.4711 acres
1 square kilometer	= 247.11 acres = 0.386 square miles
1 liter	= 1.057 quarts = 0.264 gallon
1 milligram per liter	= approximately 1 part per million
1 gram	= 0.0352 ounce

American

1 inch	= 2.54 centimeters
1 foot	= 0.3048 meter = 30.48 centimeters
1 yard	= 0.9144 meter = 91.44 centimeters
1 yard	= 3 feet = 36 inches
1 mile	= 1.609 kilometers = 1,609 meters
1 mile	= 1,760 yards = 5,280 feet
1 fathom	= 6.08 feet
1 square mile	= 2.59 square kilometers = 259 hectares = 640 acres
1 acre	= 43,560 square feet
1 ppm	= approximately 1 milligram per liter
1 pound	= 16 ounces avoirdupois
1 ounce	= 28.349 grams = 28,349 milligrams

Appendix E

Exponential Expression
of Numbers

1,000,000,000,000	$= 1 \times 10^{12}$	$=$ one trillion (American)
2,000,000,000	$= 2 \times 10^{9}$	$=$ two billion (American)
3,000,000	$= 3 \times 10^{6}$	$=$ three million (American)
400,000	$= 4 \times 10^{5}$	$=$ four hundred thousand
50,000	$= 5 \times 10^{4}$	$=$ fifty thousand
4,000	$= 6 \times 10^{3}$	$=$ six thousand
700	$= 7 \times 10^{2}$	$=$ seven hundred
80	$= 8 \times 10^{1}$	$=$ eighty
0.8	$= 8 \times 10^{-1}$	$=$ eight tenths
0.07	$= 7 \times 10^{-2}$	$=$ seven hundredths
0.006	$= 6 \times 10^{-3}$	$=$ six thousandths
0.0005	$= 5 \times 10^{-4}$	$=$ five ten-thousandths
0.00004	$= 4 \times 10^{-5}$	$=$ four hundred-thousandths
0.000003	$= 3 \times 10^{-6}$	$=$ three millionths
0.000000002	$= 2 \times 10^{-9}$	$=$ 2 billionths
0.000000000001	$= 1 \times 10^{-12}$	$=$ 1 trillionth

Glossary

Accadian (Akkadian) An ancient people of the Middle East who occupied much of Iran around present Baghdad.

accretion The process of gradual growth or addition from an external source, or the result of this process. In soils accretion is the addition of minerals.

acre foot An acre is 1/640 of a square mile or 43,560 square feet. An acre foot is that volume that covers one acre to a depth of one foot; it is equal to 43,560 cubic feet.

aerosol A system of minute particles dispersed in a gas. Particles may be solid or liquid as with smoke or perfume in air.

Agave A genus of century plant; century plants are tropical plants particularly abundant in some desert areas. They consist of a bunch of large, waxy leaves near the ground. They have been called century plants because some species mature for decades, finally bloom, and die. Century plants are related to the lilies.

aggression In ethology (the study of behavior) aggression includes not only belligerent activity, but any activity that leads to spacing, subordination, or threat.

agronomy The study of the management of both land and crops.

aldehyde An organic compound bearing a CHO radical. Aldehydes produce acids with oxidation and alcohols with reduction.

algae A group of plants including seaweeds and many small, microscopic, and even single-celled plants.

anthrax A malignant, infectious disease of hoofed animals; it can be contracted by man. It is caused by the Bacillus anthracis.

autoingesting soils Soils that expand when wet and shrink when dry. During dry periods the soils crack to a depth of two feet or more, and humus and other parts of the top soil fall down the cracks. Since cracks occur at different places through time, eventually the different soil horizons become thoroughly mixed.

Bacillus anthracis See anthrax.

bacterial digestion One of the stages in some sewage disposal systems. Bacteria are used to digest and break down sewage products as soil bacteria break down the humus in the upper part of the A horizon.

banded iron formation (BIF) Rocks rich in iron, totally Precambrian, in which layers of iron and silica were laid down so that the rocks are now colored in thin layers of dark, medium, and light reds and reddish browns, depending on the iron content.

basalt A fine grained igneous rock that (1) broke through the earth's crust to be laid down on the land surface, and (2) is composed largely of dark magnesium and iron-bearing silicate minerals.

basin Geologically, (1) a topographic depression in the earth's crust, or (2) a former depression in the earth's crust that received sediments in the geologic past.

bathymetry The topography of the seafloor. A bathymetric map is a contour map (or a map on which shading is used) showing the topography of the seafloor, usually in fathoms.

bed load That part of the sediment moving downstream by rolling or bounding along the bottom.

BIF See banded iron formation.

biomass The mass of all the living matter

growing in, on, or above a unit of ground area; usually expressed in grams per square meter.

biotope   An area inhabited by a community that is adapted to its environment.

blight   Any widespread destruction (usually a disease) of plants.

blood factor   Any chemical factor in the blood that is distinctive and can be inherited.

bomb   In volcanism, a blob of lava thrown from a volcano that hardens in the air. Such blobs usually spin, and as they harden they become elongate, often with twisted, tapering ends.

bore   Herein refers to a rush or wave of water directed by topography and momentum; e.g., a directed tsunami or tide rushing up a stream channel.

bronchitis   An inflammation of the windpipe and the major tubes leading to the lungs.

bubonic plague   A bubo is an inflammatory swelling of a lymph gland. Bubonic plague is a disease characterized by buboes. It is usually spread by fleas from small animals, particularly the rat.

calorie   A gram calorie (a small calorie) is the quantity of heat needed to raise the temperature of one gram of water one degree centigrade.

capillarity   The movement of liquid through minute passages by virtue of the attraction (surface tension) of a liquid for the containing medium.

carbon-14   A radioactive isotope of carbon with atomic weight of 14. Carbon-14 is produced when neutrons and nitrogen collide in the atmosphere. The half-life is about 5,700 years, and carbon-14 is used for dating carbonaceous materials of less than 40,000 years age.

carbonic acid   The weak acid formed with water and carbon dioxide, $H_2O + CO_2 = H_2CO_3$, which ionizes in water to $2H^+$ and CO_3^-.

cardiovascular disease   Any disease of the heart or circulatory system, including heart attacks, strokes, arteriosclerosis, atherosclerosis, etc.

catalytic converter   In air pollution, a device on an internal combustion engine that oxidizes most of the exhaust products through catalytic action of some agent; such as platinum, nickel, etc.

catalyst   A substance that, without permanently being altered itself, causes or accelerates a chemical reaction.

cellulose   An inert organic substance that is the chief constituent of the cell walls of plants, forming an essential part of paper, cotton, wood, etc.

cementation   The process whereby mineral matter is deposited in and around grains of a sediment, binding the grains together and making the rock harder and more resistant.

cesspool   A container for residential sewage. Unlike a septic tank, a cesspool has no artificial or natural processes of oxidation and neutralization.

cherrystone clam   A small, edible clam that burrows in nearshore areas and has a life span of only a few years.

chimpanzee   Probably the most advanced of nonhuman primates, largely African, mostly ground-dwelling, but sometimes feeds in trees.

cholesterol A fat-like substance, $C_{27}H_{45}OH$, found in the body, particularly in the blood stream and bile. It has been impugned as a contributor to many cardiovascular problems.

chromosome Each of several rod-, thread-, or bead-like structures that carry the hereditary substances of DNA and associated molecules.

climatic optimum A post-glacial period of higher sea level and warmer climate than now. Generally the climatic optimum extended from about 7000 years ago to 5000 years ago.

coking The process of distilling coal in an oven or enclosed space, producing a high-carbon fuel, coke. Such fuels are used in some high temperature furnaces and smelters, especially in steel manufacturing.

competent In rocks, competent refers to those rocks that are hard, resistant to erosion, and do not deform easily.

conglomerate A sedimentary rock composed largely of pieces of rock larger than 2 mm or 0.1 inch. It may or may not be cemented, but usually is.

continental climate A climate dominated by dry or boreal (also dry) air masses, usually achieved by topography that either isolates the area or that condenses moisture for precipitation before the air mass enters the particular area.

Corbicula A freshwater clam or bivalve that builds small shell banks in lakes or streams.

Coriolis effect The effect of the rotation of the earth whereby equatorial sites move more rapidly than polar sites as a result of earth rotation. Consequently, poleward-flowing air and water masses are deflected to the east.

corn blight *See* blight.

council of governments (COG) Any *ad hoc* or legislatively established, quasi-official organization made up of representatives of governments; usually refers to an organization that forms a higher category of government.

culture A particular stage or state of civilization; e.g., Roman culture. Western culture usually refers to the culture of the Christian sphere of civilization.

cyanotic babies Babies with cyanosis; that is, with a blue or lividness of the skin because of poorly oxygenated blood.

cyclone A low pressure system of the atmosphere characterized by winds blowing around the central low in a counterclockwise direction in the northern hemisphere and clockwise in the southern hemisphere.

DDT *See* dichlorodiphenyltrichloroethane.

decline curve In resource production, the curve that shows the decline in grade or amount through time.

deflation That process of erosion whereby the wind carries off unconsolidated, loose particles of sediment.

delta The low-lying plain of alluvial deposits at the mouth of a river. It takes its name from the shape of the Nile delta, which resembles the triangular shape of the Greek letter delta.

density Mass per unit of volume, where

mass may be ill-defined as the amount of matter.

depletion The process of emptying or reducing the amount of a resource to zero or to less than required.

developed nations Those nations or areas that have achieved a better than average economy and per capita income. Developed countries are the industrial nations of the world in contrast to the undeveloped countries that are not industrialized.

diabetes A disease in which the body fails to metabolize sugar. Since carbohydrates are altered to sugar in the body, starches and other carbohydrates aggravate the disease.

dichlorodiphenyltrichloroethane (DDT) A persistent chlorinated hydrocarbon pesticide banned in a few states and a few nations.

dithiocarbamate A sulfur containing carbamate pesticide. Carbamates are compounds of nitrogen and hydrogen with an organic acid radicle (COOH).

diversity In animals, the numbers of different kinds of a particular hierarchy.

dredge spoil The man-made piles of rock or soil debris produced by dredging; such as deepening harbors, or cleaning mud from stream channels, canals, irrigation ditches, etc.

drowned river A river valley that is wholly or partly submerged beneath the sea.

ecotone That area of transition and gradation between two biomes.

element One of a class of substances that cannot be chemically separated into substances of other kinds; e.g., oxygen, nitro-

gen, iron, mercury.

emphysema A disease of the lungs whereby there is abnormal extension of the alveoli (air cells of the lungs) from which oxygen is taken into the blood. This enlargement of the alveoli produces less surface area and inefficient oxygenation of the blood.

enzyme One of many complex organic substances capable of producing changes in other organic substances through catalytic action. Some enzymes are further activated or catalyzed by other substances or elements.

ester A fatty compound in animal systems formed by the interaction of an acid and an alcohol.

ethylated lead An additive to gasoline to promote smoother burning of the gasoline in the cylinders of automobiles. The additive is usually tetraethyllead $[Pb(C_2H_5)_4]$, sometimes referred to as leaded ethyl.

exothermal chemical reaction A chemical reaction that gives off heat as its product.

expanding economy An economy in which those things that are used to measure the economy, such as per capita income, gross national product, etc., are continually growing.

extensometer An instrument or gauge that is used to measure the change in distance between points.

fast breeder reactor A nuclear reactor that manufactures more fuel than it consumes. This is achieved by the addition of a neutron from a fissioning uranium-235 to a uranium-238 atom. The uranium is thus changed to plutonium-239, which can then be used as fuel for plutonium reactors.

feldspar   One of a group of minerals composed of silicate (silicon and oxygen) with sodium, potassium, or calcium, or the combinations of two of these.

fertility of soils   The degree to which a soil has those physical properties, trace minerals, and humus for optimum growth of plants. The better the combination of these factors and nutrients, the more fertile the soil.

fire flooding   In oil fields the practice of starting a fire at depth in the producing horizon and feeding it oxygen by pumping air from the surface. The fire burns through the producing horizon, warming, expanding, and pushing fluids ahead of it to producing wells. This is one of many methods of secondary recovery.

fission   In nuclear phenomena, the splitting of atoms, either naturally or by inducement.

flamboyant tree   The flame tree of the West Indies and Central America, said to have been imported. It blooms in numerous, great, red flowers, hence the name.

fungicide   Any of a large number of preparations used to destroy fungi, which are plants without chlorophyll that are either parasitic or live on dead organic matter.

fusion   The combining of the nuclei of two atoms to produce a different element; for example, the combination of deuterium plus tritium produces helium plus a neutron.

gastropod   One of those mollusks commonly with a spiral shell. Some gastropods are marine, others terrestrial.

gene   A unit of inheritance which is located at a specific site on the chromosome, and which, as it reacts with its environment and effects of other genes, develops a hereditary feature.

gene flow   The route of genes through time from generation to generation, considered both from the standpoint of increase in numbers and increase in the parts of the population in which they become established.

gene pool   The total diversity of available genes in a group of interbreeding animals, whether a total population or only a part of a population.

gene suite   The set of genes in a gene pool.

genetic base   Inheritable; implying that anything with a genetic base is hereditary.

geodetic   Pertaining to geodesy, that branch of applied mathematics concerned with the shape and area of the earth or parts of the earth.

glacial lake   Any lake produced by depositional or erosional processes of glaciers, or the damming of a stream valley with glacial ice.

glacial stage   One of the periods of glaciation as differentiated from interglacial stages during a longer period of intermittent glaciation, such as the Pleistocene.

goiter   An enlargement of the thyroid gland, producing a large swelling, up to the size of a baseball, on the sides and/or front of the neck. There are a number of causes, one of which is iodine deficiency.

grade   In evolution, a stage in the evolution of certain types. It may be achieved by closely related lineages at different times.

granite   A visibly granular igneous rock composed of feldspar, quartz, and minor amounts of silicate minerals containing iron and magnesium.

gross national product (GNP) The total production of a nation, expressed in currency; e.g., the total production of the United States exceeded one trillion dollars for the first time in 1972.

habitat The kind of area in which a particular organism lives and to which it is adapted.

half-life Of a radioactive element—the time required for one-half of the atoms in a sample of the element to disintegrate.

halogen One of a series (group 7 of the periodic table) of nonmetallic elements related to chlorine, including fluorine, bromine, and iodine.

Hebraic Referring to things Hebrew or pertaining to that group of Semites descended from Abraham.

Hellenic Herein pertaining to an ancient Greek culture, Hellenism, from about 300 B.C. to Roman acquisition.

Humboldt current A northward flowing and upwelling current of cold water along the west coast of South America. The upwelling waters provide nutrients for the teeming numbers of marine life in this area.

humus Decaying organic material, usually concentrated in or at the top of the A horizon of soils.

hydraulic mining A mining process using a jet stream of water from a nozzle to loosen the country rock containing desirable minerals.

hydrocarbon Any of a number of compounds containing only hydrogen and carbon. Sometimes used more loosely as a synonym for petroleum products.

hypertension A pathological condition, sometimes an arterial disease, of elevated blood pressure.

ice age Used loosely to include a period of glaciation on earth during which there were several episodes of continental glaciation with concurrent colder climates separated by more normal climates.

incubator An apparatus, usually a case heated by a lamp or a coil, used especially for artificially hatching eggs.

induced radiation Nuclear-fueled plants must be cooled, both to collect the heat and to prevent overheating. The contained radioactive fluids are cooled by circulating them in coils through other media. Although no radioactive material escapes, radiation products may pass through the wall of the coil and make radioactive some elements in the cooling medium. The latter is called induced radiation.

inert gases Any of those gases in the zero group of the periodic table. Their outer electron shells are saturated, and they do not actively participate in chemical reactions. They include helium, neon, argon, krypton, xenon, and radon.

ion An electrically charged atom, radical (such as, SO_4^{-2}), or molecule formed by the loss or gain of one or more electrons.

irrigation The process of providing water, in addition to natural rainfall, to increase the rate of growth and the final size and amount of agricultural produce.

isotope ratio A figure or figures comparing the relative amounts of two isotopes of the same element.

keratin   An organic substance found in horn, finger- and toenails, and skin.

lag time   That time between the initiation of a process expected to produce a result and the actual completion of the result.

laissez faire   The theory of interaction of government and economic system in which the government interferes as little as possible in economic affairs.

lapilli   Volcanic ejecta ranging from 4 to 32 millimeters in diameter.

lead time   The amount of time necessary for the planning and instigation of any program. In a petroleum exploration program, for example, a minimum of five years is required from the initiation of the program before production can be expected.

lithification   Any process or combination of processes that turns sediment into rock. Processes may be cementation, compaction, heating, etc.

magmatic   Referring to molten rock or to a source in molten rock, as in magmatic (or juvenile) water.

magnetism   The characteristic properties possessed by magnets, the attraction of certain substances (especially iron), and the formation of fields or spheres that influence the orientation of these certain substances.

melanin   The dark pigment in the bodies of men and animals, producing color in hair and sometimes in skin.

meteoric water   In contrast to juvenile or magmatic water, meteoric water is precipitated via the water cycle.

monsoonal climate   A climate of great moisture extremes, with little rainfall during part of the year and great rainfall during the remainder of the year. It is the product of winds blowing offshore part of the year (dry climate) and onshore the remainder of the year (wet climate).

montmorillonite   A clay mineral, a hydrous aluminum silicate, with weak bonds between layers of silicate. As a result, additional water can enter the sites of the weak bonds, allowing the mineral to swell and break up as the bonds are further weakened.

mustard gas   A liquid chemical warfare agent [$(CLCH_2CH_2)_2S$], called a gas because it is spread as an aerosol. It is lethal, but in lesser amounts produces burns and/or blindness.

mutagenic   Referring to substances that produce mutations in genetic systems of organisms.

Nature Conservancy   A privately financed and endowed organization totally engaged in the furtherance of ecology and actively engaged in financing the acquisition of— or actually acquiring—land that is to remain natural for future generations.

nitric acid   A corrosive liquid (HNO_3) and a powerful oxidizer.

nitrogen-fixing algae   As with bacteria, there are also algae capable of extracting nitrogen from the air. See nitrogen-fixing bacteria.

nitrogen-fixing bacteria   Bacteria that extract nitrogen from the air and convert it to nitrite (NO_2) or nitrate (NO_3). Leguminous plants have symbiotic nitrogen-fixing bacteria in their rootlets to enable them to more easily obtain nitrogen. Most plants cannot obtain nitrogen from the air directly.

Noachian flood A common name for the Biblical flood, or the flood of Deucalion, as described in Genesis, Chapters 6–8.

nucleating In the atmosphere, minute microscopic particles—separately or in an aerosol—serve to provide points attracting moisture and initiating precipitation. This is the purpose of silver iodide in cloud-seeding experiments. Any such point or particle in any medium can be described as nucleating.

Oceania Refers commonly to those islands so far out in the Pacific Ocean that they are not associated with any continent.

oceanic air masses Those warmer, moisture-laden air masses that are brought in over the continents by prevailing winds. In some temperate areas they mollify the climate and make it more uniform and equitable throughout the year.

O^{18}–O^{16} ratio Is the ratio of the oxygen-18 isotope to the oxygen-16 isotope. This ratio is approximately 1 in 500 in normal sea water.

ore deposit A deposit of desirable minerals that is economically worth extracting. Sometimes government subsidy makes a deposit economic.

organochloride pesticide One of many organic compounds which, when combined with chlorine, are lethal to one or more kinds of animals and can be used as biocides.

organotroph A form of life that feeds on organic compounds.

overbank deposits Of a stream, those deposits laid down outside of the channel.

overprint That culturally derived behavior that overrides, often only temporarily, the inherited behavior patterns.

oxidation The process of converting an element to its oxide by adding oxygen. This process removes or neutralizes excess electrons, so that oxidation can then be defined as any removal of electrons, or as changing the charge in a positive direction.

oxides Compounds of an element or a radical and oxygen.

oxygen isotope ratio See O^{18}–O^{16} ratio.

particulate pollution Generally refers to atmospheric pollution by particles instead of by gas or liquid.

parts per billion (ppb) A measure of small quantities. Mercury, for example, in Lake Huron ranges from 100 to 500 parts of mercury to one billion parts total composition.

parts per million (ppm) A measure of small quantities. In liquids one part per million is approximately the same as 1 milligram per liter.

PCB See polychlorinatedbiphenyl.

pegmatite An igneous rock that represents the final stages of crystallization from an igneous body. It has many large (often up to two feet or more in length) crystals of frequently rarer and more valuable elements than other rocks. Pegmatites usually occur in tabular or lenticular bodies in other igneous rocks.

pelecypod One of a class of mollusks that has two shells—a bivalve. Examples are clams, oysters, and scallops.

pepsinogen A substance in the body that

manufactures pepsin, which is an enzyme that helps break down proteins.

permeability   The rate of transmission of one medium through a second medium. The permeability of a rock to a liquid is measured in centimeters per second. A quite impermeable rock would be 5×10^{-7} or less.

phreatic water   That ground water below the water table.

phytoplankton   Minute floating plants, usually unicellular algae or diatoms.

picocurie   A curie is a measure of radiation received by the body. It is derived by a complicated formula which results from different degrees of harmful radiation for each radionuclide. A picocurie is one-trillionth (1×10^{-12}) of a curie.

placer deposit   A mineral deposit in alluvium or regolith from which much of the mineral wealth can be obtained by hydraulic mining or dredging.

planets   Satellites of the sun.
   inner   The inner planets are those denser planets closer to the sun—Mercury, Venus, Earth, and Mars.
   outer   The outer planets are those less dense planets farther away from the sun—Jupiter, Saturn, Uranus, Neptune, and Pluto.

polar timberline   The line between trees and tundra, or the boundary between the tundra and the conifer forest biomes. This boundary is controlled by temperature and moves north or south through time with climatic changes.

pollutant   Any addition to the ecosystem, artificial or natural, that upsets the balance of the ecosystem—or any part of the ecosystem—to a hazardous degree.

pollution   The addition of pollutants.

polychlorinatedbiphenyl (PCB)   One of a group of organochloride compounds used in industry that has two phenyl (C_6H_5) rings. Consumption is harmful to man.

positive reinforcement   The concept or process of programming the desired behavior in the young to override innate behavior so that the behavior will be socially acceptable.

ppb   *See* parts per billion.

ppm   *See* parts per million.

pressure wave   A wave that proceeds with a series of alternating pressures and relaxations parallel to the direction of propagation.

primary production   The basis of the food cycle—the conversion of solar energy to organic material via photosynthesis. Primary producers are green plants.

quahog   A large, burrowing, edible, marine clam that lives on nearshore flats, sometimes exposed at low tide; it has a life span of up to 15 years.

quartz   A hard, translucent mineral composed of silica (silicon and oxygen, SiO_2).

radiometric   Refers to measuring great lengths of geologic time through the ratios of parent to daughter-products of disintegration of radioactive elements.

radionuclide   Any isotope that is a radioactive product of fission, or the product of bombardment by fission products. There are both natural radionuclides and artificial radionuclides; the latter are induced through man-made nuclear facilities.

reclamation   Preparation for reuse, a different use, or restoration to near the natural or original condition.

recycled water   Referring specifically to sewage systems that treat the effluent until it can be returned directly to the water purification plant of the same municipality.

red beds   Naturally red sedimentary strata. The coloring is iron oxide, and as little as 0.1% is sufficient to impart a deep red to the rock.

red tides   Bloomings of particular marine microorganisms in sufficient numbers to tinge the water pink. Some organisms are toxic; others deplete the water of oxygen; both types produce large fish kills.

reduction   Chemically the opposite of oxidation (the addition of electrons), it is the changing of charge in a negative direction.

relic   Something out of the past that has lived beyond its time or usefulness; something that is out of place in time.

reserves of resources   Reserves are proved or estimated. Proved reserves are reserves known to be in the ground. Estimated reserves may range from accurate guesses, based on geological conditions similar to known resource occurrences, to phantasies based on little or no data.

residence time   The length of time of residence anywhere. In geochemical cycles it refers to the average length of time that an atom, a molecule, or a particle remains in a particular part of the ecosystem, such as the ocean, the atmosphere, the lithosphere, etc.

respiratory disease   Any one of many diseases of the lungs or passages for air leading to the lungs.

rhinitis   Inflammation of the mucous membranes of the nose.

river basin authority   A legislatively created authority for planning and controlling water usage and quality for a river system.

S^{32}–S^{34} ratio   The ratio of the sulfur-32 isotope to the sulfur-34 isotope.

septic tank   A sewage system, usually for a single residence, that takes advantage of natural processes in the soil. It consists of a holding tank and perforated lateral pipes, known as the drain field, extending out through the soil and from which the effluent seeps. The holding tank and near parts of the laterals take advantage of anaerobic soil bacteria (termed the anaerobic field) to break down the effluent, and the distal parts of the laterals take advantage of aerobic bacteria (termed the aerobic field) to oxidize and neutralize the effluent.

sesquioxide   An oxide containing 2 metallic and 3 oxygen atoms, named because each oxygen has a valence of 2 ($2 \times 3 = 6$). Alumina (Al_2O_3) and ferric oxide (Fe_2O_3) are the most abundant.

shrink-swell   Alternate shrinking and swelling with wetting and drying as in montmorillonite and other shrink-swell clays. Rocks with shrink-swell properties produce unstable ground.

sierozem   The gray desert soil.

sill   In paleogeography, a shallow threshhold separating two deeper depressions or basins.

sink   In karstic areas (areas of extensive limestone solution), a sink is a large hole in the ground, sometimes at the center of a depression, formed by solution and through which waters enter to the underground.

In geochemical cycles, a sink is a route or pathway whereby a constituent in one part of the cycle is transferred to another part; e.g., increased plant growth in nondesert areas is a sink for excess carbon dioxide, transferring it from the atmosphere to the biosphere.

slate   The metamorphic rock produced from shale or claystone.

slope   The inclination of the ground surface, usually measured in percent slope. A drop of 5 feet in 100 feet horizontal distance is a 5 percent slope.

source beds   Usually in reference to petroleum deposits. The source beds generate the oil and/or gas, which then may migrate to other beds that constitute the reservoir rock.

spoil   Man-made piles of dirt or rock waste resulting from dredging, excavation, mining, or landfill.

stalagmite   The formation on the floor of a cave that is formed by dripstone building up from the floor as mineral precipitates from water that falls from above, drop by drop.

stock pollutants   Referring to the wastes produced by different domesticated animals.

substrate   The underlying rock or stratum.

succulents   Plants with fleshy or juicy leaves; frequently used in describing the thick, waxy leaves of desert cactaceae and relatives of the lilies.

sulfuric acid   A strong, corrosive liquid and a strong oxidizer (H_2SO_4).

sulfur isotope ratio   See $S^{32}-S^{34}$ ratio.

superconductor   A material with a great many more freely movable electrons than in normal conductors. Superconductors, so far, are usually achieved only at extremely low temperatures.

tailing   Waste rock from various types of mining. Bone piles are the waste tailings from coal mines.

Tamarlane   Timur the lame, 1336–1405, Oriental conqueror and descendant of Ghengis Kahn, who reputedly stacked the skulls of one million victims in the square of Baghdad.

Tatar   The correct name for the descendants of Ghengis Kahn, usually called Mongols.

tectonic   Refers to earth movements (e.g., mountain building, uplift, depression) that are associated with changes within the earth's crust and mantle.

teratogenic   Refers to substances that produce either natural abortions or unnatural or deformed young at birth.

thermal event   Refers to proposed widespread meltings of all or major parts of the earth's crust early in earth history.

Thiobacillus   A bacterium that reduces sulfate (SO_4) to sulfide. This, among other phenomena, results in the deposition of iron sulfide (FeS_2) in recent sediments.

toxicant   A poisonous or toxic substance.

trait   A behavioral or physical feature of organisms.

"Tragedy of the Commons"   The title of a paper by Garrett Hardin, the main thesis of which states that when property becomes public property, individuals no longer feel responsibility and the property deteriorates from lack of care.

transpiration   The giving off of wastes, specifically by plants, but primarily concerned with the accompanying water which plants take out of the soil through their roots and emit through their leaves during the discharge of metabolic products.

trapping beds The impermeable strata that hold in mineral deposits, particularly hydrocarbons, and keep them from being dispersed.

turbidity currents Usually in water, currents with enough suspended sediment to be heavier than the containing fluid.

ultraviolet light The invisible light beyond the violet side of the visible spectrum; the wave length is shorter than that of visible light.

undeveloped nations Those nations or areas that are not industrialized and that have lower per capita incomes. *See* also developed nations.

viscosity A measure of the resistance to flow. More viscous mediums flow less easily.

vitamin C Ascorbic acid; a vitamin essential in brain metabolism, some digestive functions, and the metabolism of alcohol.

vitamin D Any one of several fat-soluble, antiricketic, related compounds utilized by the body in many ways.

volatile Referring to substances that evaporate rapidly.

vortical Pertaining to a whirling movement of water, air, or matter.

Western civilization The civilization characterized by Western culture. *See also* culture.

Bibliography

Chapter 1

Cloud, P., and Gibor, A. 1970. The oxygen cycle. *Scientific American* 223(3):111–23.

Coale, A. J. 1970. Man and his environment. *Science* 170:132–36.

Cole, L. C. 1958. The ecosphere. *Scientific American* 198:83–92.

Dansereau, P. 1957. Man's impact on the landscape. In *Biogeography: an ecological perspective*, ed. P. Dansereau, pp. 258–93. New York: Ronald Press.

Deevey, E. S., Jr. 1970. Mineral cycles. *Scientific American* 223(3):148–58.

Dillon, R. H. 1967. Stephen Long's Great American Desert. *Proceedings of the American Philosophical Society*, vol. 111 no. 2. (Reprinted in *Montana Magazine* 18(3):58–74.)

Douglas, W. O. 1951. *Strange lands and friendly people.* New York: Harper Brothers.

Flawn, P. T. 1966. Geology and the new conservation movement. *Science* 151: 409–12.

Gates, D. M. 1971. The flow of energy in the biosphere. *Scientific American* 225(3):89–99.

Glacken, C. J. 1967. *Traces on the Rhodian shore.* Berkeley and Los Angeles: Univ. of California Press.

Goetzmann, W. H. 1966. *Exploration and empire.* New York: Knopf.

Hardin, G. 1968. The tragedy of the Commons. *Science* 162:1243–48.

Hayden, F. V. 1867. First annual report of the United States geological survey of the territories, embracing Nebraska. General Land Office.

Herrmann, P. 1954. *Conquest by man*, trans. M. Bullock. New York: Harper & Row.

Ilic, M. 1971. O "novoj globalnoj tektonici." *Zbornik Radova* 14:9–39.

Ivey, J. B. (1969, in press). *Definition of environmental geology and purpose of the conference.* Governor's Conference on Environmental Geology. Association of Engineering Geologists, Denver Section.

Keller, W. 1956. *The Bible as history: a confirmation of the book of books,* trans. W. Neil. New York: William Morrow.

Kesteven, G. L. 1968. A policy for conservationists. *Science* 160:857–60.

Langdale-Brown, I. 1968. The relationship between soils and vegetation. In *The soil resources of tropical Africa*, ed. R. P. Moss, pp. 61–74. Cambridge: The University Press.

Lavender, D. 1968. *The Rockies.* New York: Harper & Row.

Leggett, R. F. 1968. Consequences of man's alteration of natural systems. *Texas Quarterly* 11(2):24–35.

Leith, C. K. 1935. Conservation of minerals. *Science* 82:109–17.

Leopold, L. B. 1960. *The conservation attitude.* U.S. Geological Survey Circular 414-C, pp. 15–19.

Manning, T. G. 1967. *Government in science: U.S. Geological Survey, 1867–1894.* Lexington: Univ. Press of Kentucky.

Martin, P. S. 1967. Prehistoric overkill. In *Pleistocene extinctions: a search for cause,* ed. P. S. Martin and H. E. Wright, Jr., pp. 75–120. Proceedings of the 7th Congress of the International Association for Quaternary Research. New Haven: Yale Univ. Press.

McHarg, I. L. 1970. Values, process, and form. In *The ecological conscience,* ed. R. Disch, pp. 21–36. Englewood Cliffs, New Jersey: Prentice-Hall.

Mitchell, E. R. 1971. Only people pollute. *Bulletin of the Canadian Institute of Minerology* 64(712):96–100.

Odum, E. P. 1959. *Fundamentals of ecology.* Philadelphia: W. B. Saunders.

———. 1963. *Ecology.* New York: Holt, Rinehart and Winston.

———. 1969. The strategy of ecosystem development. *Science* 164:262–70.

Pecora, W. T. 1970. Nature, not only man, degrades the environment. *West Virginia Geological Survey Newsletter* 14:6–7.

Rodin, L. E., and Basilevic, N. J. 1968. World distribution of plant biomass. In *Functioning of terrestrial ecosystems at the primary production level,* pp. 45–52. UNESCO, Proceedings of the Copenhagen Symposium.

Schnell, J. C. 1969. William Gilpin: advocate of expansion. *Montana* 19(3):30–37.

Slayter, R. O. 1969. Man's use of the environment—the need for ecological guidelines. *Australian Journal of Science* 32(4):146–53.

Smith, H. N. 1947. Rain follows the plow: the notion of increased rainfall for the Great Plains. *Huntington Library Quarterly* 10:159–94.

Speiser, A. E., ed. 1964. Genesis. *The Anchor Bible,* vol. 1. New York: World.

Sukachev, V., and Dylis, N. 1964. *Fundamentals of forest biogeocoenology,* trans. J. M. Maclennan. Edinburgh and London: Oliver & Boyd.

U.S. Department of Interior. 1968. *America's Department of Natural Resources.* Washington.

Wassink, E. C. 1968. Light conversion in photosynthesis and growth of plants. In *Functioning of terrestrial ecosystems at the primary production level,* pp. 53–66. UNESCO, Proceedings of the Copenhagen Symposium.

White, L., Jr. 1967. Historical roots of our ecological crises. *Science* 155:1203–7.

Chapter 2

American Mining Congress. 1972. *Energy and public*

policy. American Mining Congress Special Report no. 2.

Anonymous. 1968a. Updating ocean bottom mineral activity. *Ocean Industry* 3(11):39–42.

————. 1968b. Disinterest dulls U.S. oil-shale leasing. *Oil and Gas Journal* 65(53):97.

————. 1969a. Government proposals for leasing of oil shale a flop. *Engineering and Mining Journal* 170 (1):15.

————. 1969b. Japanese energy planners up a tree. *Oil and Gas Journal* 67(8):41–42.

————. 1969c. Pan Am plans major tar-sand plant. *Oil and Gas Journal* 67(1):30.

————. 1969d. U.S. reserves of crude gas, gas liquids shrinking. *Oil and Gas Journal* 67(14):78–80.

————. 1969e. USBM reports successful underground retorting of shale oil. *Engineering and Mining Journal* 170(7):90.

————. 1969f. Japan moving ahead in geothermal development. *California Division of Mines and Geology Mineral Information Service* 22(8):136–38.

————. 1969g. Project Rulison: new try at nuclear stimulation. *World Oil* 169(4):67–71.

————. 1969h. Nuclear shot ready to go at Rulison. *Oil and Gas Journal* 67(35):87, 89.

————. 1971a. News and views—energy. *International Journal of Environmental Studies* 2:170–71.

————. 1971b. Japan, new giant in the minerals industry. *Engineering and Mining Journal* 172(11):69–141.

————. 1971c. Experts see grim outlook for U.S. energy. *Oil and Gas Journal,* pp. 60–64.

————. 1972. U.S. crude imports hit 1.6 million b/d. *Oil and Gas Journal* 70(1):28–29.

Aubert, M. 1965. *Cultiver l'ocean.* Paris: Presses Universitaires de France.

Averitt, P. 1970. *Stripping—coal resources of the United States—January 1, 1970.* U.S. Geological Survey Bulletin 1322.

Bachman, W. A.. et al. 1969. Forecast for the seventies. *Oil and Gas Journal* 67(45):160–203.

Bachman, W. A. 1971. Is the U.S. vastly underestimating its oil needs? *Oil and Gas Journal* 69(8):33–35.

Barnes, J. W. 1969. Future uranium requirements. *The Mines Magazine,* Jan., pp. 20–24.

Belknap, W. B. 1970. Offshore drilling in the North Sea. In *Mining and petroleum technology,* ed. M. J. Jones. Proceedings of the 9th Commonwealth Mining and Metallurgical Congress, 1969, vol. 1, pp. 505–20. London: Institute of Mining and Metallurgy.

Bings, W. B., and Paist, D. A. 1970. Minerals from the oceans—pt. 2. *Mineral Industries Journal,* vol. 13, no. 3.

Borchers, D.; Stocken, C. G.; and Dall, A. E. 1970. Beach mining at Consolidated Diamond Mines of South West Africa, Ltd.: exploitation of the area be-

tween high- and low-water marks. In *Mining and petroleum technology,* ed. M. J. Jones. Proceedings of the 9th Commonwealth Mining and Metallurgical Congress, 1969, pp. 571–90. London: Institute of Mining and Metallurgy.

Boulding, K. E. 1967. The prospects of economic abundance. In *The control of the environment,* ed. J. D. Roslansky, pp. 39–57. Amsterdam: North-Holland.

————. 1970. *Economics as a science.* New York: McGraw-Hill.

Bullard, F. M. 1962. *Volcanoes: in history, in theory, in eruption.* Austin: Univ. of Texas Press.

Burgess, E. 1970. Geothermal power—virtually pollutionless. *Christian Science Monitor,* 2 Oct., p. 4.

Calvert, S. E., and Price, N. B. 1970. Minor metal contents of recent organic-rich sediments off South West Africa. *Nature* 227:593–95.

Carpenter, H. C., et al. 1972. Evaluation of an in-situ retorting experiment in Green River oil shale. *Journal of Petroleum Technology,* Jan., pp. 21–26.

Coale, A. J. 1970. Man and his environment. *Science* 170:132–36.

Coene, G. T. 1968. Recovery of ocean resources. *Ocean Industry* 3(11):53–58.

Coffer, H. F.; Grier, H. E.; and Aronson, H. H. 1968. The use of nuclear explosives in oil and gas production. *Earth Science Bulletin,* Mar., pp. 5–22.

Dart, R. A., and Beaumont, P. B. 1971. On a further radiocarbon date for ancient mining in southern Africa. *South African Journal of Science* 67(1):10–11.

Dieckamp, H. 1971. The fast breeder reactor: a source of abundant power for the future. *Transactions of the American Geophysical Union* 52(11):756–62.

Dunham, K. C. 1969. Practical geology and the natural environment of man—II: seas and oceans. *Quarterly Journal of the Geological Society* (London) 124:101–29.

Emigh, G. C. 1972. World phosphate reserves—are there really enough? *Engineering and Mining Journal* 173(4):90–95.

Fisher, G. 1970. Demand for metals in the seventies. *Engineering and Mining Journal* 171(11):205–6, 208.

Flawn, P. T. 1965. Ore body concept vs. cost concept. *Geotimes,* Dec. 1964-Jan. 1965, pp. 17–18.

————. 1966. *Mineral resources: geology, engineering, economics, politics, law.* Chicago: Rand McNally.

Frasche, D. F. 1962. *Mineral Resources.* National Academy of Sciences-National Research Council Publication 1000C.

Gerasimov, I. P.; Armand, D. L.; and Yefron, K. M. 1971. *Natural resources of the Soviet Union: their use and renewal,* trans. J. I. Romanowski. San Francisco: W. H. Freeman.

Gillette, R. 1971. Nuclear power in the U.S.S.R.: American visitors find surprises. *Science* 173:1003–6.

Glaser, P. E. 1968. Power from the sun: its future. *Science* 162:857–61.

Gough, W. C., and Eastland, B. J. 1971. The prospects of fusion power. *Scientific American* 224(2):56–64.

Greenberg, D. S. 1969. Uranium: three European nations plan to build centrifuge plants. *Science* 164: 53–55.

Grose, L. T. 1972. Geothermal energy: geology, exploration, and developments, pt. 2. *Mineral Industries Journal* 15(1):1–16.

Hammond, A. L. 1971. Breeder reactors: power for the future. *Science* 174:804–10.

Hampton, J. 1969. A battle rages over nuclear plant. *Christian Science Monitor,* 3 Feb., pp. 1, 14.

Hand, J. W. 1970. Planning for disposal of oil shale, chemical and mine wastes. In *Colorado Governor's Conference on Environmental Geology,* ed. J. B. Ivey, et al. Colorado Geological Survey Special Publication no. 1, pp. 33–37.

Hanna, B. 1967. Oil shale . . . public hot potato. *The Wyoming Alumni News,* Sept.-Oct., pp. 7–15.

Haun, J. D. 1972. Petroleum, public policy, and pollution. *Oil and Gas Journal* 70(17):15–16.

Herrmann, P. 1954. *Conquest by man,* trans. M. Bullock. New York: Harper & Row.

Hicks, M. E., and Woodward, T. C. 1969. Project Thunderbird. *Wyoming Geological Association 21st Annual Field Conference Guidebook,* pp. 161–63.

Hinton, C. 1958. Atomic power in Britain. *Scientific American* 198(3):29–35.

Holman, E. 1952. Our inexhaustible resources. *Bulletin of the American Association of Petroleum Geology* 36:1323–29.

Howe, E. L. 1968. Residual fuel: a key factor in world energy demand, pt. 2. *Canadian Petroleum,* Dec., pp. 41–49.

Hubert, M. K. 1962. *Energy resources: a report to the committee on natural resources.* National Academy of Sciences-National Research Council Publication 1000D.

———. 1965. *History of petroleum geology and its bearing upon present and future exploration.* Gulf Coast Association of Geological Societies.

———. 1969. Energy resources. In *Resources and man,* ed. P. E. Cloud, Jr. et al., pp. 157–242. San Francisco: W. H. Freeman.

Jaffe, F. C. 1971. Geothermal energy, a review. *Bulletin Vereinigung Schweizerischer Petroleum-Geologen und-Ingenieure* 38(93):17–40.

Katz, M. 1971. Decision-making in the production of power. *Scientific American* 224:191–200.

Keller, W. 1956. *The Bible as history: a confirmation of the book of books,* trans. W. Neil. New York: Morrow.

King, R. E. 1972. Offshore interest centers on North Sea, Far East. *World Oil* 174(6):63–66.

Kinney, G. T. 1972. Budget pushes SNG, breeder reactor. *Oil and Gas Journal* 70(5): 64–65.

Koenig, J. B. 1967. The Salton-Mexicali geothermal province. *Mineral Information Service* 20(7):75–81.

Kostuik, J. 1969. Production, reserves, and future sources of uranium. *Bulletin of the Canadian Institute of Mineralogy* 62(686):609–13.

Lambert, D. E. 1969. Increase incentives or face shortage of gas and oil. *World Oil,* Mar., pp. 11–17.

Lampietti, F. J.; Davies, W.; and Young, D. J. 1968. Prospecting for tin off Tasmania. *Mining Magazine,* Sept., p. 160–9.

Landsberg, H. H. 1970. The U.S. resource outlook: quantity and quality. In *America's changing environment,* ed. R. Revelle and H. H. Landsberg, pp. 107–30. Boston: Houghton Mifflin.

Landsberg, H. H.; Fischman, L. L.; and Fisher, J. L. 1963. *Resources in America's future: patterns of requirements and availability, 1960–2000.* Baltimore: Johns Hopkins Univ. Press.

Lane, C. E. 1968. Minerals from sea animals. *Oceanology,* Mar.-Apr., pp. 27–30.

Legget, R. F. 1972. Canadian northern pipeline: research conference. *Geotimes,* May, pp. 24–26.

Lindbergh, J. M. 1970. The untouched mines beneath the sea. *Mining in Canada,* July, pp. 12–14.

Lovering, T. S. 1969. Mineral resources from the land. In *Resources and man,* ed. P. E. Cloud, Jr. et al., pp. 109–34. San Francisco: W. H. Freeman.

Luehrmann, W. H. 1971. The high cost of offshore exploration. *Oceanology,* Oct., pp. 24–29.

Luten, D. R. 1971. The economic geography of energy. *Scientific American* 224(3): 165–75.

Mansfield, E. 1972. Contribution of R & D to economic growth in the United States. *Science* 175:477–86.

Martin, W. B., and Shaughnessy, J. 1969. Project wagon wheel. *Wyoming Geological Association 21st Annual Field Conference Guidebook,* pp. 145–52.

McDermott, J. 1969. Technology: the opiate of the intellectuals. *New York Review of Books,* 31 July, pp. 15–35.

McKelvey, V. E. 1968. Mineral potential of the submerged parts of the U.S. *Ocean Industry* 3(9):37–43.

Mel'nikov, N. V. 1972. The role of coal in the energy fuel resources in the U.S.S.R. *Canadian Mining and Metallurgical Bulletin* 65(722):77–82.

Ministry of Foreign Affairs. 1971. *Statistical survey of Japan's economy.* Economic Affairs Bureau, Ministry of Foreign Affairs, Japan.

National Commission on Materials Policy. 1972. *Towards a national materials policy: basic data and issues, an interim report.* National Commission on Materials Policy.

National Goals Research Staff. 1970. *Toward balanced growth: quantity with quality.* Washington.

Nicholls, G. D. 1968. Deep ocean sediments—a future resource? *Journal of Mines, Metals, and Fuels,* July, pp. 241–44.

Nielsen, R. G. 1960. Thermal recovery of oil. *Mineral Industries* 29(9):1, 4–8.

Nininger, R. D. 1971. Uranium reserves, future demand and the extent of the exploration problem. In *Uranium exploration geology,* pp. 3–19. Proceedings of a Panel in Vienna, International Atomic Energy Agency.

Noakes, J. E., and Harding, J. L. 1971. New techniques in seafloor mineral exploration. *Marine Technology Society Journal* 5(6):41–44.

Overall, M. P. 1968. Mining phosphorite from the sea, pt. 3: evaluation as an investment. *Ocean Industry* 3(11):51–42.

Owen, O. S. 1971. *Natural resource conservation: an ecological approach.* New York: Macmillan.

Park, C. F., Jr., with Freeman, M. C. 1969. *Affluence in jeopardy: minerals and the political economy.* San Francisco: Freeman, Cooper.

Rao, N. N. 1963. Establishment of atomic power reactors in India. *Mineral Markets* 2(5):51–54.

Rasonyi, L. 1968. Exploration and exploitation of geothermic energy in Hungary. *International Association of Hydrologists Memoirs* 8:323–26.

Reynolds, M., Jr. 1971. How successful was Project Rulison? *World Oil,* Dec., p. 54.

Risser, H. E. 1970. *Power and the environment—a potential crisis in energy supply.* Illinois State Geological Survey Environmental Geology Notes no. 44.

Risser, H. E., and Major, R. L. 1967. *Urban expansion—an opportunity and a challenge to industrial mineral producers.* Illinois State Geological Survey Environmental Geology Notes no. 15.

Rose, D. J. 1971. Controlled nuclear fusion: status and outlook. *Science* 172: 797–808.

Ross, D. A. 1972. Red Sea hot brine area revisited. *Science* 175:1455–57.

Ross, S. H. 1971. *Geothermal potential of Idaho.* Idaho Bureau of Mines and Geology Pamphlet 150.

Salotti, C. A.; Heinrich, E. W.; and Giardini, A. A. 1970. Limestone as raw material for hydrocarbon fuels. In *Sixth forum on geology of industrial minerals,* ed. W. A. Kneller, pp. 48-55. State of Michigan Department of Natural Resources.

Seaborg, G. T., and Bloom, J. L. 1970. Fast breeder reactors. *Scientific American* 223(5):13–21.

Seif, M. 1971. Fusion power: progress and problems. *Science* 173:802–3.

Shaw, D. P. 1968. Britain "makes a go" of its nuclear program. *Christian Science Monitor,* 18 Oct., p. 7.

Shigley, C. M. 1968. Seawater as raw material. *Ocean Industry* 3(11):43–46.

Sinha, N. C., and Lahiri, A. 1967. Problems of energy supply in India during the fourth five-year plan. *Journal of Mines, Metals, and Fuels,* Nov., pp. 343–48.

Skinner, B. J. 1969. *Earth resources.* Englewood Cliffs, N. J.: Prentice-Hall.

Smith, S., with Lambert, D. 1969. Gasbuggy: what's the next step? *World Oil,* Jan., pp. 10–18.

Squires, A. M. 1970. Clean power from coal. *Science* 169:821–28.

Starr, C. 1971. Energy and power. *Scientific American* 224(3):37–49.

Summers, C. M. 1971. The conversion of energy. *Scientific American* 225(3):149–60.

Taylor, J. T. M. 1970. Tin dredging off the coast of Cornwall. In *Mining and petroleum technology,* ed. M. J. Jones. Proceedings of the 9th Commonwealth Mining and Metallurgical Congress, 1969, pp. 591–627. London: Institute of Mining and Metallurgy.

Tomaszlewski, W. 1969. Superconductors may fulfill many a technical dream. *New York Times,* 23 Feb., p. 14.

Toynbee, A. J. 1947. *A study of history,* abridgement of vols. 1–6 by D. C. Somervell. New York and London: Oxford Univ. Press.

U.S. Atomic Energy Commission. 1968. *Nuclear power plants in the United States.*

U.S. Geological Survey. 1963. *Long range plan, United States Geological Survey program of resource surveys, investigations, and research: preliminary draft.*

———. 1968. *Marine geology: research beneath the sea.*

Vonder Haar, T. V., and Suomi, V. 1969. Satellite observations of the earth's radiation budget. *Science* 163: 667–69.

Wassink, E. C. 1968. Light conversion in photosynthesis and growth of plants. In *Functioning of terrestrial ecosystems at the primary production level,* pp. 53–66. UNESCO, Proceedings of the Copenhagen Symposium.

Weber, G. 1971. Petroleum and the energy crunch. *Oil and Gas Journal* 69(46):121–133.

Wenk, E., Jr. 1969. The physical resources of the ocean. *Scientific American* 221(3):166–77.

Wilson, H. M. 1969. Outlook fading for payout in deep waters of Texas. *Oil and Gas Journal* 67(14):71–74.

Chapter 3

Ackley, G. 1966. Preface to *The economics of air pollution,* ed. H. Wolozin. New York: Norton.

Allen L. H., Jr. et al. 1971. Plant response to carbon dioxide enrichment under field conditions: a simulation. *Science* 173: 256–58.

Allsopp, H. L., et al. 1962. Pb-Sr age measurements on various Swaziland granites. *Journal of Geophysical Research* 67:5307–13.

Altshuller, A. P. 1968. Analysis of organic gaseous pollutants. In *Air pollution,* 2d ed., ed. A. C. Stern, vol. 2, pp. 115–45. New York: Academic Press.

Anhaeusser, C. R. 1972. *The evolution of the early Precambrian crust of southern Africa.* Univ. of Witwatersrand Economic and Geological Research Unit Information Circular no. 70.

Anhaeusser, C. R., et al. 1968. The Barberton Mountain land: a model of the elements and evolution of Archean fold belt: annexure to Transvaal. *Geological Society of South Africa* 71:225–254. (Also available as Univ. of Witwatersrand Economic and Geological Research Unit Information Circular no. 38.)

Assaf, G., and Biscaye, P. E. 1972. Lead-212 in the urban boundary layer of New York City. *Science* 175:890–94.

Atkins, D. H. F.; Cox, R. A.; and Eggleton, A. E. J. 1972. Photochemical ozone and sulphuric acid aerosol formation in the atmosphere over southern England. *Nature* 235: 372–76.

Baehr, F. 1972. The IIEC: study of automotive emission control. *Canadian Petroleum,* May, pp. 22–26.

Battan, L. J. 1969. Climate and man. *Science* 166:536.

Bazell, R. J. 1971a. Lead poisoning: zoo animals may be the first victims. *Science* 173:130–31.

———. 1971b. Lead poisoning: combating the threat from the air. *Science* 174:574–76.

Billings, C., and Matson, W. R. 1972. Mercury emissions from coal combustion. *Science* 176:1232–33.

Black, L. P., et al. 1971. Isotope dating of very early Precambrian amphibolite facies gneisses from the Godthaab district, west Greenland. *Earth Planetary Sciences Letter* 12:245–59.

Boffey, P. M. 1968. Nerve gas: Dugway accident linked to Utah sheep kill. *Science* 162:1460–64.

Brandt, C. S., and Heck, W. W. 1968. Effects of air pollutants on vegetation. In *Air pollution,* 2d ed., ed. A.C. Stern, vol. 1, pp. 401–43. New York: Academic Press.

Breslow, R. 1972. The nature of aromatic molecules. *Scientific American* 227(2):32–40.

Brooks, C. E. P. 1949. *Climate through the ages.* New York: McGraw-Hill.

Carne, S. 1971. Quoted in R. J. F. Pinsent, "Airs, waters, and places." *International Journal of Environmental Studies* 1:237–38.

Carson, R. L. 1951. *The sea around us.* New York: Oxford Univ. Press.

Chester, R.; Elderfield, H.; and Griffin, J. J. 1971. Dust transported in the north-east and south-east trade winds in the Atlantic Ocean. Nature 223:474–76.

Chisolm, J. J. 1971. Lead poisoning. *Scientific American* 224(2):15–23.

Clemesha, B. R., and Nakamura, Y. 1972. Dust in the upper atmosphere. *Nature* 237:328–29.

Eisenbud, M., and Ehrlich, L. R. 1972. Carbon monoxide concentration trends in urban atmospheres. *Science* 176:193–94.

Essenhigh, R. H. 1971. Air pollution from combustion sources. *Earth and Mineral Sciences* (Pennsylvania State Univ.) 407(7):49–52.

Foy, C. L. 1970. *Plants and pollution.* American Society of Agronomy Special Publication 16, pp. 37–59.

Garrels, R. M. and Mackenzie, F. T. 1971. *Evolution of sedimentary rocks.* New York: Norton.

Goldich, S. S.; Hedge, C. E.; and Stern, T. W. 1970. Age of the Morton and Montevideo gneisses and related rocks, southwestern Minnesota. *Bulletin of the Geological Society of America* 81:3671–96.

Goldsmith, J. R. 1968. Effects of air pollution on human health. In *Air pollution,* 2d ed., ed. A. C. Stern, vol. 1, pp. 547–615. New York: Academic Press.

Goldsmith, J. R., and Hartman, R. 1969. Pollution medical research. *Science* 163:706–9.

Hagen-Smit, A. G. 1964. The control of air pollution. *Scientific American* 210(6):25–37.

Hutchinson, G. L. 1972. Air containing nitrogen-15 ammonia: foliar absorption by corn seedlings. *Science* 175:759–61.

Hutchinson, G. L., et al. 1972. Atmospheric ammonia: absorption by plant leaves. *Science* 175:771–72.

Incoll, L. D., and Popenoe, P. 1969. London: where smog was born. *Science* 163:339.

Inman, R. E., et al. 1971. Soil: a natural sink for carbon monoxide. *Science* 172:1229–31.

Johnson, N. M.; Reynolds, R. C.; and Likens, G. E. 1972. Atmospheric sulfur: its effect on the chemical weathering of New England. *Science* 177:514–16.

Judson, S. 1968. Erosion of the land—or what's happening to our continent. *American Scientist* 56:356–74.

Katz, M. 1968. Analysis of inorganic gaseous pollutants. In *Air pollution,* 2d ed., ed. A. C. Stern, vol. 2, pp. 53–114. New York: Academic Press.

Knight, B. C. J. C. 1936. *Bacterial nutrition; materials for a comparative physiology of bacteria.* Medical Research Council Special Report Series no. 210.

Le Roy Ladurie, E. 1971. *Times of feast, times of famine: a history of climate since the year 1000,* trans. B. Bray. Garden City, N.Y. Doubleday.

Lowry, W. P. 1967. The climate of cities. *Scientific American* 216(2):15–23.

Ludwig, J. H.; Morgan, G. B.; and McMullen, T. B. 1970. Trends in urban air quality. *Transactions of the American Geophysical Union* 51(5):468–75.

Maugh, T. H. 1972. Carbon monoxide: natural sources dwarf man's output. *Science* 177:338–39.

Oparin, A. I. 1938. *The origin of life,* trans. S. Morgulis. New York: Macmillan.

———. 1957. *The origin of life on earth,* trans. Synge.

New York: Academic Press.

Otto, H. W., and Daines, R. H. 1969. Plant injury by air pollutants: influence of humidity on stomatal apertures and plant response to ozone. *Science* 163:1209–10.

Park, C. F., Jr., with Freeman M. C. 1969. *Affluence in jeopardy: minerals and the political economy.* San Francisco: Freeman, Cooper.

Pearman, G. I., and Garratt, J. R. 1972. Global aspects of carbon dioxide. *Search* 3(3):67–73.

Pettersson, O. 1912. Climatic variations in historic and prehistoric time. *Svenska Hydrogafisk-Biologiska. Kommissionens Skrifter* 15.

Plass, G. N. 1959. Carbon dioxide and climate. *Scientific American* 201(1):41–47.

Rasool, S. I., and Schneider, S. H. 1971. Atmospheric carbon dioxide and aerosols: effects of large increases on global climate. *Science* 173:138–41.

Richards, F. R. 1965. Dissolved gases other than carbon dioxide. In *Chemical oceanography*, ed. J. P. Riley and G. Skirrow, vol. 1, pp. 197–225. New York: Academic Press.

Rubey, W. W. 1951. Geologic history of sea water: an attempt to state the problem. *Bulletin of the Geological Society of America* 62:1111–47.

Semrau, K. T. 1971. Two new processes for recovery of sulphur oxides from smelter gases. *Engineering and Mining Journal*, Apr., pp. 115–19.

Skirrow, G. 1965. The dissolved gases—carbon dioxide. In *Chemical oceanography*, ed. J. P. Riley and G. Skirrow, vol. 1, pp. 227–32. New York: Academic Press.

Stewart, R. W. 1969. The atmosphere and the ocean. *Scientific American* 221(3):76–86.

Stokinger, H. E., and Coffin, D. L. 1968. Biologic effects of air pollutants. In *Air pollution*, 2d ed., A. C. Stern, vol. 1, pp. 445–546. New York: Academic Press.

Sullivan, W. 1969. A new effort to save our environment. *New York Times* 8 June, p. 6.

Unsöld, A. O. J. 1969. Stellar abundances and the origin of the elements. *Science* 163:1015–25.

Vonder Haar, T. V., and Suomi, V. E. 1969. Satellite observations of the earth's radiation budget. *Science* 163:667–69.

Westenberg, A. A. 1972. Carbon monoxide and nitric oxide consumption in polluted air: the carbon monoxide-hydroperoxyl reaction. *Science* 177:255–56.

Wexler, H. 1952. Volcanoes and world climate. *Scientific American* 186(4):74–80.

Winkler, E. M. 1970. The importance of air pollution in the corrosion of stone and metals. *Engineering Geology* 4:327–34.

Wolozin, H., ed. 1966. *Economics of air pollution.* New York: Norton.

Yocom, J. E., and McCaldin, R. O. 1968. Effects of air pollution on materials and the economy. In *Air pollution*, 2d ed., ed. A. C. Stern, vol. 1, pp. 617–54. New York: Academic Press.

Chapter 4

Anonymous. 1968a. Oceans proposed for disposal of waste from urban centers. *New York Times.*

———. 1968b. Coast suit attacks pollution by Navy. *New York Times,* 24 Nov.

———. 1969a. Heyerdahl voices concern over pollution of open ocean. *Dallas Morning News,* 2 Aug., p. 4.

———. 1969b. W. Australia backs nuclear blasting for harbor work. *Ocean Industry* 4(2):33.

———. 1969c. Bay's calm threatened by population pressure. *Austin Statesman,* 18 Sept., p. 33.

———. 1969d. Pollution grows at Tokyo beaches. *New York Times* 18 Aug., p. 13.

———. 1969e. Beach litter irks Australians. *Christian Science Monitor,* 28 Jan.

———. 1969f. Verdict still to come in Chennel. *Oil and Gas Journal,* 3 Mar., pp. 76–77.

———. 1969g. Fresh oil leak compounds channel woes. *Oil and Gas Journal,* 3 Mar., pp. 82–83.

———. 1969h. Huge channel oil spill blows up storm. *Oil and Gas Journal,* 10 Feb., pp. 50–51.

———. 1969i. California suing for $560 million over oil leakage. *Wall Street Journal,* 19 Feb., p. 28.

———. 1969j. Oil-spill damage to marine life scant. *Oil and Gas Journal,* 17 Mar., pp. 65–66.

———. 1969k. Cook inlet tanker leaks crude oil after accident. *Oil and Gas Journal,* 10 Mar., p. 28.

———. 1969l. Oil threatening wildlife in wake or hurricane. *Dallas Morning News,* 23 Aug., p. 15.

———. 1969m. Gulf states tightening offshore rules. *Oil and Gas Journal,* 3 Mar., pp. 86–87.

———. 1971. Coliform bacteria found 80 miles from sewage dumping ground in Atlantic Ocean. *Air/Water Pollution Report,* 20 Dec., p. 517.

Bainbridge, A. E. 1963. *Nuclear geophysics.* National Academy of Sciences-National Research Council.

Beasley, T. M.; Osterberg, C. L.; and Jones, Y. M. 1969. Natural and artificial radionuclides in seafoods and marine protein concentrates. *Nature* 221:1207–9.

Blumer, M., and Sass, J. 1972. Oil pollution: persistence and degradation of spilled fuel oil. *Science* 176:1120–22.

Bourne, W. R. P., and Bogan, J. A. 1972. Polychlorinated biphenyls in North Atlantic seabirds. *Marine Pollution Bulletin* 3:171–75.

Bowen, V. T., et al. 1969. Strontium-90; concentration in surface waters of the Atlantic Ocean. *Science* 164:825–26.

Burton, J. D., and Leatherland, T. M. 1971. Mercury in a coastal marine environment. *Nature* 231:440–41.

Cahn, R., and Favre, G. H. 1969. Conservationists frame plan to save San Francisco Bay. *Christian Science Monitor,* 5 July, pp. 1, 3.

Carpenter, E. J., and Smith, K. L., Jr. 1972. Plastics on the Sargasso Sea surface. *Science* 175:1240–41.

Carson, R. L. 1951. *The sea around us.* New York: Oxford Univ. Press.

Carter, L. J. 1970. Galveston Bay: test case of an estuary in crisis. *Science* 167:1102–7.

Carthy, J. D., and Arthur, D. R., eds. 1968. *The biological effects of oil pollution in littoral communities.* Proceedings of a Symposium: Pembrade, Wales; Classy, Hampton, and Middlesex Field Study Council, London.

Chase, C. G., and Perry, E. C., Jr. 1972. The oceans: growth and oxygen isotope evolution. *Science* 177: 992–94.

Cloud, P. E., Jr. 1968. Atmospheric and hydrospheric evolution of the primitive earth. *Science* 160:729–35.
————. 1972. A working model of the primitive earth. *American Journal of Science* 272:537–48.

Connor, P. M. 1972. Distribution of cadmium, lead, and zinc in the Bristol Channel. *Marine Pollution Bulletin* 3:188–92.

Cowan, R. C. 1969. Chemist airs urgent call for ocean pollution study. *Christian Science Monitor,* 6 Jan., pp. 1, 5.

Duursma, E. K. 1972. Geochemical aspects and applications of radionuclides in the sea. In *Oceanography and marine biology: an annual review,* ed. H. Barnes, vol. 10, pp. 137–223. New York: Hafner.

Dyrssen, D. 1972. The changing chemistry of the oceans. *Ambio* 1(1):21–25.

Edwards, P. 1972. Cultured red alga to measure pollution. *Marine Pollution Bulletin* 3:184–87.

Ehrlich, P. R., and Ehrlich, A. H. 1970. The food-from-the-sea myth. *Saturday Review,* 4 Apr., pp. 53, 64.

Ellis, D. V., and Littlepage, J. L. 1972. Marine discharge of mine wastes: ecosystem effects and monitoring programs. *Canadian Mining and Metallurgical Bulletin,* Apr., pp. 45–50.

Fischer, J. L. 1969. Starfish infestation: hypothesis. *Science* 165:645.

Garrels, R. M., and MacKenzie, F. T. 1971. *Evolution of sedimentary rocks.* New York: W. W. Norton.

Garritt, D. E., et al. 1959. *Radioactive waste disposal into Atlantic and Gulf coastal waters.* National Academy of Sciences-National Research Council Publication no. 655.

Greenbaum, R. N. 1969. How San Diego cleaned up its bay. *Ocean Industry,* July, pp. 55–56.

Hardin, G. 1968. The tragedy of the Commons. *Science* 162:1243–48.

Harriss, R. C.; White, D. B.; and MacFarlane, R. B. 1970. Mercury compounds reduce photosynthesis by plankton. *Science* 170:736–37.

Hasler, A. D., and Ingersoll, B. 1968. Dwindling lakes. *Natural History* 77(9):8–20.

Hedgpeth, J. W. 1972. Atomic waste disposal in the sea. In *The careless technology,* ed. M. T. Farvar and J. P. Milton, pp. 812–28. Garden City, N.Y. Natural History Press.

Heyerdahl, T. 1971. *The Ra expeditions,* trans. Crampton. Garden City, N.Y.: Doubleday.

Hill, G. 1969. "Save" San Francisco Bay hearing on. *New York Times,* 23 Mar., p. 79.

Holden, C. 1971. Chesapeake Bay. *Science* 17:825–27.

Hood, D. W., ed. 1971. *Impingement of man on the oceans.* New York: Wiley-Interscience.

Howells, G. P. 1972. Research in estuaries. *Marine Pollution Bulletin* 3(10):152–54.

Hubbard, E. F., and Stamper, W. G. 1972. *Movement and dispersion of soluble pollutants in the northeast Cape Fear Estuary, North Carolina.* U. S. Geological Survey Water-Supply Paper 1873-E.

Hull, E. 1968. *The atom and the ocean.* Understanding the atom series. Oak Ridge, Tenn.: Atomic Energy Commission.

Imbray, J. F. 1868. *Sailing directions for the west coast of North America, pt. 1.* London: James Imbray.

Jensen, E. T. 1969. Estuarine pollution. *Oceanology,* June, pp. 32–33.

Lane, F. W., et al. 1924. *Effect of oil pollution of coast and other waters on the public health.* Public Health Service. Reprint 986.

Ludwigson, J. O. 1969. Chesapeake Bay. *Oceans,* July, pp. 6–16.

McLusky, D. S. 1971. *Ecology of estuaries.* London: Heinemann.

Menzel, D. W.; Anderson, J.; and Randtke, A. 1970. Marine phytoplankton vary in their response to chlorinated hydrocarbons. *Science* 167:1724–26.

Mikolaj, P. G. 1972. Environmental applications of the Weibull distribution function: oil pollution. *Science* 176:1019–21.

Mosser, J. L., et al. 1972. Polychlorinated biphenyls: toxicity to certain phytoplankters. *Science* 175:191–92.

Mueller, M. 1969. New canal: what about bioenvironmental research? *Science* 163:165–67.

Nelson, B. 1969. Marine Commission invokes NOAA, urges refitting of nation's Ark. *Science* 163:263–65.

Nuzzi, R. 1972. Toxicity of mercury to phytoplankton. *Nature* 237:38–40.

Odum, H. T., et al. 1963. *Experiments with engineering of marine ecosystems.* Univ. of Texas Institute of Marine Science 9:373–403.

Oglesby, R. T. 1967. Biological and physiological basis of indicator organisms and communities. In *Pollution and marine ecology,* ed. T. Olsen and F. J. Burgess, pp. 167–69. New York: Wiley-Interscience.

Pestrong, R. 1972. San Francisco Bay tidelands. *California Geology* 25(2):27–40.

Prouty, D. 1969. Wind held veto on A-blast. *Denver Post,* 4 Sept., p. 2.

Regner, A. P., and Park, R. W. A. 1972. Faecal pollution of our beaches—how serious is the situation? *Nature* 239:408–10.

Revill, L. 1969. Earthquake threat cancels A-harbor plan in Australia. *Christian Science Monitor,* 12 Apr., p. 2.

Rice, T. R., and Wolfe, D. A. 1971. Radioactivity—chemical and biological aspects. In *Impingement of man on the oceans,* ed. D. W. Hood, pp. 325–79. New York: Wiley-Interscience.

Richards, F. A. 1965. Dissolved gases other than carbon dioxide. In *Chemical oceanography,* ed. J. P. Riley and G. Skirrow, vol. 1, pp. 197–225. New York: Academic Press.

Ritchie-Calder, 1971. Oceanic farming from surface to deepest deep. *Smithsonian* 1(12):8–14.

Rubey, W. W. 1951. Geologic history of sea water: an attempt to state the problem. *Bulletin of the Geological Society of America* 62:1111–48.

Ryther, J. H., and Dunstan, W. M. 1971. Nitrogen, phosphorous, and eutrophication in the coastal marine environment. *Science* 171:1008–13.

Sawbridge, D. F., and Bell, M. A. M. 1969. Pacific shores, *Science* 164:1089.

Schindler, W. D., et al. 1972. Atmospheric carbon dioxide: its role in maintaining phytoplankton standing crops. *Science* 177:1102–94.

Schubel, J. R., and Pritchard, D. W. 1971. Chesapeake Bay: a second look. *Science* 173:943–45.

Sheffey, J. P., and Rubinoff, I. 1968. When Caribbean and Pacific waters mix. *Science* 162:1329.

Shiber, J. G., and Ramsay, B. 1972. Lead concentrations in Beirut waters. *Marine Pollution Bulletin* 3:169–71.

Skirrow, G. 1965. The dissolved gases—carbon dioxide. In *Chemical oceanography,* ed. J. P. Riley and G. Skirrow, vol. 1, pp. 227–322. New York: Academic Press.

Smith, H. W. 1971. Incidence of R+ *Escherichia coli* in coastal bathing waters of Britain. *Nature* 234:155–56.

Smith, K. L., Jr., and Teal, J. M. 1973. Deep-sea benthic community respiration: an in situ study at 1850 meters. *Science* 179:282–83.

Starr, V. P. 1956. The general circulation of the atmosphere. *Scientific American* 195:26, 40–45, December.

Stephan, E. C. 1969. Long Island's approach to coastal zone planning. *Marine Technological Society Journal* 3(6): 71–72.

Topp, R. W. 1969. Interoceanic sea-level canal: effects on the fish faunas. *Science* 165:1324–27.

Totenberg, N. 1969. Drilling attacked as oil mucks coast. *National Observer,* 10 Feb.

Turekian, K. K. 1968. *Oceans.* Englewood Cliffs, N.J.: Prentice-Hall.

Vaccaro, R. F. 1965. Inorganic nitrogen in sea water. In *Chemical oceanography,* ed. J. P. Riley and G. Skirrow, vol. 1, pp. 365–408.

Wass, M. L. 1967. Biological and physiological basis of indicator organisms and communities. In *Pollution and marine ecology,* ed. T. Olsen and F. J. Burgess, pp. 271–83. New York: Wiley-Interscience.

Wassink, E. C. 1968. Light conversion in photosynthesis and growth of plants. In *Functioning of terrestrial ecosystems at the primary production level,* pp. 53–66. UNESCO, Proceedings of the Copenhagen Symposium.

Webber, H. H. 1970. The development of maricultural technology for the pennaeid shrimp of the Gulf and Caribbean region. *Helgolander Wissenschafterliche Meeresuntersughungen* 20(1–4):455–63.

Williams, J.; Higginson, J. J.; and Rohrbough, J. D. 1968. *Sea and air: the naval environment.* Annapolis: U.S. Naval Institute.

Chapter 5

American Public Health Association. 1971. *Standard methods for the examination of water and wastewater.* 13th ed. Washington.

Bache, C. A., et al. 1972. Polychlorinated biphenyl residues: accumulation in Cayuga Lake trout with age. *Science* 177:1191–92.

Barthel, W. F., et al. 1966. Surface hydrology and pesticides. In *Pesticides and their effects on soils and water,* ed. M. E. Bloodworth et al., pp. 128–44. American Society of Agronomy Special Publication 8.

Bharadwag, O. P. 1961. The arid zone of India and Pakistan. In *A history of land use in arid regions,* ed. L. D. Stamp. UNESCO, Arid Zone Research, vol. 12, pp. 143–74.

Bibby, G. 1969. *Looking for Dilmun.* New York: Knopf.

Boswell, H., et al. 1968. *The Texas water plan.* Texas Water Development Board.

Boyko, H. 1967. Salt water agriculture. *Scientific American* 216(3):89–96.

Bradbury, J. P., and Megard, R. O. 1972. Stratigraphic record of pollution in Shagawa Lake, northeastern Minnesota. *Bulletin of the Geological Society of America* 83:2639–48.

Brier, G. W., et al. 1972. Cloud seeding experiments;

lack of bias in Florida series. *Science* 176:163–64.

Briggs, J. C., et al. 1972. Aquatic ecosystems. *Science* 176:581–83.

Carson, R. 1951. *The sea around us.* New York: Oxford Univ. Press.

Carter, L. J. 1969. Warm-water irrigation: an answer to thermal pollution. *Science* 165:478–80.

Clawson, M., et al. 1969. Desalted seawater for agriculture: is it economic? *Science* 164:1141–48.

Cronin, J. G. 1969. *Ground water in the Ogallala formation in the southern high plains of Texas and New Mexico.* U.S. Geological Survey Hydrologic Investigations, Atlas HA-330.

Csanady, G. T.; Crawford, W. R.; and Pade, B. 1971. Thermal plume study at Douglas Point, Lake Huron. In *Proceedings of the 14th Conference on Great Lakes Research,* pp. 522–34. International Association for Great Lakes Research.

Degeer, M. W. 1971. *Natural chloride pollution, Arkansas and Red River Basins.* Annals of the Oklahoma Academy of Science Publication no. 2, pp. 42–46.

Dingman, R. J., and Nuñez, J. 1969. *Hydrogeologic reconnaissance of the Canary Islands, Spain.* U.S. Geological Survey Prof. Paper 650-C, pp. C201–8.

Ellis, H. B. 1969. Dutch rinse the Rhine. *Christian Science Monitor,* 27 Aug., p. 1, 4.

Grounds, H. C., and Hartley, S. M. 1971. Alternative methods of tertiary treatment for nutrient removal. In *Proceedings of the 14th Conference on Great Lakes Research,* pp. 224–35. International Association for Great Lakes Research.

Hamilton, J. L., and Owens, W. G. 1972. Effect of urbanization on ground water levels. *Bulletin of the Association of Engineering Geologists* 9:327ff.

Hanke, S. H., and Brandt, J. J. 1972. Thermal discharges and public policy development. *Water Resources Bulletin* 8:446–58.

Hasler, A. D., and Ingersoll, B. 1968. Dwindling lakes. *Natural History,* vol. 77, no. 9.

Hem, J. D. 1972. Chemistry and occurrence of cadmium and zinc in surface water and ground water. *Water Resources Research* 8:661–79.

Herak, M., and Stringfield, V. T., eds. 1972. *Karst: important karst regions of the northern hemisphere.* Amsterdam: Elsevier.

Hill, D. N., and Bettett, M., et al. 1968. *Earthquake engineering program: progress report.* California Department of Water Resources Bulletin 116–4.

Huling, E. E., and Hollocher, T. C. 1972. Ground water contamination by road salt: steady-state concentrations in east central Massachusetts. *Science* 176: 288–90.

Hunt, C. A., and Garrells, R. M. 1972. *Water: the web of life.* New York: Norton.

Inglehart, H. H. 1967. *Occurrence and quality of ground water in Crockett County, Texas.* Texas Water Development Board Report 47.

Jackim, E., and Gentile, J. 1968. Toxins of blue-green algae: similarity to saxitoxin. *Science* 162:915–16.

James, L. 1964. The California water plan. *Geotimes,* Nov., pp. 9–11.

Jeffers, F. J. 1972. A method for minimizing effects of waste heat discharges. *International Journal of Environmental Studies* 3:321–27.

Jenkins, D. S. 1957. Fresh water from salt. *Scientific American* 196(3):37–45.

Johnson, A. I.; Moston, R. P.; and Morris, D. A. 1968. *Physical and hydrologic properties of water-bearing properties in subsiding areas of central California.* U.S. Geological Survey Prof. Paper 497–A.

Joos, L. A. 1970. Quoted in F. Kendrick, "Irrigation may stimulate new Great Plains Climate." *Austin Statesman,* 8 Jan., p. 7.

Kang, S. T. 1972. Irrigation in ancient Mesopotamia. *Water Resources Bulletin* 8:619–23.

King, D. L. 1972. Carbon as a limiting factor in lake ecology. In *Trace substances in environmental health —V,* ed. D. D. Hemphill, pp. 109–15. Columbia: Univ. of Missouri Press.

Kuiper, E. 1971. *Water resources project economics.* London: Butterworths.

Lane, R. 1969. Seattle area washes out Lake Washington. *Christian Science Monitor,* 30 Sept., p. 3.

LeGrand, H. E. 1973. Hydrological and ecological problems of karst regions. *Science* 179:859–64.

McCleskey, G. W. 1972. Problems and benefits in ground water management. *Ground Water* 10(2):2–5.

Meier, M. F. 1969. *New York Times,* 19 Oct., p. 50.

Michalowski, K. 1968. *The art of ancient Egypt,* trans. Norbert Guterman. New York: Abrams.

Mink, L. L.; Williams, R. E.; and Wallace, A. T. 1972. Effect of early day mining operations on present day water quality. *Ground Water* 10:17–26.

Moss, F. S. 1967. *The water crisis.* New York: Praeger.

Mukhergee, K. P. 1968. Origin and disposal of acidic waters from collieries and waste coal heaps, *Journal of Mines, Metals, and Fuels,* Mar., pp. 83 ff.

Nace, R. L. 1969. Quoted in the *Christian Science Monitor,* 17 Dec., p. 1.

Nair, G. U. 1967. Studies on algae with special reference to their control in potable waters. Agra University, *Journal of Research,* 16(1):175–77.

Nelson, D. M.; Romberg, G. P.; and Prepejchal, W. 1971. Radionuclide concentrations near the Big Rock Point nuclear power station. In *Proceedings of the 14th Conference on Great Lakes Research,* pp. 268–76. International Association for Great Lakes Research.

Park, C. F., Jr., with Freeman, M. C. 1969. *Affluence in*

jeopardy: minerals and the political economy. San Francisco: Freeman, Cooper.

Pearson, F. J., Jr. 1966. Ground water ages and flow rates by Carbon-14 method. Ph.D. dissertation, Univ. of Texas.

Pettyjohn, W. A. 1968. Design and construction of a dual recharge system at Minot, North Dakota. *Ground Water* 6(1):4–8.

Pettyjohn, W. A. and Fahy, V. 1968. Artificial recharge solves water problem. *Public Works,* Sept., pp. 82–87.

Popper, K.; Merson, R. L.; and Camirand, W. M. 1968. Desalination by osmosis-reverse osmosis couple. *Science* 159:1364–65.

Powell, J. W. 1969. *Down the Colorado: diary of the first trip through the Grand Canyon.* New York: Dutton.

Purushothaman, K., and Yue, C. M. 1972. Removal of phosphorous from waste water to control eutrophication. In *Trace substances in environmental health —V,* ed. D. D. Hemphill, pp. 95–108. Columbia: Univ. of Missouri Press.

Rich, V. 1972. Problems of the Caspian. *Marine Pollution Bulletin* 3:84–85.

Rosebury, T. 1969. *Life on man.* New York: Viking.

Skinner, B. J. 1969. *Earth resources.* Englewood Cliffs, N.J.: Prentice-Hall.

Snyder, A. E. 1962. Desalting water by freezing. *Scientific American* 207(6):41–47.

Spence, E. V., et al. 1948. *Progress report for the period Sept. 1, 1946-Aug. 31, 1948.* Texas Board of Water Engineers.

Spencer, J. 1970. Geological influence on regional health problems. *Texas Journal of Science* 21:459–69.

Taylor, G. C. 1965. Water, history, and the Indus Plain. *Natural History,* May, pp. 40–49.

Twenter, F. R., and Knutilla, R. L. 1972. *Water for a rapidly growing urban community—Oakland County, Michigan.* U.S. Geological Survey Water-Supply Paper 2000.

UNESCO. 1971. *Discharge of selected rivers of the world,* vol. 3.

U.S. Department of the Interior. 1970. *River of life; water: the environmental challenge.*

U.S. Geological Survey. 1968a. *Water of the world.*

———. 1968b. *Water and Industry.*

———. 1968c. *The Amazon: measuring a mighty river.*

Urrows, G. M. 1967. *Nuclear energy for desalting.* U.S. Atomic Energy Commission.

Welch, E. B. 1969. *Factors initiating phytoplankton blooms and resulting effects on dissolved oxygen in Duwamish River estuary, Seattle, Washington.* U.S. Geological Survey Water-Supply Paper 1873-A.

White, G. F., et al. 1966. *Alternatives in water management.* National Academy of Sciences-National Research Council Publication 1408.

Wigley, T. M. L. 1971. Ion pairing and water quality measurements. *Canadian Journal of Earth Sciences* 8:468–76.

Willen, T. 1972. The gradual destruction of Sweden's lakes. *Ambio* 1(1):6–14.

Wyatt, W. 1968. *Water management studies and irrigation tail water study.* High Plains Underground Water Conservation District no. 1, Lubbock, Texas.

Chapter 6

Alexander, C. S., and Prior, J. C. 1971. Holocene sedimentation rates in overbank deposits in the black bottom of the lower Ohio River, southern Illinois. *American Journal of Science* 270:361–72.

Barnes, L. 1936. Journal of Isaac McCoy for the exploring expedition of 1830. *Kansas Historical Quarterly* 5:339–77.

Campbell, C. A., et al. 1967. Factors affecting the accuracy of the carbon dating method in soil humus studies. *Soil Science* 104:81–85.

Carson, R. L. 1951. *The sea around us.* New York: Oxford Univ. Press.

Coleman, J. M., and Gagliano, S. M. 1964. Cyclic sedimentation in the Mississippi River deltaic plain. *Transactions of the Gulf Coast Association of Geologists Society* 14:67–80.

Curray, J. R. 1969. Estuaries, lagoons, tidal flats, and deltas. In *The new concepts of continental margin sedimentation,* ed. D. J. Stanley et al., Lecture 3. Philadelphia: The American Geological Institute.

Elias, M. K. 1945. Loess and its economic importance. *American Journal of Science* 243:227–30.

Fisher, W. L., et al. 1969. *Delta systems in the exploration for oil and gas: a research colloquium.* Univ. of Texas, Bureau of Economic Geology.

Food and Agriculture Council, Pakistan. 1960. *Soil erosion and its control in arid and semi-arid zones.* Proceedings of the Karachi Symposium.

Free, E. F. 1917. *The movement of soil material by the wind.* U.S. Department of Agriculture, Bureau of Soils Bulletin 68.

Gagliano, S. M., and Van Beek, J. L. 1970. *Hydrologic and geologic studies of coastal Louisiana,* vol. 1. Louisiana State Univ., Coastal Studies Institute.

Galanopoulos, A. G., and Bacon, E. 1969. *Atlantis: the truth behind the legend.* New York: Bobbs-Merrill.

Gile, L. H. 1970. Soils of the Rio Grande Valley border in southern New Mexico. *Soil Science Proceedings* 34:465–72.

Gillette, R. 1972. Stream channelization: conflict between ditchers, conservationists. *Science* 176:890–94.

Gould, H. R. 1960a. Amount of sediment. In *Comprehensive survey of sedimentation in Lake Mead, 1948–49,* ed. W. O. Smith et al., pp. 195–200. U.S. Geological Survey Prof. Paper 295.

Gould, H. R. 1960b. Sedimentation in relation to reservoir utilization. In *Comprehensive survey of sedimentation in Lake Mead, 1948–49,* ed. W. O. Smith et al., pp. 215–29. U.S. Geological Survey Prof. Paper 295.

Gross, M. G. 1972. Geologic aspects of waste solids and marine waste deposits, New York metropolitan region. *Bulletin of the Geological Society of America* 83:3163–76.

Guy, H. P. 1970. *Sediment problems in urban areas.* U.S. Geological Survey Circular 601–E.

Hayes, M. O. 1967. *Hurricanes as geological agents: case studies of hurricanes Carla, 1961, and Cindy, 1963.* Univ. of Texas, Bureau of Economic Geology Report of Investigations 61.

Johnston, W. A. 1921. *Sedimentation of the Fraser River delta.* Geological Survey of Canada Memorandum 125.

Judge, J. 1967. Florence rises from the flood. *National Geographic* 132:1–43.

Judson, S., and Ritter, D. 1964. Rates of regional denudation in the United States. *Journal of Geophysical Research* 69:3395–3401.

Kassas, M. 1972. Impact of river control schemes on the shoreline of the Nile delta. In *The careless technology,* ed. M. T. Farvar and J. P. Milton, pp. 179–88. Garden City, N.Y.: Natural History Press.

Kolb, C. R., and Van Lopik, J. R. 1966. Depositional environments of the Mississippi River deltaic plain— southeastern Louisiana. In *Deltas in their geologic framework,* ed. M. L. Shirley and J. A. Ragsdale, pp. 17–61. Houston Geological Society.

Kovda, V. A. 1961. Land use development in the arid regions of the Russian plain, the Caucasus, and Central Asia. In *A history of land use in arid regions,* ed. L. D. Stamp. UNESCO, Arid Zone Research, vol. 17, pp. 175–218.

Krieger, R. A.; Cushman, R. V.; and Thomas, N. O. 1969. *Water in Kentucky.* Univ. of Kentucky, Kentucky Geological Survey Series X.

Langbein, W. B., and Schumm, S. A. 1958. Yield of sediment in relation to mean annual precipitation. *Transactions of the American Geophysical Union* 39:1076–84.

Lockwood, Andrews, and Newman, Inc. 1967. *A new concept—water for preservation of bays and estuaries.* Texas Water Development Board Report 43.

Malde, H. E. 1968. *The catastrophic late Pleistocene, Bonneville flood in the Snake River plain, Idaho.* U.S. Geological Survey Prof. Paper 596.

Michalowski, K. 1968. *Art of ancient Egypt,* trans. Norbert Guterman. New York: Abrams.

Nicollet, I. N. 1845. *Report intended to illustrate a map of the hydrographical basin of the upper Mississippi River.* 28th Congress, 2d sess. House Document 52.

Parsons, R. B.; Balster, C. A.; and Ness, A. O. 1970. Soil development and geomorphic surfaces, Willamette Valley, Oregon. *Soil Science Proceedings* 34:485–91.

Reesman, A. L., and Godfrey, A. E. 1972. *Chemical erosion and denudation rates in middle Tennessee.* Tennessee Department of Conservation, Division of Water Resources, Water Resources Research Series no. 4.

Ritter, D. F. 1967. Rates of denudation. *Journal of Geological Education* 15:144–59.

Ruhe, R. V. 1969. *Quaternary landscapes in Iowa.* Ames: Iowa State Univ. Press.

Stall, J. B. 1966. Man's role in affecting sedimentation of streams and reservoirs. In *Proceedings of the 2d annual American Water Resources Association,* ed. K. L. Bowden, pp. 79–95.

Stanley, J. W. 1960. Significance of area, capacity, and sediment tables. In *Comprehensive survey of sedimentation in Lake Mead, 1948–49,* ed. W. O. Smith et al., pp. 83–93. U.S. Geological Survey Prof. Paper 295.

Swallow, G. C. 1866. *Preliminary report of the geological survey of Kansas.* Lawrence: John Speer.

Warnke, D. A. 1969. Beach changes at the location of landfall of hurricane Alma. *Southeastern Geology* 10(4):189–200.

Webb, H. W., et al. 1970. Road-log-storm-damaged areas in central Virginia. *Virginia Minerals* 16(1):1–10.

Whyte, R. O. 1961. Evolution of land use in southwestern Asia. In *A history of land use in arid regions,* ed. L. D. Stamp. UNESCO, Arid Zone Research, vol. 17, pp. 57–118.

Zeuner, F. E. 1959. *The Pleistocene period: its climate, chronology, and faunal successions.* London: Hutchinson Scientific and Technical.

Chapter 7

Anonymous. 1970. Disposal of radioactive waste from Windscale. *Marine Pollution Bulletin* 1:172–74.

———. 1971. Conflicting philosophies over 2–4–5–T. *Nature* 231:483.

———. 1972. *The BEIR report.* National Academy of Sciences-National Research Council-National Academy of Engineering News Report 22:2.

Bergstrom, R. E. 1968a. *Feasibility of subsurface disposal of industrial wastes in Illinois.* Illinois Geological Survey Circular 426.

———. 1968b. Feasibility criteria for subsurface waste disposal in Illinois. *Ground Water* 6(5):5–9.

Bibby, G. 1969. *Looking for Dilmun.* New York: Knopf.

Bidwell, O. W. 1967. *Soil and its relationship to maximum yield.* American Society of Agronomy Special Publication 9, pp. 37–49.

Black, A. S., and Waring, S. A. 1972. Ammonium fixation and availability in some cereal producing soils in Queensland. *Australian Journal of Soil Research* 10:197–207.

Bonnyman, J, D., et al. 1968. Concentration of caesium-137 in rainwater and milk in Australia during 1967. *Australian Journal of Science* 31:180–83.

Bonnyman, J. D., and Duggleby, J. C. 1969. Iodine-131 concentrations in Australian milk resulting from the 1968 French nuclear weapon tests in Polynesia. *Australian Journal of Science* 31:389–92.

Bray, J. R., and Jackman, R. H. 1968. Soil and climatic factors related to caesium-137 content in New Zealand milk. *New Zealand Journal of Science* 11:352–62.

Brewerton, H. V. 1969. DDT in fats of Antarctic animals. *New Zealand Journal of Science* 12:194–99.

Bridges, E. M. 1970. *World soils*. Cambridge: The University Press.

Burch, L. A. 1969. Solid waste disposal and its effect on water quality. California Vector News 16(11): 99–113.

Byerly, T. G. 1970. Nitrogen compounds used in crop production. In *Global effects of environmental pollution,* ed. S. F. Singer, pp. 104–9. New York: Springer-Verlag.

Campbell, C. A., et al. 1967a. Factors affecting the accuracy of the carbonate method in soil humus studies. *Soil Science* 104:81–85.

———. 1967b. Applicability of the carbon dating method of analysis to soil humus studies. *Soil Science* 104:217–24.

Clark, T. P. 1972. Hydrogeology, geochemistry, and public health aspects of environmental impairment at an abandoned landfill near Austin, Texas. Ph.D. dissertation, Univ. of Texas.

Colville, W. L. 1967. *Environment and maximum yield of corn*. American Society of Agronomy Special Publication 9, pp. 21–36.

Curley, A., et al. 1971. Organic mercury identified as the cause of poisoning in humans and hogs. *Science* 172:65–67.

Daniels, R. B.; Gamble, E. E.; and Cady, J. G. 1970. Some relations among coastal plain soils and geomorphic surfaces in North Carolina. *Soil Science Proceedings* 34:648ff.

Dinman, B. D. 1972. Non-concept of "no threshold" chemicals in the environment. *Science* 175:495–97.

Ehrlich, P. R., and Ehrlich, A. H. 1970. *Population, resources, environment: issues in human ecology*. 2d ed. San Francisco: W. H. Freeman.

Elias, M. K. 1945. Loess and its economic importance. *American Journal of Science* 243:227–30.

Elliott, A. M. 1957. *Zoology*. New York: Appleton-Century-Crofts.

Epstein, S. S. 1972. Toxicological and environmental implications on the use of nitrilotriacetic acid as a detergent builder—II. *International Journal of Environmental Studies* 3:13–21.

Fehrenbacher J. B ; Ray, B. W.; and Alexander, J. D. 1968. Illinois soils and factors in their development. In *The Quaternary of Illinois,* ed. R. E. Bergstrom, pp. 165–75. Univ. of Illinois, College of Agriculture Special Publication 14.

Flawn, P. T.; Turk, L. J.; and Leach, C. H. 1970. *Geological considerations in disposal of solid municipal wastes in Texas*. Univ. of Texas, Bureau of Economic Geology Geological Circular 70–2.

Galley, J. E., ed. 1968. *Subsurface disposal in geologic basins—a study of reservoir strata*. American Association of Petroleum Geologists Memoir 10.

Gazda, L. P., and Malina, J. P., Jr. 1969. *Land disposal of municipal solid wastes in selected standard metropolitan statistical areas in Texas*. Univ. of Texas, Environmental Health Engineering Research Laboratory.

Gibbs, W. J., et al. 1969. Fallout over Australia from nuclear weapons tested by France in Polynesia from July to September, 1968. *Australian Journal of Science* 31:383–88.

Gile, L. H.; Hawley, J. W.; and Grossman, R. B. 1970. *Distribution and genesis of soils and geomorphic surfaces in a desert region of southern New Mexico*. Soil Science Society of America, Soil-Geomorphology Field Conference Guidebook.

Gillette, R. 1972a. Nuclear safety. *Science* 177:771–76, 867–71, 1080–82.

———. 1972b. Radiation standards: the last word or at least a definitive one, *Science* 178:966–67, 1012.

Glasstone, S. 1971. *Public safety and underground nuclear detonations*. U.S. Atomic Energy Commission TID 25708.

Hammond, A. 1972. Fission: the pros and cons of nuclear power. *Science* 178:147–49.

Handler, P., ed. 1970. *Biology and the future of man*. London: Oxford Univ. Press.

Hanson, W. C. 1967. Cesium-137 in Alaskan lichens, caribou, and Eskimos. *Health Physics* 13:383–89. (Reprinted in *Readings in conservation ecology,* ed. G. W. Cox. New York: Appleton-Century-Crofts.)

Harvey, E. J. 1969. *Hydrologic study of a waste-disposal problem in a karst area at Springfield, Missouri*. U.S. Geological Survey Prof. Paper 600–C.

Hays, J. 1964. The relationship between laterite and land surfaces in the northern part of the Northern Territory. In *Laterites,* ed. Chowdhury, M. K. R. and A. D. Kharkwal. International Geological Congress Report of the 22d Session, section 14, pt. 14, pp. 14–28.

Hicks, M. E., and Woodward, T. C. 1969. Project Thunderbird. *Wyoming Geological Association 21st*

Annual Field Conference Guidebook, pp. 161–63.

Hillel, D. 1971. *Soil and water: physical principles and processes.* New York: Academic Press.

Hooper, A. D. L. 1969. Soils of the Adelaide-Alligator area. In *Lands of the Adelaide-Alligator area, Northern Territory,* ed. R. Story et al., pp. 93–113. Commonwealth Scientific and Industrial Research Organization, Australia, Land Research Series no. 25.

Jensen, S. 1972. The PCB story. *Ambio* 1(4):123–31.

Kalix, Z. 1968. *Utilization of fertilizer raw materials in Australia.* Economic Commission for Asia and the Far East Mineral Resources Development Series no. 32, pp. 36–39.

Karale, R. L.; Tamhane, R. V.; and Das, S. C. 1969. Soil genesis as related to parent material and climate. *Journal of the Indian Society of Soil Science* 17(2):227-39.

Kearney, P. C.; Nash, R. G.; and Isensee, A. R. 1969. Persistence of pesticide residues in soils. In *Chemical fallout,* ed. M. W. Miller and G. G. Berg, pp. 54–67. Springfield, Ill.: Charles C. Thomas.

Keller, W. D. 1957. *The principles of chemical weathering.* Rev. ed. Columbia, Mo.: Lucas Brothers.

Klein, W., and Korte, F. 1972. Conversion of pesticides under atmospheric condition and in soil. In *Trace substances in environmental health—V,* ed. D. D. Hemphill, pp. 71–80. Columbia: Univ. of Missouri Press.

Kohl, D. H.; Shearer, G. B.; and Commoner, B. 1971. Fertilizer nitrogen: contribution to nitrate in surface water in a corn belt watershed. Science 174: 1331–34.

Kovda, V. 1972. The management of soil fertility. *Nature and Resources* 8(2):2–4.

Kubiena, W. L. 1970. *Micromorphological features of soil geography.* New Brunswick, N. J.: Rutgers Univ. Press.

Lee, K. E., and Wood, T. G. 1971. *Termites and soils.* London: Academic Press.

Machta, L. 1968. Radioiodine fallout over the midwest in May. *Science* 160:64–66.

Maddox, J. 1972. Pollution and worldwide catastrophe. *Nature* 236:433–36.

Martin, J. P. 1966. Influence of pesticides on soil microbes and soil properties. In *Pesticides and their effects on soils and water,* ed. M. E. Bloodworth et al., pp. 95–108. American Society of Agronomy Special Publication 8.

Martin, W. B., and Shaughnessy, J. 1969. Project wagon wheel. *Wyoming Geological Association 21st Annual Field Conference Guidebook,* pp. 145–52.

Masuda, Y.; Kagawa, R.; and Kuratsune, M. 1972. Polychlorinated biphenyls in carbonless copying paper. *Nature* 237:41–42.

Matthews, R. A., and Franks, A. L. 1971. Cinder cone sewage disposal site at north Lake Tahoe. *California Geology* 24:183–94.

McNeil, M. 1964. Lateritic soils. *Scientific American* 211:96–102.

Mitchell, R. L. 1972. Trace elements in soils and factors that affect their availability. *Bulletin of the Geological Society of America* 83:1069–76.

Mosser, J. L.; Fisher, N. S.; and Wurster, C. F. 1972. Polychlorinated biphenyls and DDT alter species composition in mixed cultures of algae. *Science* 176:533–35.

Nash, R. G., and Woolson, E. A. 1967. Persistence of chlorinated hydrocarbon insecticides in soils. *Science* 157:924–27.

Nelson, J. L., and Haushild, W. L. 1970. Accumulation of radionuclides in bed sediments of the Columbia River between the Hanford reactors and McNary Dam. *Water Resources Research* 6:77–87.

Nielsen, J. M. 1963. Behavior of radionuclides in the Columbia River. In *Transport of radionuclides in fresh water systems,* ed. Kornegay et al., pp. 91–112. Conference, Johns Hopkins Univ. Department of Sanitary Engineering and U.S. Atomic Energy Commission.

Odum, E. P. 1971. *Fundamentals of ecology.* 3d ed. Philadelphia: W. B. Saunders.

Odum, W. E.; Woodwell, G. M.; and Wurster, C. F. 1969. DDT residues absorbed from organic detritus by fiddler crabs. *Science* 164:576–77.

Orberg, J., et al. 1972. Administration of DDT and PCB prolongs oestrous cycle in mice. *Ambio* 1:148–49.

Pape, J. C. 1970. Plaggen soils in Netherlands. *Geoderma* 4:229–55.

Peirson, D. H. 1971. Worldwide deposition of long-lived fission products from nuclear explosions. *Nature* 234:79–80.

Pocker, Y.; Beug, W. M.; and Ainardi, V. R. 1971. Carbonic anhydrase interaction with DDT, DDE, and dieldrin. *Science* 174:1336–39.

Ponomareva, V. V. 1964. *Theory of podzolization,* trans. by the Israel Program for Scientific Translations, 1969. U.S. Department of Agriculture-National Academy of Sciences.

Portland Cement Association. 1963. *PCA soil primer.* Chicago.

Quinlan, J. F. 1970. Central Kentucky karst; Réunion internationale karstologie en Languedoc-Provence, 1968. *Actes Mediterranée Études et Travaux* 7: 235–53.

Remezov, N. P., and Pogrebnyak, P. S. 1969. *Forest soil science,* trans. A. Gourevitch. U.S. Department of Commerce.

Risebrough, R. W. 1971. Chlorinated hydrocarbons. In *Impingement of man on the oceans,* ed. D. W. Hood, pp. 259–86. New York: Wiley-Interscience.

Rosebury, T. 1969. *Life on man.* New York: Viking.

Sagan, L. A. 1972. Human costs of nuclear power. *Science* 177:487–93.

Sandifer, S. H., and Keil, J. E. 1972. Pesticide exposure: association with cardiovascular factors. In *Trace substances in environmental health—V*, ed. D. D. Hemphill, pp. 329–34. Columbia: Univ. of Missouri Press.

Saunder, D. H. 1971. The minor element status of Rhodesian soils. *Rhodesia Science News* 5:304–8.

Shacklette, H. T. 1971. A U.S. Geological Survey study of elements in soils and other surficial materials in the United States. In *Trace substances in environmental health—IV*, ed. D. D. Hemphill pp. 35–45. Columbia: Univ. of Missouri Press.

Sheaffer, J. R., et al. 1967. *Introduction to flood proofing; an outline of principles and methods.* Univ. of Chicago, Center for Urban Studies.

Shimek, B. 1925. *The persistence of the prairie.* Univ. of Iowa Studies in Natural History, vol. II, no. 5.

Smit, B. 1971. The safe use of insecticides in South Africa. *South African Journal of Science* 67:485–87.

Sodergren, A., and Ulfstrand, S. 1972. DDT and PCB relocate when caged robins use fat reserves. *Ambio* 1(1):36–40.

Sowers, G. B., and Sowers, G. F. 1970. *Introductory soil mechanics and foundations.* 3d ed. London: Macmillan.

Stace, H. C. T., et al. 1968. *A handbook of Australian soils.* Glenside, So. Australia: Relim Technical Publications.

Stein, A. 1933. *On ancient Central-Asian tracks.* New York: Random House.

Stefanson, R. C. 1972. Effect on plant growth and form of nitrogen fertilizer on denitrification from four south Australia soils. *Australian Journal of Soil Research* 10:183–95.

Stephens, C. G. 1970. Laterite and silcrete in Australia: a study of the genetic relationships of laterite and silcrete and their companion materials, and their collective significance in the formation of the weathered mantle, soils, relief and drainage of the Australian continent. *Geoderma* 5:5–52.

Story, R. 1969. Summary description of the Adelaide-Alligator area. In *Lands of the Adelaide-Alligator area, Northern Territory*, ed. R. Story et al., pp. 16–23. Commonwealth Scientific and Industrial Research Organization, Australia, Land Research Series no. 25.

Stout, J. D. 1971. The distribution of soil bacteria in relation to biological activity and pedogenesis. *New Zealand Journal of Science* 14:816–50.

Thorp, J. 1968. The soil—a reflection of Quaternary environments in Illinois. In *The Quaternary of Illinois,* ed. R. E. Bergstrom, pp. 48–55. Univ. of Illinois, College of Agriculture Special Publication 14.

Tidball, R. R. 1971. Geochemical variation in Missouri soils. In *Trace elements in environmental health—IV*, ed. D. D. Hemphill, pp. 15–25. Columbia: Univ. of Missouri Press.

Turk, L. J. 1971. Nitrate contamination of ground water in Runnels County, Texas. In *Trace substances in environmental health—IV*, ed. D. D. Hemphill, pp. 25–26. Columbia: Univ. of Missouri Press.

Umeda, G. 1972. PCB poisoning in Japan. *Ambio* 1:132–34.

U.S. Soil Conservation Service. 1960. *Soil classification: a comprehensive system, 7th approximation.* U.S. Department of Agriculture.

Valeton, I. 1972. *Bauxites.* Amsterdam: Elsevier.

Van der Sluijs, 1970. Decalcification of marine clay soils connected with decalcification during silting. *Geoderma* 4:209–26.

Warner, D. L. 1968. Subsurface disposal of liquid industrial wastes by deep well injections. *American Association of Petroleum Geologists Memoir 10*, pp. 11–20.

Watson, D. G., et al. 1971. Cycling of radionuclides in Columbia River biota. In *Trace substances in environmental health—IV*, ed. D. D. Hemphill, pp. 144–157. Columbia: Univ. of Missouri Press.

Whitehead, D. C. 1970. *The role of nitrogen in grassland productivity.* Commonwealth Agricultural Bureau, Great Britain, Commonwealth Bureau of Pastures and Field Crops Bulletin 48.

Williams, M. A. J. 1969. Geomorphology of the Adelaide-Alligator area. In *Lands of the Adelaide-Alligator area, Northern Territory*, ed. R. Story et al., pp. 71–94. Commonwealth Scientific and Industrial Research Organization, Australia, Land Research Series no. 25.

World Health Organization. 1970. *Control of pesticides.* Geneva.

Wyatt, W. 1968. *Water management studies and irrigation tail water study.* High Plains Underground Water Conservation District no. 1, Lubbock, Texas.

Young, K. 1968. Early Holocene earth movements, Travis County, Texas. *Texas Journal of Science* 19:420-21.

Zeuner, F. E. 1959. *The Pleistocene period: its climate, chronology, and faunal successions.* London: Hutchinson Scientific and Technical.

Chapter 8

Abdullah, M. I., and Royle, L. G. 1972. Heavy metal content of some rivers and lakes in Wales. *Nature* 238:329–30.

Ahlberg, C. R., and Wachtmeister, C. A. 1972. Organolead compounds shown to be genetically active. *Ambio* 1(1):29–31.

Allaway, W. H. 1969. Control of the environmental

levels of selenium. In *Trace substances in environmental health—II,* ed. D. D. Hemphill, pp. 181–206. Columbia: Univ. of Missouri Press.

Allen-Price, E. D. 1960. Uneven distribution of cancer in west Devon. *Lancet,* June, pp. 1235–38.

Anast, C. S., et al. 1972. Evidence for parathyroid failure in magnesium deficiency. *Science* 177:606–8.

Banta, J., et al. 1967. Epidemiologic and demographic problems in the study of the geographic relationship of trace elements and the cardiovascular diseases. In *Relation of geology and trace elements to nutrition,* ed. H. L. Cannon and D. F. Davidson, p. 6. Geological Society of America Special Paper 90.

Barber, R. T.; Vijayakumar, A.; and Cross, F. A. 1972. Mercury concentrations in recent and ninety-eight-year-old benthopelagic fish. *Science* 178:636–38.

Bazell, R. J. 1971a. Lead poisoning: zoo animals may be the first victims. *Science* 173:130–31.

———. 1971b. Lead poisoning: combating the threat from the air. *Science* 174:574–76.

Beath, O. A. 1937. Seleniferous vegetation. In *The occurrence of selenium and seleniferous vegetation in Wyoming,* ed. S. H. Knight and O. A. Beath, pp. 29–64. Univ. of Wyoming Agricultural Experiment Station Bulletin 221.

Bernstein, D. S., et al. 1966. Prevalence of osteoporosis in high- and low-fluoride areas in North Dakota. *Journal of the American Medical Association* 198(5):499–504.

Brinster, R. L., and Cross, P. C. 1972. Effect of copper on the preimplantation mouse embryo. *Nature* 238:398–99.

Burkitt, A.; Lester, P.; and Nickless, G. 1972. Distribution of heavy metals in the vicinity of an industrial complex. *Nature* 238:327–28.

Carlisle, E. M. 1972. Silicon: an essential element for the chick. *Science* 178:619–21.

Chisolm, J. J., Jr. 1971. Lead poisoning. *Scientific American* 224(2)15–23.

Chow, T. J., and Earl, J. L. 1969. Lead aerosols in the atmosphere: increasing concentrations. *Science* 169:577–80.

Conney, A. H., and Burns, J. J. 1972. Metabolic interaction among environmental chemicals and drugs. *Science* 178:576–86.

Connor, J. J., et al. 1971. Roadside effects on trace element content of some rocks, soils, and plants of Missouri. In *Trace elements in environmental health IV,* ed. D. D. Hemphill, pp. 26–34. Columbia: Univ. of Missouri Press.

Cooper, C. R., and Jolly, W. C. 1970. Ecological effects of silver iodide and other weather modification agents: a review. *Water Resources* 6(1):88–98.

Cotzias, G. C. 1968. Importance of trace substances in environmental health as exemplified by manganese. In *Trace substances in environmental health—I,* ed. D. D. Hemphill, pp. 5–19. Columbia: Univ. of Missouri Press.

Crawford, M. D.; Gardner, M. J.; and Morris, J. N. 1968. Mortality and hardness of local water supplies. *Lancet,* April, pp. 827–31.

Crosby, N. T., et al. 1969. Estimation of steam-volatile N-nitrosamines in foods at the 1 ug/kg level. *Nature* 23:342–43.

Curley, A., et al. 1971. Organic mercury identified as the cause of poisoning in humans and hogs. *Science* 172:65–66.

Dansgaard, W. S., et al. 1969. One thousand centuries of climatic record from Camp Century on the Greenland ice sheet. *Science* 166:377–81.

Davis, G. K. 1970. Interaction of trace elements. In *Trace substances in environmental health—III,* ed. D. D. Hemphill, pp. 135–48. Columbia: Univ. of Missouri Press.

Deuel, L. E., and Swoboda, A. R. 1972. Arsenic solubility in a reduced environment. *Soil Science* 36:276–78.

Doisy, R. J., et al. 1969. Effects and metabolism of chromium normals, elderly subjects, and diabetics. In *Trace substances in environmental health—II,* ed. D. D. Hemphill, pp. 75-82. Columbia: Univ. of Missouri Press.

Duce, R. A., et al. 1972. Enrichment of heavy metals and organic compounds in the surface microlayer of Narragansett Bay, Rhode Island. *Science* 175: 161–63.

Ebashi, S. 1972. Calcium ions and muscle contraction. *Nature* 240:217–18.

Feick, G.; Horne, R. A.; and Yeaple, D. 1972. Release of mercury from contaminated freshwater sediments by the runoff of road deicing salt. *Science* 175:1142–43.

Frieden, E. 1972. The chemical elements of life. *Scientific American* 227(1):52–60.

Fristedt, B. 1969. Metal toxicity. Swedish Natural Science Research Council, Ecological Research Committee Bulletin 5, pp. 64–66.

Glass, B. 1950. Our life-sustaining land. *Texas Journal of Science* 2:287–95.

Goldwater, L. J. 1971. Mercury in the environment. *Scientific American* 224(5):15–21.

Goodman, G. T. 1969. The revegetation of derelict land contaminated with toxic heavy metals. Swedish Natural Science Research Council, Ecological Research Committee Bulletin 5, pp. 3–16.

Goodman, G. T., and Roberts, R. M. 1971. Plants and soils as indicators of metals in the air. *Nature* 231: 287–92.

Griffith, K.; Wright, E. B.; and Dormandy, T. L. 1973. Tissue zinc in malignant disease. *Nature* 241:60.

Hadjimarkos, D. M. 1971. The role of selenium in

dental caries. In *Trace substances in environmental health—IV*, ed. D. D. Hemphill, pp. 301–6. Columbia: Univ. of Missouri Press.

Hammer, D. L., et al. 1972a. Trace metals in human hair as a simple epidemiologic monitor of environmental exposure. In *Trace substances in environmental health—V*, ed. D. D. Hemphill, pp. 25–38. Columbia: Univ. of Missouri Press.

―――. 1972b. Cadmium exposure and human health effects. In *Trace substances in environmental health—V*, ed. D. D. Hemphill, pp. 269–83. Columbia: Univ. of Missouri Press.

Handler, P., ed. 1970. *Biology and the future of man.* New York and London: Oxford Univ. Press.

Hegsted, D. M. 1968. The beneficial and detrimental effects of fluoride in the environment. In *Trace substances in environmental health—I*, ed. D. D. Hemphill, pp. 105–13. Columbia: Univ. of Missouri Press.

Hem, J. D. 1970. *Study and interpretation of the chemical characteristics of natural water.* U.S. Geological Survey Water-Supply Paper 1473.

Henzel, J. H. et al., 1969. Trace elements in atherosclerosis, efficacy of zinc medication as a therapeutic modality. In *Trace substances in environmental health—II*, ed. D. D. Hemphill, pp. 83–99. Columbia: Univ. of Missouri Press.

―――. 1971. Efficacy of zinc medication as a therapeutic modality in atherosclerosis: follow-up observations on patients medicated over prolonged periods. In *Trace substances in environmental health—IV*, ed. D. D. Hemphill, pp. 336–41. Columbia: Univ. of Missouri Press.

Hemphrey, H. G., and Stratton, H. J. 1956. *Bibliography of Bureau of Mines, Health, and Safety publications, January 1947–June 1956.* U.S. Bureau of Mines Bulletin B–558.

Horne, R. A. 1972. Biological effects of chemical agents. *Science* 177:1152–53.

Hopps, H. C. 1972. Geography, geochemistry, and disease. In *Trace substances in environmental health—V*, ed. D. D. Hemphill, pp. 475–84. Columbia: Univ. of Missouri Press.

Hussein, S., et al. 1972. Mutagenic effects of TCDD on bacterial systems. *Ambio*, Feb., pp. 32–33.

Jha, K. K. 1969. Effect of vanadium and tungsten on nitrogen fixation and the growth of *Medicago sativo*. *Journal of the Indian Society of Soil Science* 17: 11–13.

Joensuu, O. I. 1971. Fossil fuels as a source of mercury pollution. *Science* 172:1027–28.

Jonasson, I. R., and Boyle, R. W. 1972. Geochemistry of mercury and origins of natural contamination of the environment. *Canadian Mining and Metallurgical Bulletin*, Jan., pp. 32–39.

Jones, R. S. 1970. *Gold content of water, plants, and animals.* U.S. Geological Survey Circular 625.

Katsuna, M. 1968. *Minamata disease.* Jumamoto Univ., Study Group of Minamata Disease.

Keller, W. D., and Smith, G. E. 1967. Ground-water contamination by dissolved nitrates. In *Relation of geology and trace elements to nutrition*, ed. H. L. Cannon and D. F. Davidson, pp. 47–59. Geological Society of America Special Paper no. 90.

Kmet, J., and Mahboubi, E. 1972. Esophageal cancer in the Caspian littoral of Iran: initial studies. *Science* 175:846–53.

Knight, S. H. 1937. Rocks and soils of Wyoming and their relations to the selenium problem. In *The occurrence of selenium and seleniferous vegetation in Wyoming,* ed. S. H. Knight and O. A. Beath, pp. 4–27. Univ. of Wyoming Agricultural Experiment Station Bulletin 221.

Kobayashi, J. 1970. *Relation between the "itai-itai" disease and the pollution of river water by cadmium from a mine.* Fifth International Water Pollution Research Conference.

―――. 1972. Air and water pollution by cadmium, lead, and zinc attributed to the largest zinc refinery in Japan. In *Trace substances in environmental health—V*, ed. D. D. Hemphill, pp. 117–28. Columbia: Univ. of Missouri Press.

Lagerwerff, J. V. 1971. Uptake of cadmium, lead, and zinc by radish from soil and air. *Soil Science* 111: 129–33.

Lakin, H. W. 1972. Selenium accumulation in soils and its absorption by plants and animals. *Bulletin of the Geological Society of America* 83:181–90.

Landner, L., and Jernelov, A. 1969. Cadmium in aquatic systems. Swedish Natural Science Research Council, Ecological Research Committee Bulletin 5, pp. 47–55.

Lauer, G. J. 1968. Some effects of metals on aquatic life. In *Proceedings of the symposium: mineral waste utilization,* ed. M. A. Schwartz, pp. 20–27. Chicago: IIT Research Institute.

McIntire, M. S., and Angle, C. R. 1972. Airlead: relation to lead in blood of black school children deficient in glucose-6-phosphate dehydrogenase. *Science* 177:520–22.

Merliss, R. R. 1971. Talc-treated rice and Japanese stomach cancer. *Science* 173:1141–42.

Mertz, W. 1968. The role of chromium in glucose metabolism. In *Trace substances in environmental health—I*, ed. D. D. Hemphill, pp. 86–95. Columbia: Univ. of Missouri Press.

―――. 1971. Health-related function of chromium. In *Environmental geochemistry in health and disease,* ed. H. L. Cannon and H. C. Hopps, pp. 197–202. Geological Society of America Memorandum 123.

Messer, H. H.; Armstrong, W. D.; and Singer, L. 1972.

Fertility impairment in mice on a low fluoride intake. *Science* 177:893–94.

Miller, G. E., et al. 1972. Mercury concentration in museum specimens of tuna and swordfish. *Science* 175:1121–22.

Mills, C. F., and Dalgarno, A. C. 1972. Copper and zinc status of ewes and lambs receiving increased dietary concentrations of cadmium. *Nature* 239: 171–73.

Moody, P. A. 1967. *Genetics of man.* New York: Norton.

National Academy of Sciences. 1970. Health and geochemical environment; what is the relationship? *National Academy of Sciences News Letter,* Apr., pp. 2–3.

Nathani, G. P., et al. 1969. Water soluble boron in irrigated medium black soils. *Journal of the Indian Society of Soil Science* 17:59–62.

Ness, G. B. 1971. Ecology of anthrax. *Science* 172: 1303–7.

Nielsen, F. H., and Higgs, D. J. 1971. Further studies involving a nickel deficiency in chicks. In *Trace substances in environmental health—IV,* ed. D. D. Hemphill, pp. 241–46. Columbia: Univ. of Missouri Press.

Pearson, I. B., and Jenner, F. A. 1971. Lithium in psychiatry. *Nature* 232:532–33.

Perry, H., Jr. 1969. Hypertension and trace metals, particularly cadmium. In *Trace substances in environmental health—II,* ed. D. D. Hemphill, pp. 101–25. Columbia: Univ. of Missouri Press.

———. 1971. Trace elements related to cardiovascular disease. In *Environmental geochemistry in health and disease,* ed. H. L. Cannon and H. C. Hopps, pp. 179–95. Geological Society of America Memorandum 123.

Pinkerton, C., et al. 1972. Cadmium content of milk and cardiovascular disease mortality. In *Trace substances in environmental health—V,* ed. D. D. Hemphill, pp. 285–92. Columbia: Univ. of Missouri Press.

Pories, W. J., et al. 1967. Trace elements and wound healing. In *Trace substances in environmental health—I,* ed. D. D. Hemphill, pp. 114–33. Columbia: Univ. of Missouri Press.

Pories, W. J.; Strain, W. H.; and Rob, C. G. 1971. Zinc deficiency in delayed healing and chronic disease. In *Environmental geochemistry in health and disease,* ed. H. L. Cotton and H. C. Hopps, pp. 73–95. Geological Society of America Memorandum 123.

Porter, W. W., II. 1972. One part per million. *Science* 177:476–77.

Rajaratnam, J. A., et al. 1971. Boron: possible role in plant metabolism. *Science* 172:1142.

Remezov, N. P., and Pogrebnyak, P. S. 1969. *Forest soil science,* trans. A. Gourevitch. U.S. Department of Agriculture and National Science Foundation.

Renshaw, G. D.; Pounds, C. A.; and Pearson, E. F. 1972. Variation in lead concentration along single hairs as measured by non-flame atomic absorption spectrophotometry. *Nature* 238:162–63.

Rife, D. C. 1967. Race and heredity. In *Race and modern science,* ed. R. E. Kuttner, pp. 141–68. New York: Social Science Press.

Ruch, R. R., et al. 1971. *Mercury content of Illinois coals.* Illinois State Geological Survey, Environmental Geology Notes no. 43.

Schalie, H. Van Der. 1969. Schistosomiasis: control in Egypt and the Sudan. *Natural History,* Feb., 78: 62–65.

Schroeder, H. A. 1965. Cadmium as a factor in hypertension. *Journal of Chronic Diseases* 18:647–56.

———. 1966. Municipal drinking water and cardiovascular death rates. *Journal of the American Medical Association* 195(2):81–85.

Schwarz, K., and Milne, D. B. 1971. Growth effects of vanadium in the rat. *Science* 174:426–28.

Shaklette, H. T. 1971. A U.S. Geological Survey study of elements in soils and other surficial materials in the United States. In *Trace substances in environmental health—IV,* ed. D. D. Hemphill, pp. 35–45. Columbia: Univ. of Missouri Press.

Shaklette, H. T., and Cuthbert, M. C. 1967. Iodine content of plant groups as influenced by variation in rock and soil types. In *Relation of geology and trace elements to nutrition,* ed. H. L. Cannon and D. F. Davidson, pp. 31–46. Geological Society of America Special Paper 90.

Shapiro, J. 1971. Arsenic and phosphate: measured by various techniques. *Science* 141:234.

Shiff, C. J. 1969. Schistosomiasis: host and parasite in Rhodesia. *Natural History,* Feb., 78:65–67.

Smith, G. E. 1970. Nitrate pollution of water supplies. In *Trace substances in environmental health—III,* ed. D. D. Hemphill, pp. 273–87. Columbia: Univ. of Missouri Press.

Smith, J. D.; Nicholson, R. A.; and Moore, P. J. 1971. Mercury in water of the tidal Thames. *Nature* 232: 393–94.

Spencer, J. M. 1970. Geological influence on regional health problems. *Texas Journal of Science* 21: 459–69.

Stoewsand, G. S., et al. 1971. Eggshell thinning in Japanese quail fed mercuric chloride. *Science* 173: 1030–31.

Suzuki, T. 1969. Neurological symptoms from concentration of mercury in the brain. In *Chemical fallout,* ed. M. W. Morton and G. G. Berg, pp. 245–57. Springfield, Ill.: Charles C. Thomas.

Swenerton, H.; Shrader, R.; and Hurley, L. S. 1969.

Zinc-deficient embryos: reduced thymidine incorporation. *Science* 166:1014–15.

Trealease, S. F., and Beath, O. A. 1949. *Selenium*. New York: published by the authors.

Turk, L. J. 1971. Nitrate contamination in ground water in Runnels County, Texas. In *Trace substances in environmental health—IV*, ed. D. D. Hemphill, pp. 25–26. Columbia: Univ. of Missouri Press.

Ulmer, D. D., and Vallee, B. L. 1969. Effects of lead on biochemical systems. In *Trace substances in environmental health—II*, ed. D. D. Hemphill, pp. 7–27. Columbia: Univ. of Missouri Press.

Vessell, E. S., et al. 1973. Hepatic drug metabolism in rats: impairment in a dirty environment. *Science* 179:896–97.

Weiss, D.; Whitten, B.; and Leddy, D. 1972. Lead content of human hair (1871–1971). *Science* 178:69–70.

Weiss, H. V.; Koide, M.; and Goldbert, E. D. 1971. Mercury in a Greenland ice sheet: evidence of recent input by man. *Science* 174:692–94.

White, J. M., and Harvey, D. R. 1972. Defective synthesis of α and β globin chains in lead poisoning. *Nature* 236:71–73.

Whitnack, G. C., and Martens, H. H. 1971. Arsenic in potable desert ground water: an analysis problem. *Science* 171:383–85.

Wolff, I. A., and Wasserman, A. E. 1972. Nitrates, nitrites, and nitrosamines. *Science* 177:15–19.

Woolson, E. A.; Axley, J. H.; and Kearney, P. C. 1971. The chemistry and phytotoxicity of arsenic in soils. *Soil Science Proceedings* 35:938–43.

Wyatt, W. 1968. *Water management studies and irrigation tail water study*. High Plains Underground Water Conservation District no. 1, Lubbock, Texas.

Young, K. 1970. Man in the geobiocoenose. In *Environmental Geology*, AGI Short Course Lecture Notes. Washington: American Geological Institute.

Zeissink, H. E. 1971. Trace element behavior in two nickeliferous laterite profiles. *Chemical Geology* 7:25–36.

Chapter 9

Adams, R. D. 1971. An immediate field survey of the San Fernando, Los Angeles, earthquake, February 9, 1971. *Bulletin of the New Zealand Society for Earthquake Engineering* 4(3):335–46.

Aki, K., et al. 1969. Near-field and far-field seismic evidences for triggering of an earthquake by the Benham explosion. *Bulletin of the Seismological Society of America* 59:2197–2207.

Ambraseys, N. N. 1962. The seismicity of Tunis. *Annali di Geofisica* 15(203):233–44.

———. 1963. The Buyin-Zara (Iran) earthquake of September, 1962: a field report. *Bulletin of the*

Seismological Society of America 53:705–40.

———. 1965a. An earthquake engineering study of the Buyin-Zahra earthquake of September 1st, 1962. *Proceedings of the 3d World Conference on Earthquake Engineering*, Paper V/A/19. Wellington, New Zealand.

———. 1965b. A note on the seismicity of the eastern Mediterranean. *Studia Geophysica et Geodaetica* 9:405–10.

———. 1966a. Seismic environment of the Skopje earthquake of July, 1963. *Revue de l'Union Internationale de Secours*, Sept.

———. 1966b. *Progress report on the seismicity of north central Iran*. Tehran University, Institute of Geophysics Publication no. 34.

———.1967. The earthquakes of 1965–1966 in the Peloponnesus, Greece: a field report. *Bulletin of the Seismological Society of America* 57(5):1025–46.

———. 1971. Value of historical records of earthquakes. *Nature* 232:375–79.

Ambraseys, N. N. and Sarma, S. K. 1967. The response of earth dams to strong earthquakes. *Geotechnique* 17:181–213.

Ambraseys, N. N., Tchalenko, J. S. 1969. The Dasht-E Bayáz (Iran) earthquake of August 31, 1968: a field report. *Bulletin of the Seismological Society of America* 59:1751–1822.

Ambraseys, N. N., and Zatopek, A. 1968. The Varto Üstükran (Anatolia) earthquake of 19 August 1966: summary of a field report. *Bulletin of the Seismological Society of America* 58:47–102.

———. 1969. The Mudurnu Valley, West Anatolia, Turkey, earthquake of 22 July 1967. *Bulletin of the Seismological Society of America* 59:521–89.

Anonymous. 1969. Earthquakes. California Division of Mines and Geology, *Mineral Information Service*, 22(5):75–81.

———. 1971a. Engineering features of the San Fernando earthquake. *California Geology* 24(11):218–19.

———. 1971b. The San Fernando earthquake: 1971. *California Geology* 24(4–5):59–74.

———. 1972. Classification of high earthquake risk buildings. *New Zealand Society for Earthquake Engineering* 5:37–46.

Barosh, J. 1969. *Use of seismic intensity data to predict the effects of earthquakes and underground nuclear explosions in various geologic settings*. U.S. Geological Survey Bulletin 1279.

Bayer, K. C. 1970. Dasht-I-Biaz, Iran earthquake. *Mines Magazine* 60(9)16–23.

Brune, J. N., and Allen, C. R. 1967. A micro-earthquake survey of the San Andreas Fault system in southern California. *Bulletin of the Seismological Society of America* 57:277–96.

Bucknam, R. C. 1969. Geologic effects of the Benham underground nuclear explosion, Nevada test site. *Bulletin of the Seismological Society of America* 59:2209–20.

Burrett, C. F. 1972. Plate tectonics and the Hercynian Orogeny. *Nature* 239:155–57.

Clark, W. B., and Hauge, C. J. 1971. The earth quakes: you can reduce the danger. *California Geology* 24(11):203–16.

Coffman, J. L., and Cloud, W. K. 1970. *United States earthquakes, 1968*. U.S. Department of Commerce, Environmental Science Services Administration.

Cook, K. L., and Smith, R. B. 1957. Seismicity in Utah, 1850 through June, 1965. *Bulletin of the Seismological Society of America* 57:689–718.

Cox, D. C., and Mink, J. F. 1963. The tsunami of 23 May 1960 in the Hawaiian Islands. *Bulletin of the Seismological Society of America* 53:1191–1209.

Eisler, J. D. 1967. Investigation of a method for determining stress accumulation at depth. *Bulletin of the Seismological Society of America* 57:891–911.

Elders, W. A., et al. 1972. Crustal spreading in southern California. *Science* 178:15–24.

Fedotov, S. A.; Gusev, A. A.; and Boldyrev, S. A. 1972. Progress of earthquake prediction in Kamchatka. *Tectonophysics* 14:279–86.

Fox, F. L. 1970. Seismic geology of the eastern United States. *Bulletin of the Association of Engineering Geologists* 7:21–43.

Galanopoulos, A. G., and Bacon, E. 1969. *Atlantis: the truth behind the legend.* Indianapolis: Bobbs-Merrill.

Gupta, H. K.; Rastogi, B. K.; and Narain, H. 1972. Some discriminatory characteristics of earthquakes near the Kariba, Kermasta, and Koyna artificial lakes. *Bulletin of the Seismological Society of America* 62:493–507.

Gumper, F. J., and Scholz, C. 1971. Microseismicity and tectonics of the Nevada seismic zone. *Bulletin of the Seismological Society of America* 61:1413–32.

Gutenberg, B., and Richter, C. F. 1954. *Seismicity of the earth and associated phenomena.* Princeton: Princeton Univ. Press.

Hagiwara, T., and Rikitake. 1967. Japanese program on earthquake prediction. *Science* 157(3790):761–68.

Hamilton, R. M., and Healy, J. H. 1969. Aftershocks of the Benham nuclear explosion. *Bulletin of the Seismological Society of America* 59:2271–81.

Hammond, A. L. 1971. Earthquake prediction and control. *Science* 173:316.

Healy, J. H., et al. 1968. The Denver earthquakes. *Science* 161:1301–10.

Heintze, C. 1968. *The circle of fire.* New York: Meredith Press.

Henley, A. D. 1966. Seismic activity near the Texas coast. *Bulletin of the Association of Engineering Geologists* 3:33–39.

Houser, F. N. 1969. Subsidence related to underground nuclear explosions, Nevada test site. *Bulletin of the Seismological Society of America* 59:2231–51.

Johnson, T.; Wu, F. T.; and Scholz, C. H. 1973. Source parameters for stick-slip and for earthquakes. *Science* 179:278–79.

Kahle, J. E., et al. 1971. Geologic surface effects of the San Fernando earthquake. *California Geology* 24:75–80.

Kaila, K. L., and Narain, H. 1971. A new approach for preparation of quantitative seismicity maps as applied to Alpide Belt-Sunda Arc and adjoining areas. *Bulletin of the Seismological Society of America* 61:1275–91.

Kazmi, A. H. 1972. Earthquakes in Pakistan. *Geonews* 2:22–40.

Keys, J. G. 1963. The tsunami of 22 May 1960 in the Somoa and Cook Islands. *Bulletin of the Seismological Society of America* 53:1211–27.

Lee, W. H. K.; Eaton, M. S.; and Brabb, E. E. 1971. The earthquake sequence near Danville, California, 1970. *Bulletin of the Seismological Society of America* 61:1771–94.

Lomnitz, C. 1970. Casualties and behavior of populations during earthquakes. *Bulletin of the Seismological Society of America* 60:1309–13.

Luce, J. V. 1969. *Lost Atlantis: new light on an old legend.* New York: McGraw-Hill.

Marshall, P. D.; Burch. R. F.; and Douglas, A. 1972. How and why to record broad band seismic signals. *Nature* 239:154–55.

Milne, W. G., and Davenport, A. G. 1969. Distribution of earthquake risk in Canada. *Bulletin of the Seismological Society of America* 59:729–54.

Mirza, M. A. 1972. Earthquake prediction research entering stage of practical application. *Geonews* 2:31–37.

National Academy of Sciences. 1971. *The great Alaska earthquake of 1964: Geology.* Washington.

Nersesov, I. 1970. Earthquake prognostication in the Soviet Union. *Bulletin of the New Zealand Society for Earthquake Engineering* 3:108–19.

Niazi, M., and Basford, J. R. 1968. Seismicity of Iranian Plateau and Hindu Kush region. *Bulletin of the Seismological Society of America* 58:417–26.

Oakeshot, G. B. 1971. The geologic setting. *California Geology* 24(4–5):69–74.

Oliver, J. 1969. Earthquake prediction: United States-Japan cooperative science program. *Science* 164:92–93.

Oxburgh, E. R. 1972. Flake tectonics and continental collisions. *Nature* 239:202–4.

Pecora, W. T. 1966. National Center for Earthquake Research, USGS. *Geotimes,* Dec. 1965-Jan. 1966, p. 13.

Pafker, G. 1969. *Tectonics of the March 27, 1964, Alaska earthquake.* U.S. Geological Survey Prof. Paper 543–I.

Rasmussen, N. 1967. Washington state earthquakes 1840 through 1965. *Bulletin of the Seismological Society of America* 57:463–76.

Renault, M. 1962. *The bull from the sea.* New York: Random House.

Saul, R. B. 1971. Effects of the San Fernando earthquake in the Oat Mountain quadrangle. *California Geology* 24:83–84.

Scholz, C. H. 1972. Crustal movements in tectonic areas. *Tectonophysics* 14:201–217.

Steinbrugge, K. V. 1969. Seismic risk to buildings and structures on filled lands in San Francisco Bay. In *Geologic and engineering aspects of San Francisco Bay fill,* ed. H. B. Goldman, pp. 103–15. California Division of Mines and Geology Special Report 97.

Steinbrugge, K. V., and Rodrigo, F. A. 1963. The Chilean earthquakes of May, 1960: a structural engineering viewpoint. *Bulletin of the Seismological Society of America* 53:225–307.

Sugawara, M. 1972. The national research centre for disaster prevention. *Nature* 240:200–201.

Tasdemiroglu, M. 1971. The 1970 Gediz earthquake in western Anatolia, Turkey. *Bulletin of the Seismological Society of America* 61:1507–27.

Terada, K. 1972. Natural disasters in Japan. *Nature* 240:197–99.

Tocher, D. 1973. Earthquakes and earthquake engineering. *Science* 179:1148.

Trifunac, M. D., and Hudson, D. E. 1971. Analysis of the Pacoima Dam accelerogram—San Fernando, California, earthquake of 1971. *Bulletin of the Seismological Society of America* 61:1393–1411.

U.S. Geological Survey. 1965. *United States earthquake program: earthquake prediction.* U.S. Geological Survey Program Issue Paper.

————. 1971. *The San Fernando, California, Earthquake of February 9, 1971.* U.S. Geological Survey Prof. Paper 733.

Vitousek, M. J. 1963. The tsunami of 22 May 1960 in French Polynesia. *Bulletin of the Seismological Society of America* 53:1229–36.

Wade, N. 1972. Earthquake research: a consequence of the pluralistic system. *Science* 178:39–43.

Wallace, R. E. 1968. Earthquake of August 19, 1966, Varto area, eastern Turkey. *Bulletin of the Seismological Society of America* 58:11–45.

Wood, J. H., and Jennings, P. C. 1971. Damage to freeway structures in the San Fernando earthquake.

Bulletin of the New Zealand Society for Earthquake Engineering 4(3):347–75.

Wood, M. D., and Allen, R. V. 1971. Anomalous microtilt preceding a local earthquake. *Bulletin of the Seismological Society of America* 61:1801–9.

Chapter 10

Anonymous. 1970. Mount Rainier Restless. *California Geology* 23:86–87.

Cadle, R. D., and Buifford, I. H. 1971. Hekla eruption clouds. *Nature* 230:573–74.

Cailleux, A. 1968. *Anatomy of the earth,* trans. J. M. Stuart. New York: McGraw-Hill.

Chesterman, C. W. 1971. Volcanism in California. *California Geology* 24:139–47.

Crandell, D. R. 1969. *Surficial geology of Mount Rainier National Park, Washington.* U.S. Geological Survey Bulletin 1288.

Fiske, R. S., and Kohanagi, R. Y. 1968. *The December, 1965, eruption of Kilauea Volcano, Hawaii.* U.S. Geological Survey Prof. Paper 607.

Johnston, J. J. S., and Mauk, F. J. 1972. Earth tides and the triggering of eruptions from Mt. Stromboli, Italy. *Nature* 239:266–67.

Keller, G. V.; Jackson, D. B., and Rapolla, A. 1972. Magnetic noise preceding the August, 1971, summit eruption of Kilauea Volcano. *Science* 175:1457–58.

Kiersch, G. A. 1965. Vaiont Reservoir disaster. *Geotimes,* May-June, pp. 9–12.

Kinoshita, W. To., et al. 1969. Kilauea Volcano: the 1967–68 summit eruption. *Science* 166:459–68.

LaCroix, A. 1908. *La Montagne Pelée après ses éruptions.* Paris: Académie de Science.

MacDonald, G. A., and Abbott, A. T. 1970. *Volcanoes in the sea: the geology of Hawaii.* Honolulu: Univ. of Hawaii Press.

Porter, S. C. 1972. Buried Caldera of Maun Kea Volcano, Hawaii. *Science* 175:1458–60.

Richter, D. H., et al. 1970. *Chronological narrative of the 1959–60 eruption of Kilauea Volcano, Hawaii.* U.S. Geological Survey Prof. Paper 537–E.

Schoo, J. H. 1969. *Hercules' labors: fact or fiction.* Chicago: Argonaut.

Tamrazyan, G. P. 1972. Peculiarities in the manifestation of gaseous-mud volcanoes. *Nature* 240:406–8.

U.S. Geological Survey. 1968. *Volcanoes of the United States.*

Chapter 11

Anonymous. 1970. Peru was hit by massive avalanche triggered by earthquake. *California Geology* 23:203–7.

Brawner, C. O., and Milligan, V., eds. 1971. *Stability in*

open pit mining. New York: Society of Mining Engineers, Em. Institute of Mining, Metallurgy, and Petroleum Engineers.

Brezinski, L. S. 1971. A review of the 1924 Kenogami Landslide. *Canadian Geotechnical Journal* 8:1–6.

Clark, J. I.; Desimon, A.; and Stepanek, M. 1971. Landslides in urban areas. *Proceedings of the 9th Annual Engineering Geology and Soils Engineering Symposium,* pp. 289–304.

Donnelly, I. 1898. *Atlantis: the antediluvian world.* 24th ed. New York: Harper Brothers.

Emery, K. O. 1969. The continental shelves. *Scientific American* 221(3):106–22.

Fairbridge, R. W. 1960. The changing level of the sea. *Scientific American* 202(5):70–79.

Foose, R. M. 1968. Surface subsidence and collapse caused by ground water withdrawal in carbonate rock areas. In *Engineering geology in country planning,* ed. Q. Zaruba et al. International Geological Congress, 23d Session, vol. 12, pp. 155–66.

Gray, H. H. 1971. *Glacial lake deposits in southern Indiana—Engineering problems and land use.* Indiana Geological Survey Report of Progress no. 30.

Hammen, T. van der, et al. 1967. Stratigraphy, climatic succession and radiocarbon dating of the last glacial in the Netherlands. *Geologie en Mijnbouw* 46(3):79–95.

Hart, M. W. 1972. Erosional remnant of landslides, west-central San Diego County, California. *Bulletin of the Association of Engineering Geologists* 9:377–99.

Hill, D. M., et al. 1968. *Earthquake engineering programs: progress report.* California Department of Water Resources Bulletin no. 116–4.

Hough, J. L. 1958. *Geology of the Great Lakes.* Urbana: Univ. of Illinois Press.

———. 1963. The prehistoric Great Lakes of North America. *American Scientist* 51(1):84–109.

Ingerson, I. M. 1941. Hydrology of the southern San Joaquin Valley, California, and its relation to important water supplies. *Transactions of the American Geophysical Union,* pt. 1, pp. 20–45.

Kerr, P. F., and Drew, I. M. 1972. Clay mobility in ridge route landslides, Castaic, California. *Bulletin of the American Association of Petroleum Geologists* 56:2168–84.

Klein, D. R. 1971. Reaction of reindeer to obstructions and disturbances. *Science* 173:393–98.

Lachenbruch, A. H. 1970. *Some estimates of the thermal effects of a heated pipeline in permafrost.* U.S. Geological Survey Circular 632.

Leggett, R. F. 1972. Duisburg Harbour lowered by controlled coal mining. *Canadian Geotechnical Journal* 9:374–83.

Longwell, C. R. 1960. Interpretation of the leveling data. In *Comprehensive survey of sedimentation in Lake Mead, 1948–49,* ed. W. O. Smith et al., pp. 33–38. U.S. Geological Survey Prof. Paper 295.

Lyell, C. 1883. *Principles of geology,* vol. I. New York: D. Appleton.

McDowell, B., and Fletcher, J. E. 1962. Avalanche. *National Geographic* 121(6):855–80.

Miller, R. E. 1966. Land subsidence in southern California. In *Engineering geology in California,* pp. 273–85. Los Angeles: Association of Engineering Geologists.

Mitchell, R. J., and Eden, W. J. 1972. Measured movements of clay slopes in the Ottawa area. *Canadian Journal of Earth Sciences* 9:1001–13.

Morgando, F. P. 1971. Surface subsidence due to coal mining, Rock Springs, Wyoming. *Proceedings of the 9th Annual Engineering Geology and Soils Engineering Symposium,* p. 189.

Morton, D. M. 1971. Seismically triggered landslides above San Fernando Valley. *California Geology* 24:81–82.

Piteau, D. R. 1972. Engineering geology considerations and approach in assessing the stability of rock slopes. *Canadian Mining and Metallurgical Bulletin* 65(721):53–100.

Poland, J. F. 1972. Land subsidence in the western states due to groundwater overdraft. *Water Resources Bulletin* 8:118–31.

Saunders, M. K., and Fookes, P. G. 1970. A review of the relationship of rock weathering and climate and its significance to foundation engineering. *Engineering Geology* 4:289–325.

Scrivenor, J. B. 1929. The mudstreams (lahars) of Gunong Kloet in Java. *Geology Magazine* 66:433–34.

Sharpe, C. F. S., and Dosch, E. F. 1942. Relation of soil creep to earthflow in the Appalachian Plateau. *Journal of Geomorphology* 5:312–24.

Sheets, M. M. 1971. Active surface faulting in the Houston area, Texas: 1971. *Houston Geological Society Bulletin* 13(7):24–33.

Siclen, D. V. 1972. Reply: The Houston fault problem. *Bulletin of the Association of Engineering Geologists* 9:69–77.

Thomason, S. 1971. The Lesueur landslide, a failure in Upper Cretaceous clay shale. *Proceedings of the 9th Annual Engineering Geology and Soils Engineering Symposium,* pp. 257–87.

Winslow, A. G., and Wood, L. A. 1959. Relation of land subsidence to ground water withdrawals in the upper Gulf Coast region, Texas. *Transactions of the American Institute of Mining, Metallurgy, and Petroleum Engineering* 214:1030–34.

Zaruba, Q., and Mencl, V. 1969. *Landslides and their control.* Amsterdam; Elsevier.

Chapter 12

Anderson, D. G. 1970. *Effects of urban development on floods in northern Virginia.* U.S. Geological Survey Water-Supply Paper 2001–C.

Benson, M. A. 1962. *Evolution of methods for evaluating the occurrence of floods.* U.S. Geological Survey Water-Supply Paper 1580–A, pp. A1–A30.

————. 1967. *Factors influencing the occurrence of floods in the humid region of diverse terrain.* U.S. Geological Survey Water-Supply Paper 1543–A.

————. 1968. Uniform flood-frequency estimating methods for federal agencies. *Water Resources Research,* vol. e, pp. 891–908.

Bodine, B. R. 1969. *Hurricane surge frequency estimated for the Gulf Coast of Texas.* U.S. Corps of Engineers, Coastal Engineering Research Center Technical Memorandum no. 26.

Bue, C. D. 1967. *Flood information for flood planning.* U.S. Geological Survey Circular 539.

Dalrymple, T. 1960. *Flood-frequency analyses.* U.S. Geological Survey Water-Supply Paper 1534–A.

Donnelly, I. 1898. *Atlantis: the antediluvian world.* 24th ed. New York: Harper Brothers.

Flierl, G. R., and Robinson, A. R. 1972. Deadly surges in the Bay of Bengal: dynamics and storm-tide tables. *Nature* 239:213–15.

Gillette, R. 1972. Stream channelization: conflict between ditchers, conservationists. *Science* 176:890–94.

Helley, E. J., and Lamarch, V. C., Jr. 1968. *December, 1964, a 400-year flood in northern California.* U.S. Geological Survey Prof. Paper 600–D, pp. D34–D37.

Hill, I. K. 1969. Runoff hydrograph as a function of rainfall excess. *Water Resources Research* 5(1):95–102.

Howard, R. A.; Matheson, J. E.; and North, D. W. 1972. The decision to seed hurricanes. *Science* 176:1191–1202.

Judge, J. 1967. Florence rises from the flood. *National Geographic* 132:1–43.

Kiersch, G. A. 1965. Vaiont Reservoir disaster. *Geotimes,* May-June, pp. 1–12.

Leopold, L. B. 1968. *Hydrology for urban land planning.* U.S. Geological Survey Circular 554.

Leopold, L. B.; Wolman, M.G.; and Miller, J. P. 1964. *Fluvial processes in geomorphology.* San Francisco: Freeman.

Martens, L. A. 1968. *Flood inundation and effects of urbanization in metropolitan Charlotte, North Carolina.* U.S. Geological Survey Water-Supply Paper 1591–C.

Matthai, H. F. 1969. *Floods of June 1965 in South Platte River Basin, Colorado.* U.S. Geological Survey Water-Supply Paper 1850–B.

May, V. J., and Allen, H. E. 1965. *Floods in Stream-*
wood Quadrangle, northeastern Illinois. U.S. Geological Survey Hydrologic Investigations, Atlas HA-203.

Michalowski, K. 1968. *Art of ancient Egypt,* trans. Norbert Guterman. New York: Abrams.

Missouri Geological Survey. 1968. Hazardous dams in Missouri. *Missouri Mineral Industry News* 8(2):16–20.

Patterson, J. L. 1963. *Floods in Texas.* Texas Water Commission Bulletin 6311.

————. 1964. *Magnitude and frequency of floods in the United States, pt. 7: lower Mississippi River basin.* U.S. Geological Survey Water-Supply Paper 1681.

————. 1965. *Magnitude and frequency of floods in the United States, pt. 8: western Gulf of Mexico basins.* U.S. Geological Survey Water-Supply Paper 1682.

Rantz, S. E. 1970. *Urban sprawl and flooding in southern California.* U.S. Geological Survey Circular 601–B.

Seaburn, G. E. 1969. *Effects of urban development on direct runoff to East Meadowbrook, Nassau County, Long Island, New York.* U.S. Geological Survey Prof. Paper 627–B.

Sheaffer, J. R.; Ellis, D. W.; and Spieker, A. M. 1970. *Flood-hazard mapping in metropolitan Chicago.* U.S. Geological Survey Circular. 601–C.

Sigafoos, R. S. 1964. *Botanical evidence of floods and flood-plain deposition.* U.S. Geological Survey Prof. Paper 485–A.

Thomas, D. M. 1964. *Height-frequency relations for New Jersey floods.* U.S. Geological Survey Prof. Paper 475–D.

Wendorf, F. 1968. The prehistory of the Nile Valley. *Science* 162:1032–33.

Chapter 13

Bandy, O. L. 1972. Variations in *Globigerina bulloides* d'Orbigny as indices of water masses. *Antarctic Journal* 7:194–95.

Binns, R. E. 1972. Flandrian strandline chronology for the British Isles and correlation of some European post-glacial strandlines. *Nature* 235:206–10.

Bradley, R. S., and Miller, G. H. 1972. Recent climatic changes and increased glacierization in the eastern Canadian Arctic. *Nature* 237:385–86.

Broecker, W. S. 1966. Absolute dating and the astronomical theory of glaciation. *Science* 151:299–304.

Brooks, C. E. P. 1949. *Climate through the ages.* New York: McGraw-Hill.

Cailleux, A. 1968. *Anatomy of the earth,* trans. J. M. Stuart. New York: Oxford Univ. Press.

Carson, R. L. 1951. *The sea around us.* New York: Oxford Univ. Press.

Churchill, W. 1956. *History of English speaking peoples. Life Magazine* 40(12):84ff.

Colinvaux, P. A. 1972. Climate and the Galapagos Islands. *Nature* 240:17–20.

Crossman, E. J., and Harington, C. R. 1970. Pleistocene Pike, *Esox lucius,* and *Esox* sp. from the Yukon Territory and Ontario. *Canadian Journal of Earth Sciences* 7:1130–38.

Dansgaard, W., et al. 1969. One thousand centuries of climatic record from Camp Century on the Greenland ice sheet. *Science* 166:377–81.

———. 1971. Climatic record revealed by the Camp Century ice core. In *The late Cenozoic glacial ages,* ed. K. K. Turekian, pp. 37–56. New Haven: Yale Univ. Press.

Deevey, E. S., Jr. 1952. Radiocarbon dating. *Scientific American* 186(2)24–26.

Denton, G. H., and Porter, S. C. 1970. Neoglaciation. *Scientific American* 222(6):101–9.

DuPlessy, J. C., et al. 1970. Continental climatic variations between 130,000 and 90,000 years B.P. *Nature* 226:631–33.

Emery, K. O. 1969. The continental shelves. *Scientific American* 221(3):107–22.

Emilani, C. 1966. Isotopic paleotemperatures. *Science* 154:851–57.

———. 1969. Interglacial high sea levels and the control of Greenland ice by the precession of the equinoxes. *Science* 166:1503–4.

———. 1971. The last interglacial: paleotemperatures and climatology. *Science* 171:571–73.

Ericson, D. B., and Wollin, G. 1962. Micropaleontology. *Scientific American* 207:96–108.

———. 1968. Pleistocene climates and chronology in deep-sea sediments. *Science* 162:1227–34.

Fairbridge, R. W. 1960. The changing level of the sea. *Scientific American* 202(5):70–79.

———. 1967. Carbonate rocks and paleoclimatology of the biogeochemical history of the planet. In *Carbonate rocks: developments in sedimentology,* ed. G. V. Chilingar, H. J. Bissel, and R. W. Fairbridge, pp. 399–432. Amsterdam: Elsevier.

———. 1970. World paleoclimatology of the Quaternary. *Revue de Géographie Physique et de Géologie Dynamique* 12(2)97–104.

Flint, R. F. 1971. *Glacial and Quaternary geology.* New York: Wiley.

Frakes, L. A. 1972. Paleoclimatology of the southern ocean. *Antarctic Journal* 7:189–90.

Frakes, L. A., and Kemp, E. M. 1972. Influence of continental positions on early Tertiary climates. *Nature* 240:97–100.

Gadd, N. R. 1971. *Pleistocene geology of the Central St. Lawrence Lowland,* with selected passages from J. W. Goldthwaite, "The St. Lawrence Lowland" (unpublished manuscript). Geological Survey of Canada Memoir 359.

Garrels, R. M., and MacKenzie, F. T. 1971. *Evolution of sedimentary rocks.* New York: Norton.

Goldthwait, R. P., et al. 1965. Pleistocene deposits of the Erie lobe. In *The Quaternary of the United States,* ed. H. E. Wright, Jr. and D. G. Frey, pp. 85–97. Princeton: Princeton Univ. Press.

Hammen, T. van der, et al. 1967. Stratigraphy, climatic succession and radiocarbon dating of the last glacial in the Netherlands. *Geologie en Mijnbouw* 46:79–95.

Holand, H. R. 1965. *Explorations in America before Columbus.* New York: Twayne.

Hough, J. L. 1958. *Geology of the Great Lakes.* Urbana: Univ. of Illinois Press.

Ingstad, H. 1969. *Westward to Vinland.* New York: St. Martin's.

Johnsen, S. J., et al. 1972. Oxygen isotope profiles through the Antarctic and Greenland ice sheets. *Nature* 235:429–34.

Kaiser, K. 1969. The climate of Europe during the Quaternary ice age. In *Quaternary geology and climate,* ed. H. E. Wright, Jr. Proceedings of the 7th Congress of the International Association for Quaternary Research, vol. 16, pp. 10–37. Washington: National Academy of Sciences.

Koenen, P. H. 1955. *Realms of water: some aspects of its cycle in nature.* New York: Wiley.

Kukla, G. J., and Kukla, H. J. 1972. Insolation regimes of interglacials. *Quaternary Research* 2:412–24.

Kurten, B. 1972. The cave bear. *Scientific American* 226(3)60–73.

LaMarche, V. C., and Mooney, H. A. 1972. Recent climatic change and development of the bristlecone pine (*P. longaeva* Bailey), Krumholz Zone, Mt. Washington, Nevada. *Arctic and Alpine Research* 4:61–72.

Lamb, H. H. 1969. The new look of climatology. *Nature* 223:1209–13.

———. 1971. Climates and circulation regimes developed over the northern hemisphere during and since the last ice age. *Palaeogeography, Palaeoclimatology, and Palaeoecology* 10:125–62.

Leopold, L. B., and Davis, K. S. 1966. *Water.* New York: Time.

Le Roy Ladurie, E. 1971. *Times of feast, times of famine: a history of climate since the year 1000,* trans. B. Bray. Garden City, N.Y.: Doubleday.

McCarthy, R. F. 1970. The Campbell strandline of glacial lake Agassiz in Walsh County, North Dakota. *The Compass* 47:147–53.

Mesolella, K. J., et al. 1969. The astronomical theory of climatic change: Barbados data. *Journal of Geology* 77:250–74.

Mitchell, J., Jr. 1972. The natural breakdown of the present interglacial and its possible intervention by human activities. *Quaternary Research* 2:436-45.

Morison, S. E. 1971. *The European discovery of America.* New York: Oxford Univ. Press.

Morner, N. 1972. Time scale and ice accumulation during the last 125,000 years as indicated by the Greenland 0^{18} curve. *Geology Magazine* 109:17–24.

Newell, N. D. 1963. Crises in the history of life. *Scientific American,* Feb., pp. 76–92.

————. 1971. The nature of the fossil record. In *Adventures in earth history,* ed. P. E. Cloud, pp. 641–64. San Francisco: W. H. Freeman.

Pannella, G.; MacClintock, C.; and Thompson, M. N. 1968. Paleontological evidence of variation in length of synodic month since late Cambrian. *Science* 162:792–96.

Pettersson, O. 1912. Climatic variations in historic and prehistoric time. *Svenska Hydrogafisk-Biologiska Kommissionens Skrifter* 15.

Plass, G. N. 1959. Carbon dioxide and climate. *Scientific American* 201:41–47.

Pons, L. J., et al. 1963. Evolution of the Netherlands coastal area during the Holocene. *Geologie en Mijnbouw* 21–2:197–208.

Raup, D. M., and Stanley, S. M. 1971. *Principles of paleontology.* San Francisco: W. H. Freeman.

Raychaudhuri, S. P. 1971. Expansion or contraction of the great Indian Desert. *Proceedings of the Indian Natural Science Academy* 36(6):331–44.

Schneider, S. H. 1972. Atmospheric particles and climate: can we evaluate the impact of man's activities? *Quaternary Research* 2:425–35.

Shaw, D. M., and Donn, W. L. 1968. Milankovitch radiation variations: a quantitative evaluation. *Science* 162:1270–72.

Strahler, A. N. 1965. *Introduction to physical geography.* New York: Wiley.

Swift, D. J. P. 1969. Outer shelf sedimentation: processes and products. In *The new concepts of continental margin sedimentation,* ed. D. J. Stanley et al. American Geological Institute Short Course Lecture Notes, Lecture 5.

Turekian, K. K., ed. 1971. *The late Cenozoic glacial ages.* New Haven: Yale Univ. Press.

Walcott, R. I. 1972. Past sea levels, eustasy and deformation of the earth. *Quaternary Research* 2:1–14.

Wells, J. W. 1963. Coral growth and geochronometry. *Nature* 197:948–950.

West, R. G. 1968. *Pleistocene geology and biology.* New York: Wiley.

Wollin, G.; Ericson, D. B.; and Ewing, M. 1971. Late Pleistocene climates recorded in Atlantic and Pacific deep-sea sediments. In *The late Cenozoic glacial ages,* ed. K. K. Turekian, pp. 199–214. New Haven: Yale Univ. Press.

Woodford, A. O. 1965. *Historical geology.* San Francisco: W. H. Freeman.

Chapter 14

Aley, T. 1969. Out of sight, out of mind—a grim fairy ·tale. *Missouri Mineral Industry News* 9(12):163–66.

Anonymous. 1972. Federal role in land-use rule at issue. *Oil and Gas Journal,* 4 Dec., pp. 40–41.

Armstrong, J. M., and Bradley, E. H., Jr. 1972. Status of state coastal zone management programs. *Marine Technological Society Journal* 6(6):7–16.

Baltimore County Planning Office. 1967. Current and Committed Development Map, Baltimore County.

Bauer, K. W. 1966. Application of soils studies in comprehensive regional planning. In *Soil surveys and land use planning,* ed. L. J. Bartelli, pp. 42–59. Soil Science Society of America and American Society of Agronomy.

Brooks, D. B. 1966. Strip mine reclamation and economic analysis. *Natural Resources Journal 6(1):* 13–44.

Carr, J. T. 1967. *Hurricanes affecting the Texas Gulf Coast.* Texas Water Development Board Report 49.

Carter, L. J. 1967. New towns: geological survey has key role in experiment. *Science* 158:752–55.

————. 1970. Galveston Bay: test case of an estuary in crisis. *Science* 167:1102–8.

Carter, W. 1970. The North Sea defeated: Holland's delta project. *Oceans* 3(1):6–15.

Christian, W. F. 1944. Land utilization in Denver. Ph.D. dissertation, Univ. of Chicago.

Conomos, T. J., et al. 1972. *A preliminary study of the effects of water circulation in the San Francisco Bay Estuary.* U.S. Geological Survey Circular 637B.

Dobrovolny, E., and Schmoll, H. R. 1968. Geology as applied to urban planning: an example from the greater Anchorage area borough, Alaska. In *Engineering geology in country planning,* ed. Q. Zaruba et al., pp. 39–56. Proceedings of the 23d Session of the International Geological Congress, 12th section.

Evans, G. 1965. Intertidal flat sediments and their deposition in the wash. *Quarterly Journal of the Geological Society* 121:209–45.

Flawn, P. T., 1966. Geology and urban development. In *Engineering geology in California,* ed. R. Lung and R. Proctor, pp. 209–13. Association of Engineering Geologists.

————. 1968. The environmental geologist and the body politic. *Geotimes,* July-Aug., pp. 14–15.

————. 1970. Subsidence due to withdrawal of fluids. In *Environmental geology,* ed. P. T. Flawn, pp. 42–49. New York: Harper & Row.

Frye, J. C. 1968. The use of geology in managing the environment. *Earth Sciences Newsletter* 4:1–5.

Grant, R. K. 1965. *Terrain, features of the Mt. Isa-Dajarra region and an assessment of their signifi-*

cance in relation to potential engineering land use. Commonwealth Scientific and Industrial Research Organization, Australia, Soil Mechanics Technical Paper no. 1.

Hayes, M. O. 1967. *Hurricanes as geological agents: case studies of hurricanes Carla, 1961, and Cindy, 1963.* Univ. of Texas, Bureau of Economic Geology Report of Investigations no. 61.

Hilpman, P. L. 1970. Urban growth and environmental geology. In *Colorado governor's conference on environmental geology,* ed. J. B. Ivey et al. Colorado Geological Survey Special Publication no. 1, pp. 16–19.

Hunter, W. R.; Tipps, C. W.; and Coover, J. R. 1966. Use of soil maps by city officials for operational planning. In *Soil surveys and land use planning,* ed. L. J. Bartelli et al., pp. 31–36. Soil Science Society of America, and American Society of Agronomy.

Ives, R. E., and Eddy, G. E. 1968. *Subsurface disposal of industrial wastes.* Interstate Oil Compact Commission.

Jahns, R. H. 1968. Geologic jeopardy. *Texas Quarterly* 11(2):69–83.

Jong, J. D. de. 1972. Man's impact on subrecent and recent geological environments of the Netherlands. *Proceedings of the 24th International Geological Congress, Symposium 1,* p. 15.

Kaye, C. A. 1967. The greater Boston urban geology project of the U.S. Geological Survey. In *Economical geology of Massachusetts,* ed. D. C. Farquhar, pp. 273–77. Amherst: Univ. of Massachusetts Graduate School.

———. 1968. *Geology and our cities.* New York Academy of Sciences, translation series 2, vol. 30, pp. 1045–51.

Kovda, V. A. 1961. Land use development in the arid regions of the Russian Plain, the Caucasus and Central Asia. In *A history of land use in arid regions,* ed. L. D. Stamp. UNESCO Arid Zone Research vol. 12, pp. 175–218.

Lamoreaux, P. E., et al. 1969. *Mineral, water, and energy resources of Wilcox County, Alabama.* Geological Survey of Alabama Information Series 40.

Lankford, R. R., and Rogers, J. W., et al. 1969. *Holocene geology of the Galveston Bay area.* Houston Geological Society.

Lea, J. P. 1972. The quality of the built environment: the problem of urban decay. *South African Journal of Science,* Feb., pp. 43–46.

Leopold, L. B. 1969. *Quantitative comparison of some aesthetic factors among rivers.* U.S. Geological Survey Circular 620.

Lockwood, Andrews, and Newman, Inc. 1967. *A new concept—water for preservation of bays and estuaries.* Texas Water Development Board Report 43.

Lutzen, E. E., and Williams, J. H. 1968. Missouri's approach to engineering geology in urban areas. *Bulletin of the Association of Engineering Geologists* 5(2):109–21.

Maugh, T. H., II. 1973. ERTS: surveying earth's resources from space. *Science* 180:49–51.

McComas, M. R. 1968. *Geology related to land use in the Hennepin region.* Illinois State Geological Survey Circular 422.

McCulloch, D. S., et al. 1972. *Some effects of fresh-water inflow on the flushing of South San Francisco Bay.* U.S. Geological Survey Circular 637A.

McGill, J. T. 1964. *Growing importance of urban geology.* U.S. Geological Survey Circular 487.

Metropolitan Washington Council of Governments. 1968. *Natural features of the Washington metropolitan area.*

Michael, E. D. 1965. *Geology and urban development.* Association of Engineering Geologists.

Mitchell, F. G. 1972. Soil deterioration associated with prehistoric agriculture in Ireland. *Proceedings of the 24th International Geological Congress, Symposium 1,* pp. 59–68.

Mohorich, L. M. 1972. The role of geologic input in urban planning. *Mountain Geology* 8:209–19.

Nelson, B. 1967. New towns: geological survey has key role in experiment. *Science* 158:752–56.

Nichols, M. 1969. Coastal engineering. *Science* 164: 590–92.

Olson, G. W. 1966. Improving soil survey interpretations through research. In *Soil surveys and land use planning,* ed. L. J. Bartelli et al. Soil Science Society of America, and American Society of Agronomy.

Pestrong, R. 1969. The role of the urban geologists in city planning. California Division of Mines and Geology, *Mineral Information Service* 21(10):151–52.

Peterson, D. H., et al. 1972. Distribution of lead and copper in surface sediments in the San Francisco Bay Estuary, California. San Francisco Bay Region environment and resources planning study, U.S. Geological Survey Miscellaneous Field Studies Map MF–323.

Poland, J. F. 1971. Land subsidence in the Santa Clara Valley, Alameda, San Mateo, and Santa Clara Counties, California. San Francisco Bay region environment and resources planning study, U.S. Geological Survey and U.S. Department of Housing and Urban Development, Technical Report no. 2 (map).

Quay, J. 1963. Lake County uses soil survey in planning its urban areas. *Soil Conservation,* Dec., pp. 99–102.

Quinlan, J. F. 1970. Central Kentucky karst. Réunion internationale karstologie en Languedoc-Provence,

1968. *Actes Méditerranée Études et Travaux* 7: 235–53.

Radbruch, D. H. 1967. Approximate location of fault traces and historic surface ruptures within the Hayward fault zone between San Pablo and Warm Springs, California. U.S. Geological Survey Miscellaneous Geological Investigations Map I–522.

———. 1968. Engineering geology in urban planning and construction in the United States. In *Engineering geology in country planning,* ed. Q. Zaruba et al., pp. 105–11. Proceedings of the 23d International Geological Congress, 12th section.

Radbruch, D. H., et al. 1966. *Tectonic creep in the Hayward fault zone, California.* U.S. Geological Survey Circular 525.

Rantz, S. E. 1971. *Suggested criteria for hydrologic design of storm-drainage facilities in the San Francisco Bay region, California.* San Francisco Bay region environment and resources planning study, U.S. Geological Survey and U.S. Department of Housing and Urban Development, Technical Report no. 3.

Rapoport, A. 1973. Some thoughts on the methodology of men-environment studies. *International Journal of Environmental Studies* 4:135–40.

Rickert, D. A., and Spieker, A. M. 1971. *Real-estate lakes.* U.S. Geological Survey Circular 601G.

Schlicker, H. G. 1967. Engineering geology—a planning tool. *The Lone Ranger,* vol. 17, no. 4.

Schon, D. A. 1967. Forecasting and technological forecasting. *Daedalus,* Summer, pp. 759–70.

School, D. W.; Craighead, F. C., Sr.; and Stuiver, M. 1969. Florida submergence curve revised: its relation to coastal sedimentation rates. *Science* 163: 562–64.

Shadman, A. 1968. Quarry site surveys in relation to country planning. In *Engineering geology in country planning,* ed. Q. Zaruba, pp. 125–32. Proceedings of the 23d International Geological Congress, 12th section.

Sheaffer, J. R., et al. 1967. *Introduction to flood proofing: an outline of principles and methods.* Univ. of Chicago, Center for Urban Studies.

Sheng, T. C. 1972. A treatment-oriented land capability classification scheme for hilly marginal lands in the humid tropics. *Journal of the Scientific Research Council of Jamaica* 3:93–112.

Smith, A., and Joseph, M. 1971. *Mato Grosso: last virgin land.* London: The Royal Society and the Royal Geographical Society.

Smith, T. W. 1966. Repair of landslides. In *Seminar on the importance of the earth sciences to the public works and building official,* ed. C. M. Scullin, pp. 205–45. Association of Engineering Geologists, Los Angeles Section.

Smith, W. C. 1968. *Geology and engineering characteristics of some surface material in McHenry County, Illinois.* Illinois Geological Survey, Environmental Geology Notes no. 19.

Stamp, L. D. 1939. Some economic aspects of coastal loss and gain. *Geography Journal* 93:497–503.

Stankowski, S. J. 1972. *Population density as an indirect indicator of urban and suburban land-surface modifications.* U.S. Geological Survey Prof. Paper 800–B.

Steinbrugge, K. V. 1968. *Earthquake hazard in the San Francisco Bay area: a continuing problem in public policy.* Berkeley: Univ. of California Institute of Governmental Studies.

Sullivan, G. D. 1965. A new science—mineral land reclamation. *Mining Engineering* (New York) 17(7): 142–44.

Trask, P. D., and Rolaton, J. W. 1951. Engineering geology of San Francisco Bay. *Bulletin of the Geological Society of America* 62:1108.

Tuthill, S. 1970. Carbonates aid ground water pollution. *The Driller,* May.

U.S. Army Coastal Engineering Center. 1964. *Land against the sea.* Miscellaneous Paper 3–64.

U.S. Department of Agriculture, et al. 1964. *Soil handbook for soil survey, metropolitan area, San Antonio, Texas.* U.S. Department of Agriculture, Soil Conservation Service, City of San Antonio, Texas, and Texas Agricultural Experiment Station.

U.S. Geological Survey. 1952. Florida quadrangle: Puerto Rico (topographic map).

———. 1972. National resource and land information program urged. U.S. Department of Interior and U.S. Geological Survey news release, 5 October.

Volmer, A. 1963. Practical geohydrology "delta project." *Verhandelingen van het Koninklijk Nederlands Geologisch Mijnbouwkundig Genootschap Serie* 21–1:171–79.

Winslow, A. G., and Wood, L. A. 1959. Relation of land subsidence to ground water withdrawals in the upper Gulf Coast Region, Texas. *Transactions of the American Institute of Mining and Metallurgical Engineers* 214:1030–34.

Withington, C. F. 1967. Geology—its role in the development and planning of metropolitan Washington. *Journal of the Washington Academy of Sciences* 57:189–99.

Wright, R. L. 1972. Some perspectives in environmental research for agricultural land-use planning in developing countries. *Geoforum* 10:15–33.

Chapter 15

Alland, A. 1967. *Evolution and human behavior.* Garden City, N.Y.: Natural History Press.

Bardach, J. 1968. *Harvest of the sea.* New York: Harper & Row.

Berg, C. O. 1971. The fly that eats the snail that spreads disease. *Smithsonian* 2(6):8–17.

Berman, E. 1970. Human nonproliferation: a political responsibility. In *Agenda for survival: the environmental crisis—2,* ed. H. W. Helfrich, Jr., pp. 15–36. New Haven: Yale Univ. Press.

Blaser, R. E., ed. 1970. *Agronomy and health.* American Society of Agronomy Special Publication 16.

Boerma, A. H. 1970. A world agricultural plan. *Scientific American* 223(2):54–63, 66–69.

Bogue, D. J. 1967. *The prospects for world population control.* Univ. of Chicago, Community and Family Study Center.

Borgstrom, G. 1969. *Too many: a study of the earth's biological limitations.* New York: Macmillan.

Boughey, A. S. 1968. *Ecology of populations.* New York: Macmillan.

Bradfield, R. 1969. Training agronomists for increasing food production in the humid tropics. In *International agronomy: training and education,* ed. J. R. Cowan and L. S. Robertson, pp. 45–63. American Society of Agronomy Special Publication 15.

Bradfield, R. B.; Jelliffee E. F. P.; and Jelliffee, D. B. 1972. Assessment of marginal malnutrition. *Nature* 235:112.

Brown, L. R. 1965. *Population growth, food needs, and production problems.* American Society of Agronomy Special Publication 6.

———. 1967. The world outlook for conventional agriculture. *Science* 158:604–11. (Reprinted in *Readings in conservation ecology,* ed. G. W. Cox. New York: Appleton-Century-Crofts.)

———. 1970a. A Nobel peace prize: developer of high-yield wheat receives award. *Science* 170:518–19.

———. 1970b. Human food production as a process in the biosphere. *Scientific American* 223(3):161–70.

———. 1971. Food supplies and the optimum level of population. In *Is there an optimum level of population?* ed. F. S. Singer, pp. 72–88. New York: McGraw-Hill.

Buffon, G. L. L., Comte de. 1797–1807. *Buffon's natural history,* translated by Barr. London.

Chandler, R. F. 1969. Improving the rice plant and its culture. *Nature* 221:1007.

Chandrasekhar, S. 1971. India: two must do. *Ecology* 1(7):6–8, 62–63.

Clarke, V. de V. 1972. Some aspects of the epidemiology of bilharziasis in Rhodesia. *Rhodesia Science News* 6:312–17.

Cook, R. C. 1962. How many people have ever lived on earth? *Population Bulletin* 18(1):1–18.

Davis, W. H., ed. 1971. *Readings in human population ecology.* Englewood Cliffs, N.J.: Prentice-Hall.

Dawson, E. Y. 1963. Ecological paradox of coastal Peru. *Natural History,* Oct., 72:32–37.

Dorn, H. F. 1962. World population growth: an international dilemma. *Science* 135:283–90. (Reprinted in *Readings in conservation ecology,* ed. G. W. Cox. New York: Appleton-Century-Crofts.)

Dornstreich, M. D.; Leopold, A. C.; and Ardrey, R. 1973. Food habits of early man: balance between hunting and gathering. *Science* 179:306–7.

Durand, J. D. 1971. The modern expansion of world population. In *Man's impact on environment,* ed. T. R. Detwyler, pp. 36–49. New York: McGraw-Hill.

Ehrlich, P. R. 1968. *The population bomb.* New York: Ballantine Books.

Ehrlich, P. R., and Ehrlich, A. H. 1972. *Population, resources, environment: issues in human ecology.* 2d ed. San Francisco: W. H. Freeman.

Endler, J. A. 1973. Gene flow and population differentiation. *Science* 179:243–50.

Eyre, S. R. 1971. Man the pest: the dim chance of survival. *New York Review of Books,* 18 Nov., pp. 18–27.

Gammon, C. 1969. The Danes scourge the seas. *Sports Illustrated* 15 Dec., pp. 28–33.

Glesinger, E. 1960. The Mediterranean project. *Scientific American* 203(1):86–103.

Gulland, J. A., and Carroz, J. E. 1968. Management of fishery resources. In *Advances in marine biology,* ed. F. S. Russell and M. Yonge, vol. 6, pp. 1–71. London: Academic Press.

Harada, T. 1970. The present status of marine fish cultivation research in Japan. *Helgolander Wissenschafterliche Meeresuntersuchungen* 20(1–4):594–601.

Hardin, G. 1971. The survival of nations and civilization. *Science* 172:1302.

Harrar, J. G. 1966. *Prospects of the world food supply.* Washington: National Academy of Sciences.

Hendricks, S. B. 1969. Food from the land. In *Resources and man,* ed. P. E. Cloud, pp. 65–85. San Francisco: W. H. Freeman.

Holden, C. 1971. Fish flour: protein supplement has yet to fulfill expectations. *Science* 173:410–12.

Holt, S. J. 1969. The food resources of the ocean. *Scientific American* 221(3):178–94.

Humphrey, H., et al. 1969. *Marine science affairs, a year of broadened participation.* Third Report of the President to the Congress on Marine Resources and Engineering Development. Washington: Government Printing Office.

Idyll, C. P. 1972. *The sea against hunger.* New York: Crowell.

Keyfitz, N. 1969. United States and world populations. In *Resources and man,* ed. P. E. Cloud, pp. 42–64. San Francisco: W. H. Freeman. (National Academy of Sciences-National Research Council Publication 1703.)

Kleiber, M. 1961. *The fire of life.* New York: Wiley.

Langer, W. L. 1964. The black death. *Scientific American* 210(2):114–19.

————. 1972. Checks on population growth: 1750–1850. *Scientific American* 226(2):92–99.

Maddox, J. 1972. Problems of predicting population. *Nature* pp. 262–72.

Martin, P. S. 1967. Prehistoric overkill. In *Pleistocene extinctions: a search for a cause,* ed. P. S. Martin and H. E. Wright, Jr., vol. 6, pp. 75–120. Proceedings of the 7th Congress of the International Association for Quaternary Research. New Haven: Yale Univ. Press.

————. 1969. Wanted: a suitable herbivore to convert 600 million acres of western scrubland to protein; all replies held in strictest confidence. Please send photo. Box AD 2000, N.H.M. *Natural History* 78(2): 35–39.

Mathieson, A. C. 1969. The promise of seaweed. *Oceanology International,* Jan.–Feb., pp. 37–39.

May, R. M. 1972. Limit cycles in predator-prey communities. *Science* 177:900–902.

Mitchell, J. B. 1972. Population, politics, and potash. *Mining Engineering* 24(5):48–51.

Murphy, R. C. 1962. The oceanic life of the Antarctic. *Scientific American* 207(3):186–210.

Odum, E. P. 1971. *Fundamentals of ecology.* 3d ed. Philadelphia: W. B. Saunders.

Pape, J. C. 1970. Plaggen soils in the Netherlands. *Geoderma* 4:229–55.

Park, C. F., Jr., with Freeman, M. C. 1969. *Affluence in jeopardy: minerals and the political economy.* San Francisco: Freeman, Cooper.

Pauling, L. 1972. Vitamin C. *Science* 177:1152.

Pequegnat, W. E. 1958. Whales, plankton, and man. *Scientific American* 198(1):84–90.

Phillips, D. A.; Torrey, J. G.; and Burris, R. H. 1971. Extending symbiotic nitrogen fixation to increase man's food supply. *Science* 174:169–71.

Pinchot, G. B. 1970. Marine farming. *Scientific American* 223(6):15–21.

Pirie, N. W. 1967. Orthodox and unorthodox methods of meeting world food needs. *Scientific American* 216(2):27–35.

Pritchard, W. R. 1967. Increasing protein food through improving animal health. In *Prospects of world food supply,* ed. G. J. Harrar, pp. 56–65. Washington: National Academy of Sciences.

Revelle, R. 1967. Population and food supplies: the edge of the knife. In *Prospects of world food supply,* ed. G. J. Harrar, pp. 24–47. Washington: National Academy of Sciences.

Ricker, W. E. 1969. Food from the sea. In *Resources and man,* ed. P. E. Cloud, pp. 87–108. San Francisco: W. H. Freeman.

Rodin, L. E., and Basilevic, N. J. 1968. World distribution of plant biomass. In *Functioning of terrestrial ecosystems at the primary production level,* pp. 45–52. UNESCO, Proceedings of the Copenhagen Symposium.

Russell, F. S., and Yonge, M., eds. *Advances in marine biology,* vol. 6. London: Academic Press.

Ryther, J. R. 1969. Photosynthesis and fish production in the sea. *Science* 166:72–74.

Scrimshaw, N. S. 1966. Applications of nutritional and food sciences to meeting world food needs. In *Prospects of the world food supply,* ed. G. J. Harrar, pp. 48–55. Washington: National Academy of Sciences.

Skinner, J. D. 1971. Productivity of the Eland: an appraisal of the last five years' research. *South African Journal of Science* 67:534–39.

Skolnick, M. H., and Cannings, C. 1972. Natural regulation of numbers in primitive human populations. *Nature* 239:287–88.

Staub, W. J., and Blase, M. G. 1971. Genetic technology and agricultural development. *Science* 173: 119–23.

Stein, Z., et al. 1972. Nutrition and mental performance. *Science* 178:708–13.

Stockwell, E. G. 1968. *Population and people.* Chicago: Quadrangle.

Thomas, J. W., and Marburger, R. G. 1971. Quantity vs. quality, pt. 1: symptoms of deer herd overpopulation. *Texas Parks and Wild Life,* Sept., pp. 17–19.

Townsend, J. 1786. A dissertation on the poor laws. In *Population, evolution, birth control,* ed. G. Hardin, pp. 29–33. San Francisco: W. H. Freeman.

U.S. Department of the Interior. 1966. *The population challenge: what it means to America.* Conservation Yearbook no. 2.

U.S. Water Resources Council. 1968. *The nation's water resources,* 7 vols. Washington: Government Printing Office.

Wassink, E. C. 1968. Light conversion in photosynthesis and growth of plants. In *Functioning of terrestrial ecosystems at the primary production level,* pp. 53–66. UNESCO, Proceedings of the Copenhagen Symposium.

Waters, J. F. 1970. *The sea-farmers.* New York: Hastings House.

Watson, A., ed. 1970. *Animal populations in relation to their food resources.* Symposium of the British Ecological Society. Oxford: Blackwell Science Publishers.

Webber, H. H. 1970. The development of maricultural technology for the pennaeid shrimp of the Gulf and Caribbean region. *Helgolander Wissenchafterliche Meeresuntersuchungen* 20(1–4):455–63.

Chapter 16

Alland, A. 1967. *Evolution and human behavior.* Garden City, N.Y.: Natural History Press.
———. 1971. *Human diversity.* New York: Columbia Univ. Press.
Ardrey, R. 1961. *African genesis.* New York: Atheneum.
———. 1966. *The territorial imperative.* New York: Atheneum.
Ashley-Montagu, M. F., ed. 1969a. *Man and aggression.* London: Oxford Univ. Press.
———. 1969b. Morris on man: a basic urge to cooperate. *Scientific Research,* 29 Sept., pp. 17–19.
Barbetti, M., and Allen, H. 1972. Prehistoric man at Lake Mungo, Australia, by 32,000 B.P. *Nature* 240:47–49.
Barnett, S. A. 1967. Rats. *Scientific American* 216(1):79–85.
Bilsborough, A. 1972. Cranial morphology of Neanderthal man. *Nature* 237:251–25.
Bowen, B. E., and Voncra, C. F. 1973. Stratigraphical relationships of the Plio-Pleistocene deposits, East Rudolf, Kenya. *Nature* 242:391–93.
Bowler, J. M.; Thoren, A. G.; and Polach, H. A. 1972. Pleistocene man in Australia: age and significance of the Mungo skeleton. *Nature* 240:49–50.
Boyden, S. 1972. Biological view of problems of urban health. *Human Biology in Oceania* 1:159–69.
Brace, C. 1970. The origin of man. *Natural History* 79(1):46–49.
Bygott, J. D. 1972. Cannibalism among wild chimpanzees. *Nature* 238:410–11.
Carpenter, C. R. 1964. Naturalistic behavior of nonhuman primates. University Park, Pa.: Pennsylvania State Univ. Press.
Chance, M. R. A., and Jolly, C. J. 1970. *Social groups of monkeys, apes, and men.* London: Cape.
Clark, W. E. LeGros. 1962. *The antecedants of man.* Chicago: Quadrangle.
———. 1965. *History of primates.* London: British Museum.
———. 1969. The crucial evidence for human evolution. In *Papers on evolution,* ed. P. R. Erlich et al., pp. 482–95. Boston: Little, Brown.
———. 1971. *The antecedants of man.* Rev. ed. Chicago: Quadrangle.
Conti, A. M. F., and Chiarelli, B. 1968. Taxonomic and phylogenetic interest of the study of serum proteins of old world primates using bidimensional electrophoresis on starch-gel. In *Taxonomy and phylogeny of old world primates with reference to the origin of man,* ed. B. Chiarelli, pp. 127–37. Torino: Rosenberg & Sellier.
Coon, C. S. 1962. *The story of man.* 2d ed. New York: Knopf.
Day, M. H. 1971. Postranian remains of *Homo erectus* from bed iv, Olduvai Gorge, Tanzania. *Nature* 232:383–87.
Devore, I., and Washburn, S. L. 1968. Baboon ecology and human evolution. In *Man in adaptation,* ed. Y. A. Cohen, pp. 93–108. Chicago: Aldine.
Dobzhansky, T. 1967. Changing man. *Science* 155:409–15.
Durnin, J. V. G. A., et al. 1973. How much food does man require? *Nature* 242:418.
Eckhardt, R. B. 1972. Population genetics and human origins. *Scientific American* 226(1):93–103.
Eibl-Eibesfeldt, I. 1961. The fighting behavior of animals. *Scientific American* 205:112–22, Dec.
Eisenberg, J. F.; Muckenhirn, N. A.; Rudran, R. 1972. The relation between ecology and social structure in primates. *Science* 176:863–73.
Gardner, R. A., and Gardner, B. T. 1969. Teaching sign language to a chimpanzee. *Science* 165:664–72.
Ghiselin, M. T. 1973. Darwin and evolutionary psychology. *Science* 179:964–68.
Ginter, E. 1973. Cholesterol: vitamin C controls its transformation to bile acids. *Science* 179:702–4.
Goodman, M. 1968. Phylogeny and taxonomy of the Catarrhine primates from immunodiffusion data. In *Taxonomy and phylogeny of old world primates with reference to the origin of man,* ed. B. Chiarelli, pp. 95–107. Torino: Rosenberg & Sellier.
Goodman, M., et al. 1971. Evolving primate genes and proteins. In *Comparative genetics in monkeys, apes and man,* ed. B. Chiarelli, pp. 153–212. London: Academic Press.
Harlow, H. F., and Kuenne, M. 1962. Social deprivation in monkeys. *Scientific American* 206(11):127–45.
Holloway, R. L. 1972. Australopithecine endocasts, brain evolution in the Hominoidea, and a model of hominid evolution. In *The functional and evolutionary biology of primates,* ed. R. Tuttle, pp. 185–203. Chicago: Aldine.
Hook, E. B. 1973. Behavioral implications of the human XYY genotype. *Science* 179:139–50.
Howell, F. C. 1969. Remains of Hominidae from Pliocene/Pleistocene formations in the lower Omo Basin, Ethiopia. *Nature* 223:1234–39.
Howells, W. W. 1960. The distribution of man. *Scientific American* 203(3):113–27.
Hyerdahl, T. 1952. *American Indians in the Pacific: the theory behind the Kon-Tiki expedition.* Chicago: Rand McNally.
Ingle, D. J. 1964. Racial differences and the future. *Science* 146:375–79.

Irving, L. 1966. Adaptations to cold. *Scientific American* 214:94–101.

Ivanoe, F. 1970. Was Virchow right about Neandertal? *Nature* 227:577–79.

Jensen, A. R. 1969. How much can we boost IQ and scholastic achievement? In *Environment, heredity, and intelligence.* Harvard Educational Review Reprint Series no. 2.

Jenkins, F. A., Jr. 1972. Chimpanzee bipedalism: cineradiographic analysis and implications for the evolution of gait. *Science* 178:877–79.

Jukes, T. H., and Holmquist, R. 1972. Evolutionary clock: nonconstancy of rate in different species. *Science* 177:530–32.

Kellogg, W. N. 1968. Communication and language in the home-raised chimpanzee. *Science* 162:423–27.

King, J. C. 1971. *The biology of race.* New York: Harcourt Brace Jovanovich.

Kurtén, B. 1972. *Not from the apes.* New York: Pantheon.

Lawick, H. van, and Lawick-Goodall, J. van. 1971. *Innocent killers.* Boston: Houghton Mifflin.

Lawick-Goodall, J. van.; Lawick, H. van; and Packer, C. 1973. Tool-use in free-living baboons in the Gombe National Park, Tanzania. *Nature* 241:212–13.

Leach, E. R. 1965. Culture and social cohesion: an anthropologist's view. *Daedalus,* Winter, pp. 24–33.

Leakey, R. E. F. Further evidence of lower Pleistocene hominids from East Rudolf, north Kenya, 1971. *Nature* 237:264–69.

Leaves, L. J., and Jinks, J. L. 1972. Insignificance of evidence for differences in heritability of IQ between races and social classes. *Nature* 240:84–88.

Loomis, W. F. 1967. Skin-pigment regulation or vitamin-D biosynthesis in man. *Science* 157:501–6.

———. 1970. Rickets. *Scientific American* 223(6):77–91.

Lorenz, K. Z. 1958. The evolution of behavior. *Scientific American* 199:67–78.

———. 1966. *On Aggression.* New York: Harcourt, Brace & World.

Lovejoy, C.; Burstein, A. H.; and Heiple, K. G. 1972. Primate phylogeny and immunological distance. *Science* 176:803–5.

Loy, J. 1971. On the primate biogram. *Science* 172:680–81.

Lumley, H. de. 1969. A Paleolithic camp at Nice. *Scientific American* 219:42–50.

Lundman, B. 1967. Race and anthropology. In *Race and modern science,* ed. R. E. Kuttner, pp. 3–26. New York: Social Science Press.

Maglio, V. J. 1972. Vertebrate faunas and chronology of hominid-bearing sediments east of Lake Rudolf, Kenya. *Nature* 239:379–84.

Marais, E. 1939. *My friends the baboons.* London: Methuen.

McCarthy, M. 1968. Letter from London: the Demo. *New York Review of Books,* 12 Dec., pp. 5–8.

Moss, G. R. 1973. Human behavior, instinct, and aggression. *Science* 179:428–30.

Morris, D., ed. 1967. *Primate ethology.* Chicago: Aldine.

———. 1969. *The human zoo.* New York: McGraw-Hill.

Mullendore, W. E., and Cooper, K. M. 1972. Effects of race on property values: the case of Dallas. *Annals of Regional Science* 6(2)61–72.

Napier, J. R., and Napier, P. H., eds. 1970. *Old world monkeys: evolution, systematics, and behavior.* New York: Academic Press.

Nei, M., and Roychoudhury, A. K. 1972. Gene differences between Caucasian, Negro, and Japanese populations. *Science* 177:434–35.

Oliver, C. P. 1967. The races of man and human genetics. In *Race and modern science,* ed. R. E. Kuttner, pp. 171–95. New York: Social Science Press.

Omoto, B. 1972. Polymorphisms and genetic affinities of the Ainu of Hokkaido. *Human Biology in Oceania* 1:278–88.

Pilbeam, D. 1968. The earliest hominids. *Nature* 219:1335–38.

———. 1970. Early hominids and cranial capacities. *Nature* 227:747–48.

Plato, C. C., and Gajdusek, D. C. 1972. Dermatoglyphics of the Asmet and Tjitak peoples of west New Guinea. *Human Biology in Oceania* 1(2):255–58.

Premack, D. 1971. Language in chimpanzee? *Science* 172:808–22.

Quarton, G. C. 1967. Deliberate efforts to control human behavior and modify personality. *Daedalus,* Summer, pp. 837–53.

Raychaudhuri, S. P. 1971. Expansion of the great Indian desert. *Proceedings of the Indian Natural Science Academy* 36(6):331–44.

Robinson, J. T. 1972. The bearing of East Rudolf fossils on early hominid systematics. *Nature 240:239–40.*

Sarich, V. M. Primate systematics with special reference to old world monkeys: a protein perspective. In *Old world monkeys,* ed. J. R. Napier and P. H. Napier, pp. 175–226. New York: Academic Press.

Sarich, V. M., and Wilson, A. C. 1973. Generation time and genomic evolution in primates. *Science* 179:1144–46.

Scrimshaw, N. S., and Gordon, E., eds. 1968. *Malnutrition, learning, and behavior.* Cambridge: M.I.T. Press.

Sheehan, E. R. G. 1968. Conversations with Konrad Lorenz. *Harper's* 236(1416):69–77.

Sherrington, C. S. 1941. *Man on his nature.* Cambridge: Cambridge Univ. Press.

Simons, E. L. 1963. Some fallacies in the study of hominid phylogeny. *Science* 141:879–89.

————. 1964. The early relatives of man. *Scientific American* 211:50–62.

————. 1965. New fossil apes from Egypt and the initial differentiation of Hominoidea. *Nature* 205:135–39.

————. 1967. The earliest apes. *Scientific American* 217(6):28–35.

————. 1969a. Miocene monkey (*Prohylobates*) from northern Egypt. *Nature* 223:687–89.

————. 1969b. Late Miocene hominid from Fort Ternan, Kenya. *Nature* 221:448–51.

————. 1969c. The earliest apes. *Scientific American* 217(6):2–9.

————. 1970. The deployment and history of old world monkeys (Cercopithecidae, Primates). In *Old world monkeys,* ed. J. R. Napier and P. H. Napier, pp. 97–137. New York: Academic Press.

Simpson, G. G. 1951. *The meaning of evolution.* New York: Mentor.

————. 1969. *Biology and man.* New York: Harcourt Brace and World.

Singh, S. D. 1969. Urban monkeys. *Scientific American* 221:108–15.

Skinner, B. F. 1948. *Walden Two.* New York: Macmillan.

————. 1966. The phylogeny and ontogeny of behavior. *Science* 153:1205–13.

Smith, S. M., and Symonds, N. S. 1973. The unexpected location of a gene conferring abnormal radiation sensitivity on Phage T4. *Nature* 241:395–96.

Speiser, E. A., ed. 1964. Genesis. *The Anchor Bible,* vol. 1. New York: World.

Stefansson, V. 1964. *Discovery.* New York: McGraw-Hill.

Stein, A. 1933. *On ancient Central-Asian tracks.* New York: Random House.

Tattersall, I. 1969. Ecology of north Indian *Ramapithecus. Nature* 221:451–52.

Teilhard de Chardin, P. 1959. *The phenomenon of man,* trans. B. Wall. New York: Harper Brothers.

Teleki, G. 1973. The omnivorous chimpanzee. *Scientific American* 228(1):32–42.

Thoma, A. 1972. On Vertesszöllös man. *Nature* 236:464–65.

Thorne, A. G., and Macumber, P. G. 1972. Discoveries of late Pleistocene man at Kow Swamp, Australia. *Nature* 238:316–19.

Tobias, P. V. 1972. Progress. Progress and problems in the study of early man in sub-saharan Africa. In *The functional and evolutionary biology of primates,* ed. R. Tuttle, pp. 63–91. Chicago: Aldine.

Tuttle, R., ed. 1972. *The functional and evolutionary biology of primates.* Chicago: Aldine.

Washburn, S. L. 1969. The analysis of primate evolution with particular reference to the origin of man. In *Papers on evolution,* ed. P. R. Ehrlich et al., pp. 523–40. Boston: Little, Brown.

Watson, J. D. 1968. *The double helix.* New York: Atheneum.

Weckler, J. E. 1957. Neanderthal man. *Scientific American* 197(6):89–95.

Wiener, A. S., and Moor-Jankowski, M. 1971. Blood groups of non-human primates and their relationship to the blood groups of man. In *Comparative genetics of monkeys, apes, and man,* ed. B. Chiarelli, pp. 71–95. London: Academic Press.

Wilson, R. S. 1972. Twins: early mental development. *Science* 175:914–17.

Yamazaki, T., and Naruyama, T. 1972. Evidence for the neutral hypothesis of protein polymorphism. *Science* 178:56–57.

Zwell, M. 1972. On the supposed "*Kenyapithecus africanus*" mandible. *Nature* 240:236–39.

Index

Photo Credits

Art Credits

2–4 From Brian J. Skinner, *Earth Resources,* © Copyright 1969 by Prentice-Hall, Inc., Englewood Cliffs, New Jersey. By permission of Prentice-Hall, Inc., Englewood Cliffs, New Jersey.

2–6 From *Resources and Man: A Study and Recommendations* by the Committee on Resources and Man of the Division of Earth Sciences, National Academy of Sciences–National Research Council, with the cooperation of the Division of Biology and Agriculture. W. H. Freeman and Company. Copyright © 1969.

2–8 Modified from Charles F. Park, Jr. 1968. *Affluence in jeopardy: Minerals and the political economy.* San Francisco: Freeman, Cooper & Co., Fig. 9–3, p. 195.

2–9 Data from M. King Hubbert, "Energy resources." In *Resources and Man,* Publication 1703, Committee on Resources and Man, National Academy of Sciences–National Research Council, W. H. Freeman and Company, San Francisco, 1969. By permission of the National Academy of Sciences and M. King Hubbert.

2–11a From M. King Hubbert. 1956. Nuclear energy and the fossil fuels. In *Drilling and production practice.* Washington, D.C.: American Petroleum Institute, Fig. 21.

2–11b From M. King Hubbert. 1962. *Energy resources: A report to the Committee on Natural Resources.* Washington, D.C.: National Academy of Sciences–National Research Council Publication 1000–D, Fig. 47, p. 83. By permission of the National Academy of Sciences and M. King Hubbert.

2–12 From M. King Hubbert. 1962. *Energy resources: A report to the Committee on Natural Resources.* Washington, D.C.: National Academy of Sciences–National Research Council Publication 1000–D, Fig. 41, p. 75. By permission of the National Academy of Sciences and M. King Hubbert.

2–13 From *Oil and the Middle East,* with permission of Exxon Corporation. (Map based on 1969 data.)

2–15 Modified from H. F. Coffer, *et al.* 1968. The use of nuclear explosives in oil and gas production. *Earth Science Bulletin,* March, Fig. 8, p. 16.

2–16 From *World Oil.* 1969. Project Rulison: New try at nuclear stimulation. *World Oil* 169(4):69.

2–19 Modified from W. A. Bachman, *et al.* 1969 Forecast for the seventies. *Oil & Gas Journal* 67(45):162–164.

2–20 Modified from W. A. Bachman, *et al.* 1969.

Forecast for the seventies. *Oil & Gas Journal* 67(45):162–164.

2–21 Data from the tables in W. A. Bachman, *et al.* 1969. Forecast for the seventies. *Oil & Gas Journal* 67(45):162–164.

3–3 Data from Clair C. Patterson and Joseph D. Salvia. 1968. Lead in the modern environment: How much is natural? *Environment* 10(3):77, Table 1. © 1968, Committee for Environmental Information.

3–4 From Richard H. Wagner. 1971. *Environment and Man.* New York: W. W. Norton, Fig. 10.10, p. 190. Copyright © 1971 by W. W. Norton & Company, Inc.

3–5 From Emmanuel LeRoy Ladurie. 1971. *Times of feast—Times of famine: A history of climate since the year 1000.* Garden City, N.Y.: Doubleday & Company, Inc., p. 83. © Flammarion, 1967, Copyright © 1971, Doubleday & Company, Inc.

4–1 From *The Dynamics of the Earth: An Introduction to Physical Geology* by Edgar Winston Spencer. Copyright © 1972 by Thomas Y. Crowell Company, Inc., with permission of the publisher.

4–2 Data from Seymour Tilson. 1966. The ocean. *Science and Technology,* February (50):27.

4–3 Modified from Seymour Tilson. 1966. The ocean. *Science and Technology,* February (50):31.

4–4 Modified from Charles F. Wurster, Jr. 1968. DDT reduces photosynthesis by marine plankton. *Science* 159:1474.

4–6 From J. F. Santos and J. D. Stoner. 1972. *Physical, chemical, and biological aspects of the Duwamish River Estuary, King County, Washington: 1963–1967.* U.S. Geological Survey Water Supply Paper 1873–C.

4–7 From J. F. Santos and J. D. Stoner. 1972. *Physical, chemical, and biological aspects of the Duwamish River Estuary, King County, Washington: 1963–1967.* U.S. Geological Survey Water Supply Paper 1873–C.

5–3 From Robert J. Dingman and José Nuñez. 1969. *Hydrogeologic reconnaissance of the Canary Islands, Spain.* U.S. Geological Survey Prof. Paper 650–C.

5–7 From G. M. Urrows. 1967. *Nuclear energy for desalting.* Washington, D.C.: U.S. Atomic Energy Commission.

5–13 From "The Aging Great Lakes" by Charles F. Powers and Andrew Robertson. Copyright © 1966 by Scientific American, Inc. All rights reserved.

5–14 Reproduced with permission from "Dwindling

Lakes'' by Arthur D. Hasler and Bruce Ingersoll, *Natural History Magazine,* November 1968. Copyright © The American Museum of Natural History, 1968.

5–15 Modified from J. R. Irwin and R. B. Morton. 1969. *Hydrogeologic information on the Glorietta Sandstone and the Ogallala Formation in the Oklahoma Panhandle and the adjoining areas as related to underground waste disposal.* U.S. Geological Survey Circular 630, p. 14.

5–16 From ''The Aging Great Lakes'' by Charles F. Powers and Andrew Robertson. Copyright © 1966 by Scientific American, Inc. All rights reserved.

5–17 From ''Thermal Pollution and Aquatic Life'' by John R. Clark. Copyright © 1969 by Scientific American, Inc. All rights reserved.

6–1 From Donald Leet and Sheldon Judson, *Physical Geology,* 3rd edition, © Copyright 1954, 1958, 1965 by Prentice-Hall, Inc., Englewood Cliffs, New Jersey. By permission of Prentice-Hall, Inc., Englewood Cliffs, New Jersey. Redrawn from Henry Gannett. 1901. *Profiles of Rivers in the United States.* Washington, D.C.: U.S. Geological Survey Water Supply Paper 44.

6–4 From P. H. Kuenen. 1955. *Realms of water: Some aspects of its cycle in nature.* New York: John Wiley, Fig. 149, p. 266.

6–5 From Sheldon Judson. 1968. Erosion of the land—Or what's happening to our continents? *American Scientist* 56:362, Fig. 6.

6–9 From Sheldon Judson. 1968. Erosion of the land—Or what's happening to our continents? *American Scientist* 56:364, Fig. 7.

6–12 From Sheldon Judson. 1968. Erosion of the land—Or what's happening to our continents? *American Scientist* 56:366, Fig. 9.

6–13 After A. N. Strahler. 1973. *Introduction to Physical Geography,* 3rd edition. New York: John Wiley. Copyright © 1973 by John Wiley & Sons, Inc.

6–19 From A. E. J. Engel. 1969. Time and the earth. *American Scientist* 57:479, Fig. 12.

7–2 Modified from E. M. Bridges, *World Soils,* p. 18. © 1970 Cambridge University Press.

7–3 Modified from E. M. Bridges, *World Soils,* p. 19. © 1970 Cambridge University Press.

7–4 Modified from W. D. Keller. 1969. *Chemistry in introductory geology.* Columbia, Mo.: Lucas Bros., Fig. 12, p. 54.

7–5 Modified from James Thorp. 1968. The soil—A reflection of Quaternary environments in Illinois. In Robert E. Bergstrom, ed., *The Quaternary of Illinois.* Urbana: Univ. of Illinois College of Agriculture Spec. Publ. 14, Figs. 1, 2, 3, p. 51.

7–8 From *Population, Resources, Environment: Issues in Human Ecology,* 2nd ed., by Paul R. Ehrlich and Anne H. Ehrlich. W. H. Freeman and Company. Copyright © 1972.

7–9 From *Population, Resources, Environment: Issues in Human Ecology,* 2nd ed., by Paul R. Ehrlich and Anne H. Ehrlich. W. H. Freeman and Company. Copyright © 1972.

7–10 Data from Ralph G. Nash and Edwin A. Woolson. 1967. Persistence of chlorinated hydrocarbon insecticides in soils. *Science* 157:924–925, Figs. 1 and 2.

7–11 (left) From Robert D. Teeters and Annabelle Desmond. 1960. Population growth and economic development in the ECAFE region, Part 2. *Population Bulletin* 16(2):39, Fig. 4. *And* (right) from Robert C. Cook. 1962. The Common Market. *Population Bulletin* 18(4):84, Fig. 4.

7–13 Reprinted from *Subsurface disposal in geologic basins—A study of reservoir strata,* published by The American Association of Petroleum Geologists, 1968.

7–15 From C. L. Comar. 1966. *Fallout from nuclear tests.* Washington, D.C.: U.S. Atomic Energy Commission, Division of Technical Information, Understanding the Atom series, Fig. 7, p. 25.

8–1 Modified from Gerald S. Schatz, Health and geochemical environment—What is the relationship? National Academy of Sciences *News Report,* April 1970. (Adapted from maps by Armed Forces Institute of Pathology.)

8–2 From Jane F. Brody. 1970. 25 years of fluoride cuts tooth decay in Newburgh. *The New York Times,* 3 May, p. 64L. © 1970 by The New York Times Company. Reprinted by permission.

8–3 From Gerald S. Schatz, Health and geochemical environment—What is the relationship? National Academy of Sciences *News Report,* April 1970. (Adapted from maps by Armed Forces Institute of Pathology.)

8–4 Data from Hansford T. Shaklette. 1971. *Elemental composition of surficial materials in the conterminous United States.* U.S. Geological Survey Prof. Paper 574-D, Fig. 6.

8–5 Data from Hansford T. Shaklette. 1971. *Elemental composition of surficial materials in the conterminous United States.* U.S. Geological Survey Prof. Paper 574-D, Fig. 9.

8–6 From R. A. Horne. 1972. Biological effects of chemical agents. *Science* 177:1152–1153.

8–7 From George K. Davis. 1970. Interaction of

trace elements. In *Trace substances in environmental health*–III. Delbert D. Hemphill, ed., University of Missouri, Columbia, p. 141.

8–8   From H. Mitchell Perry, Jr. 1969. Hypertension and trace metals, particularly cadmium. In *Trace substances in environmental health*–II. Delbert D. Hemphill, ed., University of Missouri, Columbia, p. 107.

8–9   From Jean M. Spencer. 1970. Geological influence on regional health problems. *Texas Journal of Science* 21:461, Fig. 1.

9–1   From *California Geology*. 1969. Active faults of California. *California Geology* 22(5):81.

9–3   From M. D. Trifunac and D. E. Hudson. 1971. Analysis of the Pacoima Dam accelerogram—San Fernando, California, earthquake of 1971. *Seismological Society of America Bulletin* 61(5):1405, Fig. 12.

9–5   From Aubrey D. Henley. 1966. Seismic activity near the Texas coast. *Association of Engineering Geologists* 3:34.

9–6   Modified from Karl V. Steinbrugge. 1969. Seismic risk to buildings and structures on filled lands in San Francisco Bay. In Harold B. Goldman, ed., *Geologic and engineering aspects of San Francisco Bay fill*. California Division of Mines and Geology Special Report 97, p. 107.

9–8   From Gordon B. Oakeshott. 1971. The geologic setting. *California Geology* 24(4–5):69–74.

9–9   From Roger Greensfelder. 1971. Seismologic and crustal movement investigations of the San Fernando earthquake. *California Geology* 24(4–5):62–68.

9–10   By permission of Hawthorn Books, Inc. from *The Circle of Fire* by Carl Heintze. Illustration copyright © 1968 by Meredith Corporation. All rights reserved.

9–11   From "Tsunamis" by Joseph Bernstein. Copyright © 1954 by Scientific American, Inc. All rights reserved.

9–12   From David M. Evans. 1966. Man-made earthquakes in Denver. *Geotimes,* May-June, Fig. 4, p. 13. Copyright © 1966 American Geological Institute.

9–13   From Karl V. Steinbrugge, *Earthquake Hazard in the San Francisco Bay Area* (Berkeley: Institute of Governmental Studies, University of California, 1968). Reproduced by permission of the Institute of Governmental Studies.

10–1   From John Victor Luce. 1969. *Lost Atlantis: A new light on an old legend.* New York: McGraw-Hill, Fig. 8, p. 75.

10–2   From John Victor Luce. 1969. *Lost Atlantis: A new light on an old legend.* New York: McGraw-Hill, Fig. 9, p. 77.

10–6   From Richard S. Fiske and Willie T. Kinoshita. 1969. Inflation of Kilauea Volcano prior to its 1967–1968 eruption. *Science* 165:343. Source: U.S. Geological Survey.

10–7   From Richard S. Fiske and Robert Y. Kohanagi. 1968. *The December 1965 eruption of Kilauea Volcano, Hawaii.* U.S. Geological Survey Prof. Paper 607, p. 3.

10–10 From John Victor Luce. 1969. *Lost Atlantis: A new light on an old legend.* New York: McGraw-Hill, Figs. 12 and 13, p. 85.

11–1   From Arthur B. Cleaves. 1961. *Landslide Investigations: A field handbook for use in highway location and design.* Washington, D.C.: U.S. Department of Commerce, U.S. Bureau of Public Roads.

11–2   From C. F. S. Sharpe and E. F. Dosch. 1942. Relation of soil creep to earthflow in the Appalachian Plateau. *Journal of Geomorphology* 5:316.

11–5   From "Quick Clay" by Paul F. Kerr. Copyright © 1963 by Scientific American, Inc. All rights reserved.

11–7   Modified from George A. Kiersch. 1965. Vaiont Reservoir disaster. *Geotimes,* May-June, p. 10.

11–8   Modified from George A. Kiersch. 1965. Vaiont Reservoir disaster. *Geotimes,* May-June, p. 12.

11–10 Modified from Ronald L. Shreve. 1966. Sherman landslide, Alaska. *Science* 154:1640.

11–12 From Richard C. Leach. 1968. *The problem and correction of landslides in West Virginia.* West Virginia Geological Survey Circular 10, Fig. 9.

11–13 From Arthur B. Cleaves. 1961. *Landslide Investigations: A field handbook for use in highway location and design.* Washington, D.C.: U.S. Department of Commerce, U.S. Bureau of Public Roads.

11–15 From Arthur B. Cleaves. 1961. *Landslide Investigations: A field handbook for use in highway location and design.* Washington, D.C.: U.S. Department of Commerce, U.S. Bureau of Public Roads.

11–16 From André Cailleux. 1968. *Anatomy of the Earth,* translated by J. Moody Stuart. New York: McGraw-Hill, Fig. 42, p. 134. © André Cailleux, 1968. Translation © George Weidenfeld & Nicolson Limited, 1968. Used with permission of McGraw-Hill Book Company.

11–17 Modified from R. Köster. 1968. Postglacial sea-level changes in the western Baltic region in relation to world-wide eustatic movements. In Roger G. Morrison and Herbert E. Wright, Jr., eds., *Means of correlation of Quaternary successions.* Proceedings of the VII Congress of the International Association for Quaternary

13–12 From *Quaternary Geology and Climate,* Volume 16 of the Proceedings of the VII Congress of the International Association for Quaternary Research, Publication ISBN 0–309–01701–7, National Academy of Sciences–National Research Council, Washington, D.C., 1969.

13–13 From N. R. Gadd. 1971. Pleistocene geology of the central St. Lawrence Lowland. Geological Survey of Canada Memoir 359, Fig. 10, p. 80.

13–14 From "Micropaleontology" by David B. Ericson and Goesta Wollin. Copyright © 1962 by Scientific American, Inc. All rights reserved.

13–15 From "Micropaleontology" by David B. Ericson and Goesta Wollin. Copyright © 1962 by Scientific American, Inc. All rights reserved.

13–16 From K. J. Mesolella, *et al.* 1969. The astronomical theory of climatic change: Barbados data. *Journal of Geology* 77:261, Fig. 7. © 1969 by The University of Chicago. With permission of The University of Chicago Press.

13–17 From W. S. Dansgaard, *et al.* 1969. One thousand centuries of climatic record from Camp Century on the Greenland ice sheet. *Science* 166:380, Fig. 5.

13–18 From Rhodes W. Fairbridge. 1967. Carbonate rocks and paleoclimatology of the biogeochemical history of the planet. In G. V. Chilingar, *et al.,* eds., *Carbonate rocks: Developments in sedimentology.* Amsterdam: Elsevier Publishing Co., Fig. 1, p. 401.

13–19 From "The Changing Level of the Sea" by Rhodes W. Fairbridge. Copyright © 1960 by Scientific American, Inc. All rights reserved.

13–20 From Olav Liestöl. 1960. Glaciers of the present day. In Olaf Holtedahl, ed., *Geology of Norway.* Norges Geol. Undersokelse 208, Fig. 160, p. 485.

13–21 From "Neoglaciation" by George H. Denton and Stephen C. Porter. Copyright © 1970 by Scientific American, Inc. All rights reserved.

13–22 Reprinted by permission of Yale University Press from *The Vinland Map and the Tartar Relation,* by R. A. Skelton, Thomas E. Marston and George D. Painter. Copyright © 1965 by Yale University.

13–23 From K. W. Butzer. 1961. Climatic change in arid regions since the Pliocene. In L. Dudley Stamp, ed., *A history of land use in arid regions.* Paris: UNESCO, Fig. 3.

13–24 From C. E. P. Brooks. 1949. *Climate through the ages.* London: Ernest Benn, Fig. 39, p. 361.

13–25 From Louis M. Thompson. 1964. *Our recent high yields—How much due to weather?* Madison, Wisc.: American Society of Agronomy, Spec. Publ. no. 4, Figs. 1 and 2.

13–26 From S. J. Johnsen, *et al.* 1970. Climatic oscillations 1200–2000 A.D. *Nature* 227:482, Fig. 1.

14–4 From W. L. Fisher, *et al.* 1972. *Environmental geologic atlas of the Texas coastal zone: Galveston-Houston area.* Austin: Univ. of Texas, Bureau of Economic Geology.

14–5 From W. L. Fisher, *et al.* 1972. *Environmental geologic atlas of the Texas coastal zone: Galveston-Houston area.* Austin: Univ. of Texas, Bureau of Economic Geology.

15–1 From Arthur S. Boughey. 1968. *Ecology of population.* New York: Macmillan, Fig. 2.10, p. 38. © Copyright, Arthur S. Boughey, 1968.

15–2 From Alexander Alland. 1967. *Evolution and human behavior.* Garden City, N.Y.: Natural History Press, Fig. 4, p. 18. Copyright © 1967 by Alexander Alland, Jr. With permission of Doubleday & Company, Inc.

15–3 From "The Black Death" by William L. Langer. Copyright © 1964 by Scientific American, Inc. All rights reserved.

15–4 Data from Robert C. Cook. 1962. How many people have ever lived on earth? *Population Bulletin* 18(1):5, Fig. 1.

15–5 From *Resources and Man: A Study and Recommendations* by the Committee on Resources and Man of the Division of Earth Sciences, National Academy of Sciences–National Research Council, with the cooperation of the Division of Biology and Agriculture. W. H. Freeman and Company. Copyright © 1969.

15–6 From Anthony J. Wiener. 1969. *Technology and the human environment.* Hearing before the U.S. Senate subcommittee on intergovernmental relations of the Committee on Governmental Operations. Washington, D.C.: U.S. Senate, Fig. 1, p. 58.

15–7 From *Resources and Man: A Study and Recommendations* by the Committee on Resources and Man of the Division of Earth Sciences, National Academy of Sciences–National Research Council, with the cooperation of the Division of Biology and Agriculture. W. H. Freeman and Company. Copyright © 1969.

15–9 From Hubert H. Humphrey, *et al.* 1969. *Marine science affairs, a year of broadened participation.* Third report of the President to the Congress on marine resources and engineering development. Washington, D.C.: U.S. Government Printing Office, Fig. 11–7, p. 28.

15–10 From *Resources and Man: A Study and Recommendations* by the Committee on Resources and Man of the Division of Earth Sciences, National Academy of Sciences–National Research Council, with the cooperation of the Division